Wesley Stoker Barker Woolhouse

Measures, Weights, & Moneys of all Nations

And an Analysis of the Christian, Hebrew, and Mahometan Calendars. Seventh

Edition

Wesley Stoker Barker Woolhouse

Measures, Weights, & Moneys of all Nations
And an Analysis of the Christian, Hebrew, and Mahometan Calendars. Seventh Edition

ISBN/EAN: 9783337165123

Hergestellt in Europa, USA, Kanada, Australien, Japan

Cover: Foto ©ninafisch / pixelio.de

Weitere Bücher finden Sie auf **www.hansebooks.com**

MEASURES, WEIGHTS, & MONEYS

OF ALL NATIONS

AND AN ANALYSIS OF THE CHRISTIAN, HEBREW,
AND MAHOMETAN CALENDARS

BY

W. S. B. WOOLHOUSE, F.R.A.S., F.S.S.

ETC.

Seventh Edition, carefully Revised and Enlarged

LONDON
CROSBY LOCKWOOD AND SON
7, STATIONERS' HALL COURT, LUDGATE HILL
1890

PREFACE.

The measures, weights, and moneys established throughout the world are so diversified in their comparative values and systematic relations that a correct classification of them is a task of greater magnitude than would commonly be supposed. In the present work no labour has been spared to ascertain, in every case, the best attainable information, and the various details have been arranged with especial regard to facility of reference. With this object the materials appertaining to each locality are uniformly tabulated in the same order, and opposite to each separate quantity or value the English equivalent is distinctly exhibited, so as to obviate as much as possible the necessity of any calculation.

The comprehensive principles which influence the fluctuations of exchange are also briefly stated, and correct rules are given for computing the sterling value of coins and bullion.

The tables for the conversion of the standard linear and square measures of one country into those of another were before published in another form, and had recently become out of print. They are here enlarged and more conveniently arranged, and their utility will be fully appreciated by those who may have occasion to consult the architectural and other works of the Continent.

The second part of the volume relates to the measurement of time, and comprises a detailed investigation of the Christian, Hebrew, and Mahometan Calendars, with formulæ, tables, and practical rules for performing the various calculations. We have been more specially induced to go at length into these subjects, as they are imperfectly treated in chronological works generally.

PREFACE TO THE SEVENTH EDITION.

In revising this work for the seventh edition opportunity has been taken to make as many suitable emendations and additions as were possible. By means of the valuable Reports of the Board of Trade under the Weights and Measures Act, 1878, the author has been enabled to give, in an Appendix, the most reliable account of the Weights and Measures of China and Japan, which were formerly involved in much obscurity. Full and authentic information is given on the subject of Wire and Plate Gauges, and more especially in relation to the new Imperial Standard Wire Gauge which, in 1883, was legalised by an Order in Council. The chief particulars are also given of the new Weights and Measures Act, 1889 ; and the Appendix furthermore supplies a variety of miscellaneous and interesting information on other matters.

The utmost facility for detailed reference is also secured by the addition of a full and complete Index of the entire contents of the volume.

London, 12*th August*, 1890.

CONTENTS.

MEASURES, WEIGHTS, AND MONEYS.

STANDARDS.

	PAGE
MEASURES AND WEIGHTS	1
METRICAL SYSTEM OF FRANCE	26
COINS	28
STANDARD WEIGHTS OF BRITISH COINS	35
PROJECTED DECIMAL SYSTEM OF COINAGE	36

COMPUTATION OF COINS AND BULLION.

ASSAYS	39
RULES FOR GOLD AND SILVER COINS	40
TABLE OF GOLD AND SILVER COINS	44

GENERAL PRINCIPLES OF EXCHANGE . . . 45

MEASURES, WEIGHTS, AND MONEYS.

GREAT BRITAIN	50
OTHER PLACES, ACCORDING TO AN ALPHABETICAL ARRANGEMENT	57

GENERAL TABLES 119

I. CONVERSION OF STANDARD LINEAR MEASURES	120
II. ,, ,, SQUARE ,,	130
III. ITINERARY OR ROAD MEASURES	140

MEASURES OF TIME.

THE GREGORIAN CALENDAR 145

DIFFERENCE OF STYLE	148
DOMINICAL LETTER	149
TABLE SHOWING THE DAY OF THE WEEK	155
CYCLE OF THE SUN	ib.
GOLDEN NUMBER	157
EPACT	160
NUMBER OF DIRECTION	164
EASTER-DAY	167
INDICTION	169
DIONYSIAN PERIOD	ib.
JULIAN PERIOD	170
MOON'S AGE	ib.
MOVABLE FEASTS	ib.

CONTENTS.

	PAGE
LAW TERMS	172
LAW SITTINGS	176
UNIVERSITY TERMS	177
ELEMENTS OF GREGORIAN CALENDAR (1800 TO 2100)	178
HEBREW CALENDAR	187
TABLE OF MONTHS AND CALCULATION OF DATES	189
TABLES FOR DETERMINING THE LENGTH OF A JEWISH YEAR AND THE DAY OF THE WEEK ON WHICH IT COMMENCES	193
PRINCIPAL DAYS OF THE JEWISH CALENDAR	194
TABLE OF HEBREW YEARS UP TO THE YEAR 2072 OF THE CHRISTIAN ERA	195
MAHOMETAN CALENDAR	198
TABLES AND RULES FOR DETERMINING THE DAY OF THE WEEK AND THE DATE ON WHICH ANY YEAR OF THE HEGIRA COMMENCES	199
TABLE OF MAHOMETAN YEARS UP TO THE YEAR 2047 OF THE CHRISTIAN ERA	203
TABLE OF THE PRINCIPAL EPOCHS	206
CAMBRIDGE UNIVERSITY TERMS, *after* 1883	207
EASTER-DAY, FOUND FROM THE GOLDEN NUMBER AND THE DOMINICAL LETTER ONLY	208

APPENDIX.

IMPERIAL WEIGHTS AND MEASURES IN FORCE AT PLACES MENTIONED	211
METRIC SYSTEM LEGALLY ADOPTED IN GERMANY	ib.
CHINA.—WEIGHTS AND MEASURES LEGALLY IN USE	212
JAPAN.—WEIGHTS AND MEASURES LEGALLY IN USE	213
IMPERIAL STANDARD YARD, COPY OF	214
INDIAN SEER, VALUE OF	ib.
ILLEGAL DENOMINATIONS OF WEIGHT	ib.
WIRE AND SHEET GAUGES	215
IMPERIAL STANDARD WIRE GAUGE, TABLES I., II., III., IV.	217
STUBS' IRON WIRE GAUGE, TABLE V.	221
GAUGE FOR STEEL WIRE AND PINION WIRE, TABLE VI.	222
GAUGE FOR SHEET AND HOOP IRON, TABLE VII.	223
WEIGHTS AND MEASURES ACT, 1889	224
AMOUNTS OF ERROR ALLOWED IN WEIGHTS, ETC.	226
WEAR OBSERVED IN WEIGHTS OF SOVEREIGNS	228
ASCERTAINED STANDARDS FOR ADJUSTING BANKERS' WEIGHTS	ib.
INDEX	229

MEASURES, WEIGHTS, AND MONEYS

OF

ALL NATIONS.

STANDARDS.

MEASURES AND WEIGHTS.

MEASURES are of three kinds, viz. measures of LENGTH, called Linear or Long measure; measures of SURFACE, called Square or Superficial measure; and measures of CAPACITY, called Solid or Cubic measure.

WEIGHT may be defined to be the quantity of ponderous matter contained in a solid or fluid substance, taking into account its bulk and the density or specific gravity of its parts, and is determined by the same being balanced in an accurate scale against some known or acknowledged weights placed in the opposite scale.

STANDARDS are carefully-constructed measures or weights of acknowledged authority, by which others are tested or adjusted.

Measures and Weights are of indispensable utility, and are continually employed both in commercial and scientific pursuits. For the latter of these, minute accuracy is particularly essential. There is, however, more difficulty than would at first be supposed, in establishing and preserving correct, uniform, and invariable standards of weights and measures, and a vast amount of scientific research, ingenuity, and labour has been expended upon its accomplishment.

The origin of measures of length is to be found in parts of the human body. Their values, roughly estimated, as well as their names, establish this beyond a doubt. The foot, the digit, the palm, the span, the cubit, the nail, the arm, &c., are in all languages derived from the same source; and, in the popular view of measurement, they do not considerably differ in length. It is also unquestionable that in former times, when authenticated measures were not so easily to be obtained, the hands, arms, and feet were much more frequently used than they are at present, when every workman, however humble, is in possession of a measure.

Taking a well-proportioned man, the fathom is reckoned to equal his height or stature; the girth, or the pace, $\frac{1}{2}$ of his stature; the cubit, or measurement from the elbow to the ends of the extended fingers, $\frac{1}{4}$; the foot, $\frac{1}{6}$; the span, $\frac{1}{8}$; and the breadth of the palm, $\frac{1}{24}$.

The statute 17 Edward II. (A.D. 1324) provides that three barley-corns, round and dry, make an inch, 12 inches a foot, &c. But it is so difficult to know how much of the sharp end of a barley-corn must be cut or worn away before it becomes what was called "round," that this mode of measuring by the *lengths* of barley-corns is very indefinite.

The complete table of the sixteenth century is as follows:—
The *breadth* of 4 barley-corns make a digit, or finger-breadth; 4 digits make a palm (measured across the middle joints of the fingers); 4 palms are a foot; $1\frac{1}{2}$ foot is a cubit; 10 palms or $2\frac{1}{2}$ feet are a step (gressus); 2 steps, or 5 feet, are a pace (passus); 10 feet are a perch; 125 paces are an Italic stadium; 8 stadia, or 1000 paces, are an Italic mile; 4 Italic miles are a German mile; and 5 Italic miles are a Swiss mile. From this table it would appear that the foot was considerably less even than the ancient Roman foot of 11·6 English inches; the average human foot certainly has not that length.

In the year 1742, the Royal Society had a standard yard constructed, from a minute comparison of the standard ells or yards of the reigns of Henry VII. and Elizabeth, kept at the Exchequer.

In the year 1758, a select Committee of the House of Commons was appointed to inquire into the state of English weights and measures, and were assisted in their researches by several eminent mechanists, among whom may be mentioned Mr. Bird, a celebrated optician, and Mr. Harris, the Assay Master of the Mint. This Committee prepared with great accuracy two standards, viz. the yard and the pound troy, which were afterwards carefully preserved and justly considered of the highest authority. The standard yard was copied from that of the Royal Society, and having been examined by the Committee was reported to be equal to the standard yard and marked as such.

Since that period no alteration has been made in these standards, though much attention has been paid to the subject, both in and out of Parliament, especially since the adoption of the metrical system of France.

In 1816, in consequence of an address from the House of Commons to the Prince Regent, his Royal Highness appointed a Commission, composed of Sir J. Banks, Sir G. Clerk, Davies Gilbert, Wollaston, Young, and Kater, to consider the subject of English weights and measures; to determine the length of the pendulum vibrating seconds in the latitude of London; and to settle the proportion between the long measures of England and France.

In the first Report, made in June, 1819, no alteration is proposed to be made in the English standards of long measure and of weight, those established by the Committee of 1758 having been found quite accurate.

The second Report, made in 1820, contains the final determination of the Commissioners on the standard of long measure, the length of the second's pendulum, and of that of the metre. The following are the concluding words of this decision:—

" We prefer the Parliamentary standard executed by Mr.

Bird, in 1760, both as being laid down in the most accurate manner, and as the best agreeing with the most extensive comparisons, which have been hitherto executed by various observers, and circulated throughout Europe; and in particular with the scale employed by the late Sir George Shuckburgh.

" We have, therefore, now to propose that this standard be considered as the foundation of all legal weights and measures; and that it be declared that the length of the pendulum vibrating seconds in a vacuum, on the level of the sea, in London, is 39·1393 inches, and that of the French metre, 39·37079 inches, the English standards being employed at 62° of Fahrenheit."

The third Report of the Royal Commission, made in 1821, contains a confirmation of the other two Reports, with respect to measures of length: and as to weights, the Parliamentary troy pound of 1758 is recommended to remain unaltered, and the pound avoirdupois to continue at 7000 troy grains. It is also announced, by new experiments, that a cubic inch of distilled water at 62° is 252·722 grains of the standard pound of 1758, when weighed in a vacuum.

The House of Commons again appointed a Committee in 1821, to which these Reports were submitted: this Committee agreed with the Commissioners, and a bill was introduced in 1823. A petition from the Chamber of Commerce at Glasgow to the House of Lords occasioned an investigation, when Dr. Kelly, as one of the witnesses before the Committee, called attention to certain difficulties and imperfections in the pendulum experiments, and expressed an opinion, supported by few others at the time, though now generally received, that "nature seems to refuse invariable standards; for, as science advances, difficulties are found to multiply, or at least they become more perceptible, and some appear insuperable." The House of Lords adjourned the question over till 1824; when the act 5 Geo. IV. c. 74, was passed. This act came into operation on January 1, 1826, and made no change in the lineal and superficial measures, nor did it alter either the troy or avoirdupois weights previously established; but the measures of capacity underwent considerable change. The old Ale, Wine, and Dry measures were formerly the three authorized measures of capacity. The old Wine gallon contained 231 cubic inches; the Corn gallon, 268·8; and the old Ale gallon, 282. These measures were altered to the Imperial Gallon, containing 277·274 cubic inches.

The act also states that the pendulum vibrating seconds of mean time in the latitude of London in a vacuum at the level of the sea is 39·1393 inches of the standard; and that the cubic inch of distilled water, weighed in air by brass weights, at 62° of Fahrenheit, the barometer being at 30 inches, is equal to 252·458 grains, according to the brass weight or troy pound of 1758, declared to remain the original and genuine standard measure of weight.

Under this system, a gallon of water weighing 70,000 grains, or 10 lbs., it follows that

"A pint of pure water
Weighs a pound and a quarter;"

where, of course, reference is made to imperial measures and the avoirdupois pound.

The Houses of Parliament were burnt in 1834, and it is remarkable that 5 and 6 Wm. IV. c. 63, which passed after the fire, takes no notice of the destruction of the standards, but refers to them as still in existence.

It was, however, fortunate that in the year 1832 the council of the Royal Astronomical Society caused a scale of one yard to be constructed for themselves, and obtained permission of the Speaker of the House of Commons to adjust and compare it with Bird's Imperial Standard of 1760. This was accomplished in 1834 by an extensive set of delicate experiments, ably conducted by Mr. Baily and Lieutenant Murphy (since deceased); and, after the subsequent loss of the Imperial Standard, it may now, perhaps, be referred to as the only measure of length from which the standard scale of Great Britain can be satisfactorily deduced. The Astronomical Society's scale was also compared with the Royal Society's scale of 1742, having two scales in it marked E and Exch.; a scale called Aubert's, the prototype of one which was used in the Indian survey by Lambton; one which had been used by Sir G. Shuckburgh; one belonging to the town of Aberdeen; one belonging to Mr. T. Jones; and four new ones made after the model of the Society's scale—one for the Danish government, one for the Russian government, one for Mr. Baily, and one retained for himself by Mr. Simms the constructor. The middle yard of the Astronomical Society's scale being taken as 36 inches, the various scales, according to the mean of many observations, were found to be as follows:—

Scale.	Standard Portion.	Mean Inches of Ast. Soc. Scale.
Astron. Society	Centre yard	36·000000
Danish	Ditto	35·999758
Russian	Ditto	36·000050
Simms's	Ditto	35·999903
Baily's	Ditto	35·999949
Aberdeen	Ditto	35·998615
Jones's	Ditto	35·999802
Aubert's	0 in. — 36 in.	35·998447
Shuckburgh	0 in. — 36 in.	36·000185
Ditto	10 in. — 46 in.	35·999921
Royal Society	Line "E."	36·001473
Ditto	Line "Exch."	35·993684
Imperial standard of Bird's, of 1760, afterwards destroyed in 1834 by the burning of the Houses of Parliament		35 999624

Thus the Astronomical Society's standard being ·000376 longer than the Imperial standard, and the standard temperature being 62°, it follows that the length of the former standard, observed at 62°, and diminished by ·000376 of an inch, will give the true standard of the law.

In the year 1838, the Government appointed a Commission "to consider the steps to be taken for the Restoration of the Standards of Weight and Measure, which had been destroyed and lost by the burning of the Houses of Parliament." The members of this commission were Messrs. Airy, Baily, Bethune, Davies Gilbert, Herschel, Lefevre, Lubbock, Peacock, and Sheepshanks; and the following is extracted from their valuable report made in 1841:—

"We are of opinion that the definition contained in the act 5 Geo. IV. c. 74, ss. 1 and 4, by which the standard yard and pound are declared to be respectively a certain brass rod and a certain brass weight therein specified, is the best which it is possible to adopt.

"Since the passing of the said act, it has been ascertained that several elements of reduction of the pendulum experiments therein referred to, are doubtful or erroneous. Thus, the reduction to the level of the sea was doubtful, the reduction for the weight of air was erroneous, the specific gravity of the pendulum was erroneously stated, the faults of the agate plates introduced some degree of doubt, and sensible errors were introduced in the operation of comparing the length of the pendulum with Shuckburgh's scale, used as the representative of the legal standard. It is evident, therefore, that the course prescribed by the act would not necessarily reproduce the length of the original yard. It appears also, that the determination of the weight of a cubic inch of distilled water is yet doubtful, the greatest difference among the best English, French, Austrian, Swedish, and Russian determinations being about $\frac{1}{1200}$ of the whole weight, whereas the mere operation of weighing may be performed to the accuracy of $\frac{1}{1000000}$ of the whole weight.

"Several measures, however, exist, which were most accurately compared with the former standard yard (in particular the Royal Astronomical Society's Scale, described in their Memoirs, vol. ix., and the iron bars belonging to the Board of Ordnance, in the custody of Colonel Colby); and several metallic weights exist which were most accurately compared with the former standard pound: and by the use of these the values of the original standards can be respectively restored without sensible error. And we are fully persuaded that, with reasonable precautions, it will always be possible to provide for the accurate restoration of standards, by means of material copies which have been carefully compared with them, more securely than by reference to any experiments referring to natural constants.

"From the evidence of persons best able to judge of the comparative use of troy weight and avoirdupois weight, the proportion does not exceed one set of troy weights to many thousand sets of avoirdupois. The statements of medical men and those of persons concerned in the trade of bullion, show that even to them the troy pound is useless. The avoirdupois pound, on the other hand, is universally known through this kingdom; and moreover, being now made equal to 7000 grains, it is well adapted to subdivision by the decimal scale,—an object which we think ought never to be placed out of view in considering the changes (in other respects producing no inconvenience) which may be made in the weights and measures of the country. We feel it our duty, therefore, to recommend that the *avoirdupois* pound be adopted instead of the *troy* pound as the standard of weight. With regard to the standard of length, we do not feel the necessity of proposing any change.

"That two modes of estimating weight should coexist, is undoubtedly an evil; its bad effects are greatly increased by the identity of the names used in the different scales for describing weights of very different values. Thus we have the *pound* in the avoirdupois scale, and the *pound* in the troy or apothecaries' scale, the former being *greater* than the latter in the proportion of 7000 to 5760; we have the *ounce* in the avoirdupois scale, and the *ounce* in the troy or apothecaries' scale, the former being *less* than the latter in the proportion of 7000 to 7680; we have the *dram* in the avoirdupois scale, and the *drachm* in the apothecaries' scale, the former being *less* than the latter in the proportion of 7000 to 15,360. In examining into the actual uses of these several denominations, we see at once that it is impossible to abrogate the avoirdupois pound and ounce, which are used so extensively by persons of every class, that they must be considered as being emphatically the British weights. The avoirdupois dram, however, does not appear to be used at all. The troy pound, as we have already mentioned, appears to be wholly useless: it is not used in contracts for gold and silver, or in medical prescriptions; and we are not aware of any obstacle whatever to its entire abolition, except the existence of certain printed tables for the reduction of assays of the precious metals, in which the denominations of the larger weights are expressed by multiples of troy pounds. We propose, with the view of removing the confusion caused by the existence of this pound, and at the same time of respecting the private interests which (though to a very inconsiderable extent) are concerned in the change, that the Government should compute and print, for the use of bullion merchants and assayers, a new edition of these tables, in which the larger weights shall be expressed by decimal multiples of the troy ounce, to the entire exclusion of the troy pound. We are inclined to think that the troy ounce itself could not be abolished at present without some difficulty; we think it right, however, that the persons using it should be imperatively required to describe it in such a manner that no confusion with the avoirdupois ounce can possibly occur. The remaining weights of the troy and apothecaries' scales may, for the present, be tolerated (for certain substances only), as leading to no ambiguity. Still we think it desirable, that measures should now be taken which may ultimately tend to the removal of the troy scale; and remarking both the convenience of a decimal scale of subdivision of the avoirdupois pound, and the general willingness of bullion merchants to adopt a decimal scale, and remarking also, that by descending in such a scale we arrive at a small weight (7 grains), bearing a simple relation to the grain on which the troy weight is based, we propose that the Government should use its influence for the introduction of such decimal scale.

"We beg leave to invite the attention of the Government to the advantage and the facility of establishing in this country a decimal system of coinage. In our opinion, no single change, which it is in the power of a Government to effect in our monetary system, would be felt by all classes as equally beneficial with this, when the temporary inconveniences attending the change had passed away. The facility consists in the ease of interposing between the sovereign (or pound) and the shilling, a new coin equivalent to two shillings (to be called by a distinctive name); of considering the farthing (which now passes as the $\frac{1}{960}$th part of the pound) as the $\frac{1}{1000}$th part of that unit; of establishing a coin of value equal to $\frac{1}{100}$th part of the pound; and of circulating, besides the principal members of a decimal coinage, other coins of values bearing a simple relation to them, including coins of the same value as the present shilling and sixpence. We do not feel ourselves at liberty further to enter into this subject; but we have felt it imperative on us to advert to it, because no circumstance whatever would contribute so much to the introduction of decimal scale in weights and measures, in those respects in which it is really useful, as the establishment of a decimal coinage."

The report contains also the following recommendations:—
"That the standard of length be defined by the whole length of a certain piece of metal or other durable substance, supported in a certain manner, at a certain temperature; or, by the distance between two points or lines engraved upon the surface of a certain piece of metal or other durable substance, supported in a certain manner, at a certain temperature: but that the standard be in no way defined by reference to any natural basis, such as the length of a degree of meridian on the earth's surface in an assigned latitude, or the length of the pendulum vibrating seconds in a specified place.

"That the length of the new Parliamentary standard be one yard; there appearing no sufficient reason for departing from the length hitherto adopted for the standard.

"That the name *milyard*, or some other to be fixed by act of Parliament, be recognized as describing the measure of 1000 yards, without the necessity of further definition.

"That the standard of weight be defined by a certain piece of metal or other durable substance.

"That the standard of capacity be defined by the capacity which, under certain circumstances of the barometer and thermometer, contains a certain weight of distilled water, but that it be in no way defined by reference to the standard of length. That, nevertheless, the contents of the standard of capacity, as expressed in units and fractions of the cubical measure dependent on the standard of length, be stated by way of recital, as the best determination made by scientific men which has come to the knowledge of the Legislature, and as permitted for use when it is impracticable to refer to trial by the weight of distilled water.

"That no standard of capacity be constructed; the definition of the gallon as 'the capacity which contains 10 pounds' weight of distilled water weighed in air at the temperature of 62° Fahrenheit, the barometer being at 30 inches,' as specified in the act 5 Geo. IV., being still retained.

"That where it shall be impracticable to ascertain the contents of any vessel by the weight of distilled water which it contains, or by pouring water into it from a standard measure, it be permitted to ascertain the contents by gauging (the gallon being assumed to be 277·274 cubic inches), or by pouring seed into it from a standard measure.

"That a sufficient number of weights of multiples of grains be constructed (we would recommend 10 sets of 10, 100 and 1000 grains); and that their relative errors be found by comparison among themselves, and their absolute errors by comparison with the copies of the old troy pound of 5760 grains.

"That by the use of these, a platinum weight of 7000 grains be constructed; and that this be declared the Parliamentary standard of weight, by the name of the pound weight; the distinctive word 'avoirdupois' being hereafter omitted.

"That the avoirdupois dram be no longer recognized in any contract.

"That the troy pound be no longer recognized; and that the word 'pound,' or any letters or symbols commonly used to denote the pound, as applied to a weight, be always interpreted to mean the pound of 7000 grains (formerly called the avoirdupois pound).

"That the word 'ounce' be always interpreted to mean $\frac{1}{16}$th part of the pound, except it be described as the 'troy ounce.'

"That the name *millet*, or some other to be fixed by act of Parliament, be recognized as describing the thousandth part of the pound, without the necessity of further definition.

"That the only legal weights above one pound, be weights of multiples of 1 pound not exceeding 10 pounds; and weights of 10 pounds and its multiples, not exceeding 100 pounds.

"That the name *centner*, or some other to be fixed by act of Parliament, be recognized as describing the weight of 100 pounds, without the necessity of further definition.

"That the Exchequer standards of weight be 1 lb. and several multiples thereof, not exceeding 10 lbs.; 10 lbs. and several multiples thereof, not exceeding 100 lbs.; 100 lbs.; but no weight of 14 lbs. or any multiple of 14 lbs. except 70 lbs. Also the tenth, hundredth, and thousandth part of the pound, and several multiples of them. Also 1 troy ounce, 10 troy ounces, 100 troy ounces, and several multiples of each; 1000 troy ounces; but no weight of 12 troy ounces or any of its multiples (except those included in the decimal scale above described). Also 1 pennyweight, and several multiples of it. Also 1 grain, 10 grains, 100 grains, 1000 grains, 10,000 grains, and several multiples of each.

"That the Exchequer standards of weight be used in the trial of weights brought for examination in the same manner as at present; and that no greater error than $\frac{1}{30,000}$th part of the quantity weighed be tolerated.

"We would recommend that the influence of the Government be employed to introduce the use of the decimal subdivisions of the acre; of which the first step is actually given by the square land-chain, and the others are contained in the numerical expression obtained in the first multiplication for finding the area of a piece of ground.

"Before leaving the subject of length-measures, we beg strongly to call the attention of Government to the importance of encouraging the use of the decimal scale, and especially of sanctioning its use where custom has already adopted it. We beg particularly to indicate the decimal subdivision of the foot (which is even now engraved on foot rules and levelling staves), as one extensively used in the practice of engineers, and one which we would recommend for the recognition of Government in every case."

Although the preceding recommendations of the Commission have not been carried out they are here inserted, as the subject is very important and worthy of being generally understood.

It is probable that the varieties of gallons arose from the varieties of pounds, since the original definition of the gallon depended upon the pound, and there is a close relation not only between the old gallons and the weights, but even between the different versions of the old gallons and the weights. There was a gallon of 282 cubic inches in the Exchequer as a standard; there was one of $272\frac{1}{4}$ inches in common use; there was one of 231 inches in common use; and there was one of 224 inches in the Guildhall. Now 282 and 232 are, as near as integers can represent it, in the proportion of the pound avoirdupois to the pound troy; and $272\frac{1}{4}$ and 224 are as nearly in the same proportion. It is unlikely that this should have been accidental.

The imperial weights and measures now in use are fixed by the act 5 Geo. IV. c. 74, of which the following is an abstract.

Abstract of an Act of Parliament, 5 Geo. IV. c. 74, passed June 17, 1824, "for ascertaining and establishing Uniformity of Weights and Measures," which came into operation on the 1st of January, 1826.

This is an act declaratory of the accuracy and legality of the existing standards, both of long measure and weight; but it orders the abolition of all measures of capacity for wine, ale, corn, coals, &c., and the establishment of one only in their stead, which is to be called "Imperial Measure."

1. The standard yard is declared to be the distance between the centres of the two points on the gold studs in the straight brass rod now in the custody of the Clerk of the House of Commons, whereon is engraved "Standard Yard, 1760," the brass being at the temperature of 62° by Fahrenheit's thermometer. It is to be called "the Imperial Standard Yard."

2. The dimensions for measuring land are unaltered: they are the statute measure, of which the acre contains 4840 square yards.

3. The yard, if lost, defaced, or otherwise injured, may be restored by comparing it with the pendulum vibrating seconds of mean time, in the latitude of London, in a vacuum on the level of the sea, the yard being in the proportion of 36 inches to 39·1393 of the pendulum.

4. The standard pound is declared to be the standard brass weight of one pound troy weight made in the year 1758, and now in the custody of the Clerk of the House of Commons, and it is denominated "the Imperial Troy Pound."

5. If the imperial pound be lost, defaced, or otherwise

injured, it shall be restored by comparison with a cubic inch of distilled water, weighed in air by brass weights, at the temperature of 62° of Fahrenheit's thermometer, the barometer being at 30 inches. Such cubic inch of water is equal to 252·458 grains, the standard troy pound being 5760 such grains; and the avoirdupois pound 7000 such grains troy. All operations of restoring or correcting standards are to be made under the directions of the Lord High Treasurer, or the Commissioners of His Majesty's Treasury, or any three of them for the time being.

6. The standard measure of capacity, as well for liquids as for dry goods, not measured by heaped * measure, shall be the gallon, containing 10 lbs. avoirdupois weight of distilled water, weighed in air at the temperature of 62° of Fahrenheit's thermometer, the barometer being at 30 inches; and such brass measure shall be "the Imperial Standard Gallon," and is declared to be the unit and only standard measure of capacity from which all other measures of capacity for all sorts of liquids, as well as for dry goods not measured by heaped * measure, shall be derived; and that all measures shall be taken in parts and multiples, the quart, pint, peck, bushel, and quarter continuing in the same proportion as heretofore for dry measure.

7. That the standard measure of capacity for coals, culm, lime, fish, potatoes, or fruit, and all other goods and things commonly sold by heaped * measure, shall be the imperial bushel, containing 80 lbs. avoirdupois of water as aforesaid; the same being made round, with a plain and even bottom, and being 19½ inches from outside to outside.

8. That coals and other goods sold by heaped * measure shall be duly heaped up in the said bushel in the form of a cone, such cone to be of the height of at least 6 inches; and the outside of the bushel to be the extremity of the base of

---

* Heaped measures have been abolished since 1st Jan. 1835 (4, 5, and 6 Will. IV. c. 49 and 63).

such cone; and that three bushels shall be a sack, and twelve such sacks a chaldron.

9. That for articles not sold by heaped * measure, such as corn, pulse, &c., the same shall be stricken with a round stick or roller, straight, and of the same diameter from end to end.

10. That this law of imperial measure is not to extend to Ireland for any articles hitherto sold by weight.

11. That copies and models of the standard of length, weight, and measure aforesaid, are to be made and verified within three months after passing the act, under the direction of the Lords of the Treasury; and that such copies or models shall be deposited in the office of the Chamberlain of the Exchequer at Westminster; and that copies shall be sent to the Lord Mayor of London, and the Chief Magistrate of Edinburgh and of Dublin, and to such other places or persons as the Lord High Treasurer or Commissioners of the Treasury may from time to time direct.

12. That His Majesty's justices of the peace, in every county of the British Empire, or every town or place, being a county within itself, shall, within six months after passing the act, purchase a model of each of the standards aforesaid, with their parts and multiples; and that such shall be compared and verified with the models deposited at the Exchequer, on payment of the usual fees; and that such verified copies shall be placed for custody and inspection with such persons as the magistrates shall choose to appoint; and that the same shall be produced by the keepers thereof, upon reasonable notice, the persons requiring such production paying the customary charges for the same.

13. The expenses of procuring models for magistrates, counties, &c., are to be raised by the usual modes of taxation.

14. That when reference cannot be easily had to verified copies of the standard measures of capacity, it may be lawful for any justice of the peace, or magistrate having jurisdiction, to ascertain the content of a measure of capacity, by direct

reference to the weight of pure or rain water which such measure is capable of containing; 10 lbs. avoirdupois weight of such water, at the temperature of 62° by Fahrenheit's thermometer, being the standard gallon ascertained by the act, the same being in bulk equal to 277·274 cubic inches.

15. That all contracts for sale, &c., by weight or measure, shall be according to the imperial standard, when no special agreement shall be made to the contrary; and in all cases where any special agreement shall be made, with reference to any weight or measure established by local custom, the proportion which every such local weight or measure shall bear to any of the said standard weights and measures, shall be expressly declared and specified, or otherwise such agreement shall be null and void.

16. That existing measures may be used, being marked so as to show the proportions which they have to the imperial measures; but that after the 1st of May, 1825, no person shall be permitted to make any weights or measures, otherwise than according to the provisions of the new act.

17. That for ascertaining rents, &c., payable in grain or malt in England or Ireland, the amount is to be ascertained according to the standard by this act established, by a jury of 12 substantial freeholders.

18. That for ascertaining rents, &c., payable in grain or malt in Scotland, such rents shall be determined according to the new standard, by such juries as strike the fiar prices of grain.

19. That tables of equalization shall be made and constructed under the Commissioners of the Treasury, showing the proportions between the weights and measures heretofore in use and those now established.

20. That tables shall be also constructed for the collection of the customs and excise, under the direction of the said Commissioners of the Treasury.

21. The present act may be enforced in England and

Scotland by all the regulations and penalties contained in the following statutes, except such parts of the said statutes as may be repealed or altered by this present act, viz. 29 Geo. II. c. 25; 31 Geo. II. c. 17; 35 Geo. III. c. 102; 55 Geo. III. c. 43.

22. The present act may be enforced in Ireland by all the regulations and penalties contained in the following statutes, except such parts of the said statutes as are repealed or altered by this act, viz. 4 Anne (*I.*); 11 Geo. II. (*I.*); 25 Geo. II. (*I.*); 27 Geo. III. (*I.*); 28 Geo. III. (*I.*)

23. The repeal of numerous laws is declared in this article; some of uncertain date before the reign of Edward the Third, and many since that period. These are chiefly statutes which fixed the weight and measure of certain kinds of goods, such as wool, cheese, salt, wine, beer, fish, fruit, &c.; and also the denominations which determine their quantity, as the sack, wey, load, tun, hogshead, barrel, &c. For the particulars of these statutes (which are now repealed either wholly or in part), recourse must be had to the originals, as referred to in the margin of the present act, and which amount to about 60 statutes.

24. That this act shall not extend to affect or alter the rights of the Dean and High Steward of Westminster, to appoint proper officers to sign and seal all weights and measures used in the said city and the liberties thereof.

25. That gaugeable liquors brought into the port of the city of London shall be gauged as heretofore by the Lord Mayor or his deputies; but the contents shall be ascertained by the standard measure directed by this act.

26. This act shall not extend to prohibit or diminish the right of the Lord Mayor and Commonalty of the city of London, concerning the office of gauger of any gaugeable liquors imported within the city of London, or the liberties thereof.

Numerical relations appertaining to the foregoing Act for equalizing Measures.

Weight of a cubic inch of distilled water, in a vacuum at the temperature 62° = 252·722 grains.
Consequently, a cubic foot = 62·3862 lbs. avoirdupois.
Weight of a cubic inch of distilled water in air at 62° of temperature with a mean height of the barometer = 252·458 grains.
Consequently, a cubic foot = 62·3206 lbs. avoirdupois.
And an ounce of water = 1·73298 cubic inches.
Cubic inches in the imperial gallon = 277·274.
Diameter of the cylinder, containing a gallon at one inch in depth = 18·78933 inches.

Specific Gravity of Water at different temperatures, that at 62° being taken as unity.

70°	0·99913	61°	1·00010	52°	1·00076	43°	1·00109
69	0·99925	60	1·00019	51	1·00082	42	1·00111
68	0·99936	59	1·00027	50	1·00087	41	1·00112
67	0·99947	58	1·00035	49	1·00091	40	1·00113
66	0·99958	57	1·00043	48	1·00095	39	1·00113
65	0·99969	56	1·00050	47	1·00099	38	1·00113
64	0·99980	55	1·00057	46	1·00102	37	1·00112
63	0·99990	54	1·00064	45	1·00105	36	1·00111
62	1·00000	53	1·00070	44	1·00107	35	1·00109

The difference of temperatures between 62° and 39°, where water attains its greatest density, will vary the bulk of a gallon of water rather less than the third of a cubic inch.

And assuming from the mean of numerous estimates the expansion of brass 0·00001044 for each degree of Fahrenheit's thermometer, the difference of temperatures from 62° to 39° will vary the content of a brass gallon measure just one-fifth of a cubic inch.

It appears that the specific gravity of clear water from the Thames exceeds that of distilled water at the mean temperature, in the proportion of 1·0006 to 1, making a difference of about one-sixth of a cubic inch on a gallon.

Rain water does not differ from distilled water, so as to require any allowance for common purposes.

Comparisons of Old and New Measures.

The foregoing calculation of the diameter of a cylinder, which contains 1 gallon for every inch in depth, will be found useful in constructing both corn and coal bushels on the new plan of imperial measure.

Thus the corn bushel, with the diameter 18·78933, and 8 inches deep, will answer to 2218·192 cubic inches, the imperial bushel; being about $\frac{1}{32}$ part more than the Winchester bushel, which is 2150·42 cubic inches.

The new coal bushel, with the above diameter and depth, and heaped as directed in Art. 8, the rim being about $\frac{3}{5}$ of an inch thick, and the diameter $19\frac{1}{2}$ inches from outside to outside, will measure 2816·459 cubic inches, which is only $1\frac{1}{4}$ inch more than the present coal bushel, viz. 2814·9 cubic inches.

The proportion between the old and new wine measures is very nearly as 5 to 6. Thus 5 imperial gallons equal 6 wine gallons and about $\frac{1}{4500}$ of a gallon over.

The proportion between the old and new ale measures is about as 60 to 59.

The following table will show the relative contents more accurately, both in measure and in weight, the latter having been computed according to the principles stated in the act.

Table showing the Contents of the different Gallons, both in Measure and Weight.

	Cubic Inches.	Avoirdupois Weight.			Troy Weight.			
		lb.	oz.	dr.	lb.	oz.	dwt.	gr.
Imperial Gallon . .	277·274	10	0	0	12	1	16	16
Corn Gallon . . .	268·8	9	10	$1\frac{3}{4}$	11	9	7	12
Wine Gallon . . .	231	8	5	$6\frac{1}{4}$	10	1	9	22
Ale Gallon . . .	282	10	2	$11\frac{1}{2}$	12	4	6	8

These results will be found useful in comparing different vessels where gauging cannot be relied on; and although they are computed according to the conditions of temperature, &c., as stated in Art. 6, yet the proportions will answer with sufficient correctness for all common purposes of business, with any kind of fresh water; but where great accuracy is required, it may be determined at any other temperature by means of the table given on page 19. Thus to find the weight of the wine gallon at 56° Fahrenheit, multiply its weight at 62°, viz. 8·3311 by 1·0005, and the result will be 8·3353, the required weight.

The reduction of the different measures may be easily computed by means of *factors* or multipliers, given in the following table:—

Table of Factors, for converting Old and New Measures.

	By Decimals.			By Fractions.		
	Corn Measure.	Wine Measure.	Ale Measure.	Corn Measure.	Wine Measure.	Ale Measure.
To convert Old Measures into New .	·96943	·83311	1·01704	$\frac{33}{34}$	$\frac{5}{6}$	$\frac{60}{59}$
To convert New Measures into Old .	1·03153	1·20032	·98324	$\frac{33}{32}$	$\frac{6}{5}$	$\frac{59}{60}$

EXAMPLE I.—Reduce 63 gallons wine measure to imperial measure.

$63 \times ·83311 = 52·486$; or $63 \times \frac{5}{6} = 52\frac{1}{2}$ imperial gallons nearly.

EXAMPLE II.—Reduce 8 bushels imperial measure to Winchester measure.

$8 \times 1·03153 = 8·25224$; or $8 \times \frac{33}{32} = 8\frac{1}{4}$ Winchester bushels nearly.

It should be observed, that the computations by the fractions are only approximative, but they will be found sufficiently correct for most purposes.

In 1870 the Standards Commission, in connection with the Standard Weights and Measures Department of the Board of Trade, made the following important recommendation in their Fourth Report:—

"That mural or public standards of length be securely fixed for public use in all populous towns, and the expenses connected with them be defrayed from the local funds. The local authorities to be responsible for keeping them in good order, and the local inspectors to report to them periodically on their condition. These public standards not to be valid for convictions of false or defective measures."

In pursuance of this recommendation, measures of 100 feet, and of the chain of 66 feet, with their subdivisions, were laid down on the north side of Trafalgar Square, and these were legalised as secondary standards of length by Her Majesty's Order in Council, dated 27th June, 1876, wherein it is pointed out that by the 1st section of "The Standards of Weights, Measures, and Coinage Act, 1866," the custody of the imperial standards of length and of weight, &c., was transferred to the Board of Trade. The following condensed account, appertaining to the laying down of these secondary standards, has been extracted from an Appendix to the Annual Report, 1875-76, of H. W. Chisholm, Warden of the Standards.

The exhibition of public and mural standards of length in populous places throughout the country, was considered by the Commission for Restoration of the Standards to be one of the best auxiliary means for attaining and maintaining uniformity in the measures of length used by the public, and was recommended by the Commission in their Reports of 1841 and 1854. Similar recommendations were made by the Standards Commission appointed in 1867; and the mode in which they proposed such recommendations should be carried out by the Standards Department in laying down public standards of the chains of 100 feet and of 66 feet, and erecting public standards of the yard and its principal parts.

In selecting a site for such public standards, the Commission expressed their opinion that they should be placed—

Where they are easily accessible to the public who may wish to compare their measures;

Where such comparisons may be made without being disturbed by passers or idle gazers;

Where sufficient protection from wilful damage may at all times be secured;

And where there is a long and level plot of ground upon which the longer standard measures may be laid down, with an adjacent wall for the mural standards.

It was not until the spring of 1875 that the site of the north side of Trafalgar Square was obtained from the Office of Works, through the intervention of the Astronomer Royal. A working plan for the construction of these public standards was laid before the Standards Commission by the Astronomer Royal, which had been prepared by his son, Mr. Wilfrid Airy, an able practical civil engineer, whose valuable services were retained for making the working drawings, as well as for superintending the work of constructing the platform and laying down the standards. Mr. W. Airy also gave his assistance to the officers of the Standards Department in the work of comparing the standard chains when laid down, and in superintending the subsequent adjustment of some of the bronze blocks on which the defining lines were cut. All the blocks for defining the several measures are of bronze. Those inlaid in the granite are slightly conical, and are secured by cement in somewhat larger conical holes in the granite. The mural standard measures are also of bronze, securely fastened by screws to a solid frame. This frame, and those containing the descriptive tablets, are of bronze, and are secured to the granite wall by screws, the heads of which have since been removed to the level of the adjoining metal.

The mode in which the metal blocks bearing the defining lines of the standard measure of 100 feet were laid down, was by first marking out the several lengths by means of 10 of the pine 10-feet measuring-rods of the Standards Department, having subdivisions of 1 foot each marked upon them. Square holes were next cut in the granite in a marked-out straight line to receive the metal blocks, and were made sufficiently large to admit of a slight adjustment of the blocks in the direction of the length of the measure. When all was prepared, the square metal blocks for the 100-feet chain were placed in the several holes, and were embedded in cement, being secured on each side by an iron wedge. Each length was then carefully measured with the pine measuring-rods, and every requisite adjustment made, after which the surface of the cement was levelled and finished off.

A similar course was pursued with the whole length of the 66-feet chain, and with half of the whole length, the measurements being made with six of the 10-feet pine measuring-rods, and with two of the similar 3-feet measuring-rods of the Standards Department. The intermediate lengths of 10 links, each 7·92 inches, were laid down by using two 10-link measuring-rods of Messrs. Troughton & Simms, the intervals of each half-chain being adjusted so as to be as nearly equal as possible.

The definitive comparisons of the two measures of length were made with the two standard 10-feet measuring-bars, and the 6-feet bar constructed expressly for verifying continuous length measurements of this kind.

The errors of these two standard bars at 62° Fahr. have been determined, as follows:—

Standard 10-feet end-bar No. 1 . . . ·006739 in excess.
" " " No. 2 . . . ·005919 "
The two bars . . . . . . ·012658 "
And 100 feet measured with them . . ·063290 "

The standard 6-feet steel end-bar is similar in its construction to the two 10-feet standard bars. Its standard length at 62° F. has been determined to be ·00691 inch in excess.

The mean results of two series of minute comparisons may be taken as the determined standard length at 62° F. of the several 10-feet subdivisions of the standard 100-feet measure from the zero block, viz. :—

Ascertained Errors of the several 10-feet Subdivisions of the Standard 100-feet Measure from the Zero Block, measured to the middle of the defining line :—

0—10 feet − ·007 inch.	0—60 feet + ·007 inch.
20 ,, − ·019 ,,	70 ,, + ·011 ,,
30 ,, − ·022 ,,	80 ,, + ·021 ,,
40 ,, − .015 ,,	90 ,, + ·017 ,,
50 ,, − ·008 ,,	100 ,, − ·008 ,,

All these errors are within the breadth of the several defining lines.

The surveyors' chain of 66 feet is divided into 100 links, each being 7·92 inches in length. The standard 66 feet now laid down is divided into 10 lengths of 10 links each. The only practical mode of verifying all these lengths was to compare the whole distance of 66 feet with the standard 10-feet steel bars carried successively along it for 60 feet, and completing the measurement with the standard 6-feet bar. The mean result of two series of minute observations may be taken as the determined standard length, viz.—

Ascertained error of standard chain of 66 feet at 62° F. = − ·019 inch.

A comparison was also made of each of the 10-link divisions of the 66-feet chain, with the two pine measuring-rods of 10 links each of Mr. Simms, in order to test any relative difference between them, and in this manner to verify the length of each, with reference to the relation of the sum of the 10 links to the whole length of the 66-feet chain. It was thus ascertained that the whole distance of 66 feet was accurately divided into 10 equal parts of 10 links each, within half the breadth of the defining lines.

The several mural standards of length, consisting of bed measures of the imperial yard of 3 feet, of 2 feet, of 1 foot, and of the line divisions of these measures, including those of 1 inch divided into tenths, have been accurately compared, and found to have no sensible error at the normal temperature of 62° F.

In France the new Metric System of measures, now completely established, was introduced in 1799, and is called metrical, as derived from the measurement of the earth. Its fundamental measure, the *metre*, is presumed to be the ten-millionth part of a meridian line drawn from the pole to the equator, and is 39·37079 English inches. Taking the one-hundredth of the metre, we obtain the *centimetre;* and for the standard of weight the *gramme* is a cubic centimetre of distilled water at the temperature of maximum density, the same being ·0022054 of an English avoirdupois pound, or 15·438 English grains. All the multiples and subdivisions of the current coins, as well as of every measure and weight, are decimal, full details of which will be found under the article "France." The advantage of such a system, when once established, is so great, that all who are capable of appreciating its merits look forward with interest to the introduction of a similar one into Great Britain. It has already been adopted in Austria, Belgium, Greece, Holland, Germany, Italy, Norway, Portugal, Spain, Sweden, and Switzerland.

The Weights and Measures Act, 1878 (41 & 42 Vict. cap. 49), came into operation on the 1st of January, 1879. By it all previous Acts are for the most part repealed. It is enacted,—

That the Board of Trade shall have all such powers and perform all such duties relative to standards of measure and weight, and to weights and measures, as are by any Act or otherwise vested in or imposed on the Treasury, or the

Comptroller-General of the Exchequer, or the Warden of the Standards. It shall be the duty of the Board of Trade to conduct all such comparisons, verifications, and other operations with reference to standards of measure and weight, in aid of scientific researches, or otherwise, as the Board of Trade from time to time thinks expedient, and to make from time to time a report to Parliament on their proceedings and business under this Act.

The Board of Trade shall cause to be compared with the Board of Trade standards all copies of any of those standards which are submitted for the purpose by any local authority, and have been used, or are intended to be used, as local standards, and, if approved, shall cause them to be stamped as verified in such manner as to show the date of such verification.

A contract or dealing shall not be invalid on the ground that the weights or measures used therein are weights or measures of the metric system, or on the ground that decimal subdivisions of imperial weights and measures, whether metric or otherwise, are used.

The Board of Trade also possess accurate copies of metric standards, and may, if they think fit, cause to be compared with the metric standards in their custody, and verified, all metric weights and measures, which are submitted to them for the purpose, and are of such construction as may be from time to time directed, and which the Board of Trade are satisfied are intended to be used for the purpose of science or of manufacture, and not for the purpose of trade within the meaning of this Act.

The following is a list of the standards in the custody of the Board of Trade at the passing of this Act:—

Measures of Length.

Feet: 100, 66, 16½, 10, 6, 5, 4, 3, 2, 1.

And 1 inch divided into $\begin{cases} 12 \text{ duodecimal} \\ 10 \text{ decimal} \\ 16 \text{ binary} \end{cases}$ equal parts.

STANDARDS.

Measures of Capacity.
Bushel, half-bushel, peck,
Gallon, half-gallon,
Quart, pint, half-pint,
Gill, half-gill, quarter-gill.

Measures used in the Sale of Drugs.
Fluid ounces: 4, 3, 2, 1.
„ drachms: 4, 3, 2, 1.
Minims: 30, 20, 10, 5, 4, 3, 2, 1.

Avoirdupois Weights.
Pounds: 56, 28, 14, 7, 4, 2, 1. Ounces: 8, 4, 2, 1.
Drams: 8, 4, 2, 1, ½. Grains: 240, 120, 72, 48, 24.

Troy Bullion Weights.
Ounces: 500, 400, 300, 200, 100, 50, 40, 30, 20, 10, 5, 4, 3, 2, 1.
Decimals: ·5, ·4, ·3, ·2, ·1, ·05, ·04, ·03, ·02, ·01, ·005, ·004, ·003, ·002, 001.

Decimal Grain Weights.
Grains: 4,000, 2,000, 1,000, 500, 300, 200, 100, 50, 30, 20, 10, 5, 3, 2, 1, ·5, ·3, ·2, ·1, ·05, ·03, ·02, ·01.

Metric Measures and Weights.
1. Length.   Metres: 2, 1, ·1, ·01, ·001.
2. Weights. Kilograms: 20, 10, 5, 2, 1.
   Grams: 500, 200, 100, 50, 20, 10, 5, 2, 1.
   Decigrams: 5, 2, 1.   Milligrams: 5, 2, 1, ·05.
3. Capacity. Litres: 20, 10, 5, 2, 1, ·5, ·2, ·1, ·05, ·02, ·01, ·005, ·002, ·001.

COINS.

Coins are pieces of metal, mostly of a round and flat shape, stamped by authority with certain impressions, which are intended to give them a legal and current value, and also to serve as a guarantee for their weight and fineness.

Gold and silver are the principal metals of which coins are made, being found the fittest for that purpose, both on account of their qualities and their scarcity.

Copper and billon are likewise used, but always for coins of inferior value.

The proportional value of gold and silver is variable; for

although they are generally considered as equivalents of other property, and standard measures of value by which commodities are bought, sold, and estimated, yet, being themselves also saleable articles, they are liable to constant fluctuation in price, as exchanged for each other, as well as with respect to all property.

Pure gold and silver are invariable in their qualities, from whatever mines they are produced. In their fine state they are considered too flexible to make coins fit for general wear; and hence the practice of mixing with them a certain proportion of harder metal, which is called *alloy*.

The coinage of William the Conqueror was after the plan established by Charlemagne, in France, in the eighth century, and is supposed to be derived from the Romans, with respect to dividing the pound into 20 shillings and the shilling into 12 pence.

The Saxon pound weight was adopted by King William, and was called the moneyer's pound; and from it 20 shillings were coined, which made $21\frac{1}{3}$ to the pound troy. This number was increased in succeeding reigns until the year 1665 (18th Charles II.), at which time it was settled at 62 shillings, and so continued until the year 1816, when it was altered to 66 shillings, its present rate.

In the early coinages the silver penny or sterling was minted with a deep cross. When it was broken into two parts, each was called a *halfpenny*, and when into four, each part was called a *fourth-thing*, or farthing. Larger silver pieces of fourpence were also coined, which were called *greats*, or *groats*, and also *grosses*. There were besides, silver halfpence and farthings minted; but no shillings until the reign of Henry VII. (1504), nor copper coins until the reign of Charles II. (1665.)

As to gold coins, the first after the Norman Conquest, according to Snelling, was struck by order of Henry III. in the year 1257. It weighed two silver pence, passed for twenty pence, and was called the *gold pennie*. The same author observes, "that the king tried this expedient of coining gold through necessity, and that the city of London made a representation against this measure."

The next gold coinage in England was in the year 1344, when the florin was struck, which took its name from *Florence*, where it had been first minted in 1252. It was afterwards coined in most countries of Europe. In Germany and Holland it was called the *gulden*, on account of its having been originally gold. The florin, however, has been long a silver coin, and also a money of account.

The above coins are supposed to have been of pure gold; but those minted in the subsequent reigns down to that of Henry VIII. were 23 carats $3\frac{1}{2}$ grains fine, with $\frac{1}{2}$ grain of alloy. This was called the old standard to distinguish it from the new or the present standard, which was first called *crown gold*, as being minted into crown pieces in 1527.

The principal gold coins of the old standard were nobles of 6s. 8d. each, with halves and quarters: the latter were called *farthing nobles*. There were also marks of 13s. 4d., angels of 10 shillings, and sovereigns of 20 shillings each. Sovereigns were first minted by Henry VII., and were frequently altered during the four subsequent reigns; but in the 2nd of James I. they were fixed at 22 carats, at which fineness all gold coins have since been minted. The 20 shilling pieces first coined at this rate were called *unites*, and $33\frac{1}{2}$ pieces were struck from the pound troy; but in the reign of Charles II. (1666) a new coinage of $44\frac{1}{2}$ to the lb. was minted, and these were called *guineas*, on account of the country from which the gold was originally brought. The guinea varied in its current price from 20 shillings up to 30, until the year 1717, when, by the recommendation of Sir Isaac Newton, it was fixed at 21 shillings, its present rate.

The system of both metals being standard measures of value, which they were in virtue of each being a legal tender to any amount, was the source of much disorder; for as their market prices were always subject to variation, one kind of coin had a constant tendency to drive the other out of circulation.

To remedy this great inconvenience, our present monetary system was established in 1816, at which time, as gold was the metal in which the principal payments were made in England, the following law was enacted:—" That gold coins shall be in future the sole standard measure of value, and legal tender of payment, without any limitation of amount, and that silver coins shall be a legal tender for the limited amount of forty shillings only, at any one time."

In the same year a new coinage of 20 shilling pieces, called *sovereigns*, was minted, in due proportion to the guinea, viz. $46\frac{29}{40}$ sovereigns to the pound troy; and an extensive silver coinage also took place, at the new rate of 66 instead of 62 shillings to the troy pound, which affords a profit or *seignorage* of $6\frac{4}{31}$ per cent., but its actual amount must always depend on the market price of the metal.

The total existing quantity of gold when compared with that of silver is estimated to be nearly as 1 to 50, and the relative value of gold to silver is as about 16 to 1; consequently the value of the general silver currency of the world as compared with the gold currency is about 3 to 1. In Great Britain however, gold being the only legal tender for sums above forty shillings, the metallic currency is essentially gold, and the silver and copper coins are only introduced as auxiliary tokens for the purpose of effecting the fractional and smaller payments. The circulation of silver coin, about 13,000,000*l.* sterling including our colonies, is therefore of inconsiderable amount when compared with that of other countries.

The following table, extracted from Kelly's Universal Cambist, exhibits the history of English coins in a condensed form. It should be observed that the last column is calculated according to the Mint proportions which overrate the metallic value of silver.

Table showing the Alterations English Coins have undergone with respect to Weight and Fineness, and also the Comparative Value of Gold and Silver, from the Reign of WILLIAM THE CONQUEROR *to that of* GEORGE IV.

Date.	Reign.	SILVER.		GOLD.		Comparative Value of fine Gold and Silver.
		Fineness of Silver Coins.	Pound Troy of such Silver coined into	Fineness of Gold Coins.	Pound Troy of such Gold coined into	
		oz. dwt.	£ s. d.	car. gr.	£ s. d.	Gold. Silver.
1066	William I.	11 2	1 1 4	... ...	... ...	... ...
1280	8 Edward I.	— —	1 1 4	... ...	... ...	... ...
1344	18 Edward III.	— —	1 1 6	23 3½	14 0 10	1 to 12·584
1349	23 ————	— —	1 3 0	— —	14 18 8	1 — 11·571
1356	30 ————	— —	1 6 8	— —	16 0 0	1 — 11·158
1421	9 Henry V.	— —	1 12 0	— —	17 16 0	1 — 10·331
1464	4 Edward IV.	— —	2 0 0	— —	22 4 6	1 — 10·331
1465	5 ————	— —	2 0 0	— —	24 0 0	1 — 11·158
1470	49 Henry VI.	— —	2 0 0	— —	24 0 0	1 — 11·158
1482	22 Edward IV.	— —	2 0 0	... ...	24 0 0	1 — 11·158
1509	1 Henry VIII.	— —	2 0 0	... ...	24 0 0	1 — 11·158
1527	18 ————	— —	2 2 8	22 0	24 0 0	1 — 11·268
1543	34 ————	10 0	2 8 0	23 0	28 16 0	1 — 10·434
1545	36 ————	6 0	2 8 0	22 0	30 0 0	1 — 6·818
1546	37 ————	4 0	2 8 0	20 0	30 0 0	1 — 5·000
1547	1 Edward VI.	4 0	2 8 0	20 0	30 0 0	1 — 5 000
1549	3 ————	6 0	3 12 0	22 0	34 0 0	1 — 5·151
1551	5 ————	5 0	3 12 0	23 3½	34 0 0	1 — 11·000
1552	6 ————	11 1	3 0 0	22 0	36 0 0	1 — 11·050
1553	1 Mary.	11 0	3 0 0	23 3½	36 0 0	1 — 11·057
1560	2 Elizabeth.	11 2	3 0 0	22 0	36 0 0	1 — 11·100
1600	43 ————	— —	3 2 0	23 3½	36 10 0	1 — 10·904
1604	2 James I.	— —	3 2 0	22 0	33 10 0	1 — 12·109
1626	2 Charles I.	— —	3 2 0	— —	41 0 0	1 — 13 346
1666	18 Charles II.	— —	3 2 0	— —	44 10 0	1 — 14·485
1717	3 George I.	— —	3 2 0	— —	46 14 6	1 — 15 209
1816	56 George III.	— —	3 6 0	— —	46 14 6	1 — 14·287
1821	2 George IV.	— —	3 6 0	— —	46 14 6	1 — 14·287

By the above table it appears that silver coins have been diminished in value, during the last 500 years, in the ratio of 99 to 32, and gold coins nearly as 3¼ to 1.

In all regular governments, there has been a standard for coins fixed by law; that is, a certain proportion between the quantity of pure metal and its alloy. Thus, the established legal standard for gold in England is $\frac{22}{24}$ or $\frac{11}{12}$; that is, eleven parts of pure metal, and one of alloy. The fineness of gold is generally expressed in carats; the whole weight being supposed to be divided into 24 equal parts or carats, 22 of which are of pure metal and 2 of alloy: and hence, the English standard gold is said to be 22 carats fine; and the carat is divided into 4 parts, called grains: but these proportions differ in other countries. Experiments have shown that the proportions of the British gold standard give the combination of the two metals[1] which possesses the greatest degree of hardness.

The carat is the 24th part of the pound troy, or 10 pennyweights; and therefore the carat grain is $2\frac{1}{2}$ pennyweights or 60 grains.

The English standard for silver is $\frac{222}{240}$ or $\frac{37}{40}$; it is ex-

---

[1] Wrought gold has two legal standards; one is 22 carats, the same as the coin, and the other is 18 carats. The latter commenced in 1798, and is used chiefly in the manufacture of watch-cases and rings.

Wrought silver has also two legal standards; one the same as the coin, and the other 8 dwts. better, that is 11 oz. 10 dwts. The latter, called new-sterling, is seldom used.

All articles manufactured of gold and silver, except watch-cases, have to be taken to the Assay Office of the district, and if found of legal quality are stamped thus:—

The Hall Mark, showing the district where manufactured, or the hall where assayed, is at *Birmingham*, an anchor; *Chester*, three wheatsheaves, or a dagger; *Dublin*, a harp or figure of Hibernia; *Edinburgh*, a thistle, or castle and lion; *Exeter*, a castle with two wings; *Glasgow*, a tree, and a salmon with a ring in its mouth; LONDON, a leopard's head; *Newcastle-on-Tyne*, three castles; *Sheffield*, a crown; *York*, five lions and a cross.

The Standard Mark for gold of 22 carats, and silver of 11 oz. 2 dwts., is for England a lion passant; for Edinburgh, a thistle; for Glasgow, a lion rampant; for Ireland, a harp crowned. Gold of 18 carats fine, a crown and the figures 18. Silver of the new standard, figure of Britannia.

The Duty Mark is the head of the Sovereign, and indicates the duty has been paid.

The Date Mark is a letter of the alphabet, which is changed every year; it differs however in different companies. The Goldsmiths' company of London have used the following: from 1716 to 1755, Roman capital letters; 1756 to 1775, small Roman letters; 1776 to 1795, old English letters; 1796 to 1815, Roman capital letters, A to U; 1816 to 1835, small Roman letters a to u; 1836 to 1855, old English letters a to b. In 1856 a new Date Mark will be issued. (I and J are always regarded as one letter.)

pressed in troy ounces and pennyweights, that is, 11 oz. 2 dwt. of pure, and 18 dwt. of alloy, making together 1 pound troy.

The alloy of silver is mostly copper, and that of gold both silver and copper; but in the computation of coins the alloy is never reckoned of any value, being always allowed in order to save the trouble and expense that would be incurred in refining the metals to their highest degree of purity.

Besides this standard fineness of coins, there is also a legal weight fixed according to the mint regulation or rate of coinage of each country. Thus, in England a pound troy of standard gold is coined into $44\frac{1}{2}$ guineas, or $46\frac{22}{40}$ sovereigns, and a pound of standard silver into 66 shillings, with divisions and multiples in proportion; and hence, the mint price of standard gold is $3l.$ $17s.$ $10\frac{1}{2}d.$ per ounce, and that of standard silver, 66 pence per ounce. Before the year 1816, silver was coined at the rate of 62 pence per ounce; and this is sometimes reckoned the standard price in the valuation of foreign silver coins.

Copper money is coined in the proportion of 24 pence to the pound avoirdupois. Thus the penny should weigh $10\frac{2}{3}$ drams or $291\frac{2}{3}$ grains, and the other pieces in proportion.

Silver coin is a legal tender for the limited amount of 40 shillings; and copper to the amount of 12 pence.

According to the mint regulations of most countries, there is an allowance for deviation from the standard weight and fineness of coins, which is called the *remedy of the mint*. In some places the remedy is allowed in the weight, in others in the fineness; but mostly in both weight and fineness. It is considered generally as an allowance for the fallibility of workmanship. In some mints, however, it is made a source of emolument; and where governments issue coins at a rate above their intrinsic value, or the market price of the metal, the profit thus made is called *seignorage*, and charges for mint expenses are called *brassage*.

The *remedy of the mint*, according to the law of 1815, for gold coins is 12 grains per lb. in the weight, and $\frac{1}{16}$ of a carat in the fineness; and for silver coins 1 dwt. per lb. in the weight, and the same in the fineness. The remedy for copper coins is $\frac{1}{40}$ of the weight.

The following tables exhibit the British imperial standard weights and fineness of the various coins; those marked † are not now in ordinary circulation :—

British Gold Coins.

Number of pieces in the Troy pound.	Standard Weight of each piece.	Fine Gold in each piece.	Alloy in each piece.
	gr.	gr.	gr.
93$\frac{9}{20}$ Half sovereigns	61·637	56·500	5·137
46$\frac{20}{40}$ Sovereigns	123·274	113·001	10·273
†92$\frac{20}{80}$ £2 pieces	246·548	226·003	20·545
† 9$\frac{89}{200}$ £5 ,,	616·372	565·008	51·364
†178 ½ guineas	· 32·3595	29·6629	2·6966
†133½ ⅗ ,,	43·1460	39·5505	3·5955
† 89 ½ ,,	64·7190	59·3258	5·3932
† 44 Guineas	129·4382	118·6516	10·7866
† 22¼ 2 guineas	258·8764	237·3034	21·5730
† 8½ 5 guineas	647·1910	593·2584	53·9326

Gold Standard $\begin{cases} 22 \text{ carats, or } \frac{11}{12} \text{ fine gold} \\ 2 \quad ,, \quad \frac{1}{12} \text{ alloy} \end{cases}$ per pound troy.

British Silver Coins.

Number of pieces in the Troy pound.	Standard Weight of each piece.	Fine Silver in each piece.	Alloy in each piece.
	gr.	gr.	gr.
†792 Pence	7·2727	6·7273	0·5455
†396 Twopences	14·5455	13·4545	1·0909
264 Threepences	21·8182	20·1818	1·6364
198 Fourpences	29·0909	26·9091	2·1818
132 Sixpences	43·6364	40·3636	3·2727
66 Shillings	87·2727	80·7273	6·5455
33 Florins	174·5455	161·4545	13·0909
26$\frac{2}{5}$ Halfcrowns	218·1818	201·8182	16·3636
13$\frac{1}{5}$ Crowns	436·3636	403·6364	32·7273

Silver Standard $\begin{cases} 11 \text{ oz. 2 dwt. or } \frac{37}{40} \text{ fine silver} \\ 18 \quad ,, \quad \frac{3}{40} \text{ alloy} \end{cases}$ per pound troy.

British Bronze Coins.

An avoirdupois pound of copper was formerly coined into twenty-four pence, or forty-eight halfpence, or ninety-six farthings. In 1860 these were replaced by a bronze coinage, of less weight, composed of ninety-five parts of copper, four of tin, and one of zinc. A pound of bronze (7,000 grains) is coined into forty-eight pence, or eighty halfpence, or one hundred and sixty farthings.

The halfpence measure exactly an inch in diameter, so that, in case of emergency, twelve of them placed in a line might be used as a substitute for a foot-rule.

These bronze coins are also available as convenient letter-weights, observing that

$$\left.\begin{array}{r}\text{3 pence}\\ \text{5 halfpence}\\ \text{or 10 farthings}\end{array}\right\}\text{go to the ounce.}$$

The silver coins in circulation are considered only as tokens, payable by the government, and pass for more than their metallic value as compared with gold. Precaution is taken that it shall not be worth while to melt the silver coin into bullion, and it is so nearly worth its current value that imitation would not be ventured on so small a profit. The government will always receive back its tokens however worn they may be, provided they be not wilfully defaced or fraudulently reduced. But gold, being the sole standard measure of value, and legal tender of payment, circulates as a commodity; and hence the necessity of government receiving it at value on its return to the mint, and making a deduction for loss of weight when the same exceeds the remedy of the mint. The wear and tear of the gold coinage is such, that very nearly 3 per cent. of the whole circulation goes out annually; and the quantity which will suffice to throw a sovereign out of circulation is about one-fourth of a grain.

In 1853 a Select Committee was appointed to take into consideration, and report to Parliament, the practicability and advantages, or otherwise, that would arise from adopting a decimal system of coinage. The practical substance of the Report is conveyed in the following extracts.

"The first question to be decided is, what shall be the unit of the new system of coinage; and your Committee have no hesitation in recommending the present pound sterling. Considering that the pound is the present standard, and therefore associated with all our ideas of money value, and that it is the basis on which all our exchange transactions with the whole world rest, it appears to your Committee that any alteration of it would lead to infinite complication and embarrassment in our commercial dealings; in addition to which it fortunately happens, that its retention would afford the means of introducing the decimal system with the minimum of change. Its tenth part already exists in the shape of the florin or two-shilling piece, while an alteration of four per cent. in the present farthing will serve to convert that coin into the lowest step of the decimal scale which it is necessary to represent by means of an actual coin, viz. the thousandth part of a pound. To this lowest denomination your Committee propose, in order to mark its relation to the unit of value, to give the name of mil. The addition of a coin to be called a cent., of the value of 10 mils, and equal to the hundredth part of the pound, or the tenth part of the florin, would serve to complete the list of coins necessary to represent the monies of account, which would accordingly be pounds, florins, cents, and mils.

"As respects the coins, it will be necessary to withdraw from circulation certain of the coins at present in use, and to substitute in their place other coins, having reference to the decimal scale, before the decimal system can be considered as fully developed. Your Committee contemplate the retention under any circumstances, of the present sovereign (1000 mils), half-sovereign (500 mils), florin (100 mils), and shilling (50 mils, or 5 cents). The present sixpence, under the denomination of 25 mils, might be retained, and the crown, or piece of 250 mils, of which few are in circulation, need not be withdrawn. On the other hand, it will be desirable to withdraw the halfcrown, and the threepenny and fourpenny pieces, which are inconsistent with the decimal scale.

"With regard to the coins not in actual existence, but which it will be necessary eventually to introduce, it appears to your Committee, from the evidence taken by them on the subject, that copper coins of 1, 2, and 5 mils, and silver coins of 20 and 10 mils, will be required, to which should be added such others as experience may show to be desirable. It is important, however, to bear in mind, that the smaller the number of the coins with which it is practicable to effect purchases and exchanges, the better.

"Your Committee recommend that all the silver coins hereafter coined should have their value in mils marked upon them, in order that the public might, at the earliest possible period, associate the idea of the system with their different pecuniary transactions. They further recommend that all the copper coins that may be issued under the decimal system, should also have their value in mils similarly marked upon them. They believe that the necessary inconvenience attending a transition state will be far more than compensated by the great and permanent benefits which the change will confer upon the Public of this country, and of which the advantages will be participated in to a still greater extent by future generations."

Such is the decimal system of coinage recommended by the Committee, after having taken voluminous evidence upon the subject, and we must here accord to the report our full and unreserved approval, not only as regards the general principle, but in all its details. It must eventually be adopted, and it is indeed convenient that so little modification in existing coins will be requisite to bring it into practical operation. The radical coins in circulation will be pounds, florins, cents, and mils; but it will not be absolutely essential to treat them as separate moneys of account. There will be no necessity to employ any other denominations in accounts than the pound and mil—that is, the pound and its decimals. For further details on this subject we must refer to Dr. Bowring's treatise on "The Decimal System," which also contains well-executed wood-cut illustrations of English, French, Greek, and Roman coins, and an elaborate and interesting chapter on the numerals of different nations.

COMPUTATION OF COINS AND BULLION.

The value of bullion, or of a gold or silver coin, depends entirely upon the quantity of pure metal which, by a process called its assay, it is found to contain. The alloy which enters into its composition, is not estimated in the value, being always disregarded, and allowed as a compensation for the expense of refining.

Assays of gold are weighed in carats and carat grains, the carat being a nominal weight containing 4 carat grains. The carat weight is usually considered to be the 24th part of the pound troy = 10 dwts. = 240 grains troy; according to which the carat grain = 60 grains troy. It is not, however, requisite to give to it any fixed or absolute weight, as it is only used to determine the proportions of pure metal and alloy.

Assays of silver are weighed in ounces and pennyweights, the ounce being a nominal weight containing 20 pennyweights, and so far these weights are analogous to the troy table.

The British standard of gold is 22 carats fine ; that is to say, 24 carats weight of the metal should contain just 22 carats of fine or pure gold, the same being $\frac{11}{12}$ of the total weight [1].

The British standard of silver is 11 oz. 2 dwts. fine ; that is, 12 oz. weight of the metal should contain just 11 oz. 2 dwts. of fine or pure silver, being $\frac{222}{240}$ or $\frac{37}{40}$ of the total weight [1].

The reports of English assayers are made after comparing the ascertained weights with these standards, the difference being usually called the betterness or worseness of the metal, as the case may be. Thus gold found to be 23 carats 2 grains fine, is reported " better 1 carat 2 grains ;" and gold of 20 carats 2 grains is reported " worse 1 carat 2 grains." Also pure silver, 12 oz. fine, would be reported

---

[1] The British standards are also the mintage standards of the gold and silver coins of the realm.

"better 18 dwts.;" and silver ascertained to be 9 oz. 14 dwts. fine would be reported "worse 1 oz. 8 dwts."

Gold. Let w denote the total weight of a piece of gold or gold coin, expressed in troy grains; and g the number of carat grains in the fineness of the metal, as obtained from the assay report.

Then, as the British standard of fineness contains 88 carat grains, the proportion of standard gold contained in the metal is expressed by the fraction $\frac{g}{88}$. Consequently the actual weight, in troy grains, of standard gold contained in the piece $= \frac{wg}{88}$.

The same, in troy ounces, $= \frac{wg}{88} \times \frac{1}{480} = \frac{wg}{42240}$.

Now, the Mint value of a troy ounce of standard gold is $3l.$ $17s.$ $10\frac{1}{2}d.$, which, expressed in fractions of the pound sterling, $= \frac{623}{160}$.

Therefore the value of the piece of gold or coin, expressed in pounds sterling $=$

$$\frac{wg}{88} \times \frac{1}{480} \times \frac{623}{160} = \frac{6 \cdot 23}{67584} wg = \frac{wg}{10848}.$$

The same, in shillings sterling, $= \frac{10wg}{5424} = \frac{\cdot 059}{32} wg$.

From the preceding the following practical rules are deduced:

(1.) *From an assay report on gold metal to find the fineness.*

Rule. Put down 22 carats and the report underneath it: add the report if "better," or subtract it if "worse," and the sum or difference will be the "carats fine."

(2.) *From an assay report and the weight of a piece of gold, or gold coin, to find the quantity of fine or pure gold it contains.*

Rule. Find the fineness by (1), and reduce the same to

carat grains at the rate of 4 grains to the carat. Multiply the full weight of the metal by this number of carat grains, and divide by 96. The result will be the "fine weight."

(3.) To find the quantity of *standard* gold contained in the piece.

Rule. Proceed as in (2); only divide by 88 instead of 96, and the result will be the "standard weight."

Otherwise, multiply the weight of the piece by the number of carat grains in the assay report, "better or worse;" divide by 88, and add the quotient to the full weight, if "better," or subtract it if "worse."

(4.) To find the *value* of a piece of gold, or gold coin, in sterling money.

Rule. From the assay report determine the fineness by (1); reduce it at the rate of 4 grains to the carat, and multiply the number so found into the total weight of the metal, or coin, expressed in troy grains. Then divide the product by 10848, and the result will be the required value expressed in pounds sterling; or, otherwise, multiply the product by 59, divide by 4 and by 8, and finally cut off three decimals, and the result will be the required value expressed in shillings sterling.

The value, in shillings sterling, of a piece of gold, or gold coin, may be otherwise obtained from the calculated number of grains of fine gold it contains, by multiplying the same by ·177.

Silver. Let w be the weight of a piece of silver, or silver coin, expressed in troy grains; p the number of pennyweights contained in its degree of fineness.

Then, as the British standard of fineness contains 222 dwts., the proportion of standard silver is $\frac{p}{222}$, and the actual quantity of standard silver, in troy grains, $= \frac{wp}{222}$.

The same in troy ounces $= \frac{wp}{222} \times \frac{1}{480} = \frac{wp}{106560}$.

Thus, taking [1] 5s. or 60d. as the value of a troy ounce of standard silver, the value of the piece of silver or coin, expressed in pence sterling =

$$\frac{wp}{222} \times \frac{1}{480} \times 60 = \frac{wp}{1776}.$$

Hence the following practical rules for computations of silver:—

(1.) From an assay report on silver metal to find the *fineness*.

Rule. Put down 11 oz. 2 dwts., and the report underneath it; add the report if "better," or subtract it if "worse," and the sum or difference will be the "fineness."

(2.) From an assay report and the weight of a piece of silver, or silver coin, to find the quantity of *fine* or *pure* silver it contains.

Rule. Find the fineness by (1), and reduce the same to dwts., at the rate of 20 dwts. to the ounce. Multiply the full weight of the metal by this number of dwts., and divide by 240, and the quotient will be the "fine weight" required.

(3.) To find the quantity of *standard* silver.

Rule. Proceed as in (2); only divide by 222 instead of 240, and the result will be the "standard weight."

Otherwise, multiply the weight of the piece by the number of dwts. in the assay report, "better" or "worse:" divide by 222, and add the quotient to the full weight, if "better," or subtract it if "worse.":

(4.) To find the *value* of a piece of silver, or silver coin.

Rule. From the assay report ascertain the fineness by (1); reduce it to dwts., at the rate of 20 dwts. to the ounce, and multiply the number so found into the total weight of the metal, or coin, expressed in troy grains. Divide the product by 1776 (or, if preferred, by 4, 4, and 111), and

---

[1] The Mint price of standard silver, which was formerly 62d., is now 66d. per ounce; but this is above the average market value, which is considered to be about 60d., the price now usually adopted in the valuation of coins.

COINS AND BULLION. 43

the result will be the required value expressed in pence sterling.

The value in pence sterling may be also obtained from the calculated number of grains of pure silver, by multiplying the same by 10, and then dividing by 74; or it may be obtained from the number of grains of standard weight, by simply dividing by 8.

Note.—In France and Holland assays are made on the decimal system, the *proportion* of fine metal estimated in thousandth parts of the whole weight being called MILLIÈMES. By some assay calculators the same proportion put down in hundredth parts, which, of course, expresses the pure metal as a per centage, is called the TOUCH.

Both the Bank and the Mint now receive decimal reports, which are both simple and convenient, and will ultimately supersede the unnecessary cumbrous system of carats and carat grains.

To calculate the fine weight from a decimal report it is only requisite to multiply the total weight by the millièmes and then to point off three decimals.

To obtain the value of coin from the fine weight:—

$$\left.\begin{matrix}\text{Silver}\\ \text{Gold}\\ \text{,,}\end{matrix}\right\} \text{fine grains} \times \left\{\begin{matrix}\frac{10}{11}\\ 2\frac{1}{8}\\ 0{\cdot}177\end{matrix}\right\} = \text{value in} \left\{\begin{matrix}\text{pence}\\ \text{,,}\\ \text{shillings}\end{matrix}\right\} \text{sterling.}$$

For bullion:—

$$\left.\begin{matrix}\text{Silver}\\ \text{Gold}\end{matrix}\right\} \text{fine ounces} \times \left\{\begin{matrix}\frac{10}{37}\\ 4\frac{1}{4}\end{matrix}\right\} = \text{value in pounds sterling.}$$

With the gold, if worth while, deduct 1*d.* for every 8*l.* of value from this last calculation.

To show by actual examples the practical application of the foregoing rules, the several calculations have been made with respect to various coins, and the results of these calculations are exhibited in the last four columns of the following table. These results are determined from the data given in the two preceding columns, viz. the Assay and the Weight.

ASSAYS, WEIGHTS, AND VALUES OF GOLD AND SILVER COINS.

Coin.		Assay.	Weight.	Fineness.	Fine weight, or pure metal.	Standard weight.	Value in sterling.	
GOLD.		car. gr.	grains.	car. gr.	grains.	grains.	s.	d.
Austria........	Ducat..................	B 1 2¾	54	23 2¾	53·30	58·14	9	5·2
Baden	Ducat..................	B 1 2¾	47½	23 2¾	46·90	51·17	8	3·6
England	Sovereign	Standard	123¼	22 0	112·98	123·25	20	0
France..........	Napoleon, or 20 franc piece.........	W 0 1¾	99½	21 2⅓	89·40	97·52	15	9·9
Hanover	Ducat..................	B 1 3½	53⅝	23 3½	53·33	58·18	9	5·3
,,	Gold Florin	W 3 0½	50	18 3½	39·32	42·90	6	11·5
Holland	Ducat..................	B 1 2¼	53¾	23 2¼	52·77	57·57	9	4·1
Milan	Sequin	B 1 3	53¾	23 3	53·19	58·03	9	5·0
Naples	Oncetta, or 3 ducat piece (1818)	B 1 3½	58½	23 3½	57·95	63·21	10	3·1
Prussia.........	Frederick (1800)....	W 0 2	103	21 2	92·27	100·66	16	4·0
Russia	Ducat (1796)..........	B 1 2½	54	23 2½	53·16	57·99	9	4·9
,,	Imperial (1801)......	B 1 2¼	185½	23 2¼	181·87	198·41	32	2·3
Spain	Doubloon (1772)....	W 0 2¼	416¾	21 1¾	372·03	405·85	65	10·2
,,	Pistole (1801)........	W 1 1	104¼	20 3	90·13	98·33	15	11·4
Sweden	Ducat..................	B 1 2	53	23 2	51·90	56·61	9	2·2
United States,	Eagle	W 0 0½	270	21 3½	246·09	268·47	43	6·7
SILVER.		oz. dwt.	dwt. gr.	oz. dwt.	grains.	grains.	s.	d.
Austria.........	Rixdollar (1800) ...	W 1 5	18 1	9 17	355·4	384·2	4	0
,,	Copfstuck, or 20 kreutzer piece ...	W 4 3	4 6½	6 19	59·4	64 2	0	8
East Indies...	Sicca Rupee, coined at Calcutta by the East India Company	B 0 13	7 11½	11 15	175·8	190·0	1	11·8
,,	Rupee of later coinage	B 0 4½	7 8½	11 6½	166·6	180·1	1	10·5
,,	Company's Rupee now in circulation	W 0 2	7 12	11 0	165·0	178·4	1	10·3
England	Shilling, or half-florin	Standard	3 21	11 2	86·0	93 0	0	11·6
France	Franc (1818).........	W 0 7	3 5½	10 15	69·4	75 1	0	9·4
Hamburgh ...	Rixdollar specie ...	W 0 10	18 18	10 12	397·5	429·7	4	5·7
Holland	Florin, or Guilder.	W 0 4½	6 18	10 17½	146·3	158·7	1	7·8
Milan	Lira.....................	W 4 10	4 0	6 12	52·8	57·1	0	7·1
Naples	Ducat..................	W 1 0	14 15	10 2	295·4	319·4	3	3·9
Netherlands ..	Florin (1816).........	W 0 7½	6 22	10 14½	148·4	160·4	1	8
Portugal	New Crusado (1809)	W 0 4	9 3	10 18	198·9	215·1	2	2·9
Prussia.........	Rixdollar (Convention)	W 1 3	18 1	9 19	359·0	388·1	4	0·5
Russia	Ruble (1805).........	W 0 16	13 12	10 6	278·1	300·6	3	1 6
,,	10 Copec piece (1802)	W 0 13	1 8½	10 9	28·3	30·6	0	3·8
Sardinia (Piedmont),	5 Franc piece (1801)	W 0 8	16 1½	10 14	343·7	371·6	3	10·5
Sicily...........	Scudo	W 1 4	17 14	9 18	348·1	376·4	3	11·0
Spain...........	Dollar..................	W 0 8	17 8	10 14	370·9	401·0	4	2·1
Sweden.........	Rixdollar	W 0 14⅓	18 17	10 7⅔	388·5	420·0	4	4·5
Tuscany	Lira (1803)...........	B 0 7	2 8	11 9	53·4	57·8	0	7·2
United States,	Dollar..................	W 0 8½	17 8	10 13½	370·1	400·1	4	2·0
Venice	Ducato	W 1 5	14 6	9 17	280·7	303·5	3	1·9

GENERAL PRINCIPLES OF EXCHANGE.

A Bill of Exchange is a written order addressed by one person to another, directing the latter to pay on account of the former to some third person, or his order, or to the order of the person drawing the bill, a certain sum of money at a time therein specified. It is a mercantile contract, in which four persons are mostly concerned, viz. :—

The *drawer* and *seller* of the bill, who receives the value.

The *drawee*, his debtor, upon whom the bill is drawn. He is called the *acceptor*, when he writes his acceptance across the bill, and thereby engages to pay it when it becomes due.

The *payee*, or person to whom it is ordered to be paid, and who may, by indorsement, pass it to any other person.

The *buyer*, who gives value for the bill. Mercantile payments are, for the most part, made in bills of exchange, which generally pass from hand to hand, like any other circulating medium, until due. The person who at any time has a bill in his possession is called the *holder*, the payee being the holder in the first instance. When the holder of a bill disposes of it, he writes his name on the back, which is called *indorsing*. Any person may indorse a bill, and every indorser, as well as the acceptor or payee, is a security for the bill, and liable to be sued for payment.

Some bills are drawn at sight; others at a certain number of days, or months, after date, or after sight, and some at *usance*, which is meant to express the customary or usual term between different places. Days of grace are a certain number of days granted to the acceptor or payee, after the term of the bill is expired.

Inland exchange is the remittance of bills to places in the same country, by which means debts are discharged more conveniently than by cash remittances. Thus reciprocal debts, of equal amount, due between persons in different parts of the country, may be discharged without remitting specie, and such an operation is recommended by general convenience; but when the debts are unequal, the debtor place must pay its balance, either by transmitting cash or bills; and as the latter mode is generally preferred, an increased demand for bills must be the consequence, which enhances their price, as it would that of any other article of sale or purchase.

This is the principle of exchange, and it is exemplified in the premium paid for inland bills on London. The metropolis is the grand emporium of commerce that supplies other places in the kingdom with foreign merchandise, and being also the seat of government to which the revenue is transmitted, and the residence of wealthy landlords, whose rents must be remitted to them from the country, it has generally a large balance of debt in its favour; and as this balance is chiefly paid in bills, a demand for them is created, and a premium is the consequence. The premium on inland bills is generally commuted for time; that is, for a certain number of days after date, or after sight, which date, or term, varies according to the demand and other circumstances.

Foreign exchange is essentially the same as that of inland, with respect to settling accounts by a transfer of claims, and also by the premium or price of bills being regulated by the proportion between the demand and the supply; but the mode of paying the premium for foreign bills is different, and the operation of adjustment is more complicated, owing to the introduction of the comparative values of different moneys, since different countries have different coins, different in denomination, in weight, and consequently in value.

In foreign exchange, one place always gives another a fixed sum, or piece, of money for a variable price, expressed by other coins; the former is called the *certain price*, and the latter the *uncertain price*. Thus London is said to give to Paris the certain for the uncertain when the pound sterling is made exchangeable for a variable number of francs; and to Spain the uncertain for the certain when a variable number of pence sterling is exchangeable for the dollar of exchange. The uncertain price, as quoted at any time, is called the *rate*, or *course of exchange*.

When the demand in London for bills on Paris is great, a less number of francs is given for the pound sterling, and *vice versâ*. Again, if the course of exchange between London and Paris be 25 francs for the pound sterling, and if this number of francs contains the same quantity of pure silver as 20 shillings sterling, then the exchange is considered at par; but if Paris should give a higher price, the exchange is said to be against France, and in favour of England. This is the general mode of judging whether the exchange is favourable or unfavourable, though it is not always that on which merchants act or speculate.

The *intrinsic par* of exchange is the value of the money of one country compared with that of another, with respect both to weight and fineness according to accredited assays. It is, in effect, the metallic par; for though the moneys of exchange are many of them imaginary, their value is always deducible from that of the coins they represent, or to which they have an established relation.

The *commercial par* is the comparative value of the moneys of different countries, according to the weight, fineness, and market prices of the metals.

Thus two sums of different countries are *intrinsically* at par when they *contain* an equal quantity of the same kind of pure metal; and two sums of different countries are *commercially* at par when they can *purchase* an equal quantity of the same kind of pure metal.

The intrinsic par of exchange may be computed from gold or from silver coins. As a general rule, the measure of value should be of that metal in which the principal payments are made; and, therefore, in some countries the par should be computed from gold, and in others from silver, according to the kind of money in which bills of exchange are paid. It is, however, obvious that the intrinsic par of exchange can be determined only between places which pay their bills in the same kind of metal. Even the same metal must differ considerably between two countries where one possesses mines, and supplies the other with materials of coinage, as between Spain and France, or between Portugal and England. The difference in all such cases is usually estimated according to the expenses of transporting the precious metals; and thus, from the intrinsic par and the various charges and prices, the commercial equivalence is computed.

The fluctuations of exchange are occasioned by various circumstances, both political and commercial. A greater or less demand for money in a stated place at a particular time may increase or diminish its commercial value without reference to its intrinsic value. The principal cause of fluctuation is generally stated to be the *balance of trade*, by which is meant the difference between the commercial exports and imports of one country with respect to another. The demand for bills of exchange arises out of the necessity of paying for importations. The supply arises out of the practice of drawing for the amount of exportations. If the supply and the demand be equal, if for every pound's worth of goods imported there be a pound's worth of exported goods to be drawn for, there will be no real exchange; that is, the real exchange, however much the nominal exchange may alter, will be at par. When, however, the importations are not equal to the exportations, exchange can no longer remain at par. An excess of importation would cause the exchange to advance against the importing country, and *vice versâ*. The exchange may, however, be unfavourable to a country when the balance of trade is greatly in its favour; for the demand for bills must chiefly depend on the balance of such debts as come into immediate liquidation, that is to say, on the balance of payments. Besides, it does not follow that large exports are always successful, or quick in their returns; and even should this be the case, the balance of payments may be still unfavourable from political causes, such as foreign loans, subsidies, expeditions, or colonial establishments.

When any legal changes take place in the coinage or currency of a country, the exchange will of course vary, so as to keep pace or correspond with such alterations. The same remark is applicable to the debasement of coin through clipping and wear. This, however, cannot in either case be considered as an absolute change in the price of bills, but only in the money or medium through which they are bought or sold.

In times of peace, the course of exchange seldom remains long unfavourable to a country, at least beyond the expenses that might be incurred by the transportation of the precious metals; for bullion is considered the universal currency of merchants, and exchange gives it circulation, and thus tends to maintain the level of money throughout the commercial world. An unfavourable rate of exchange also operates as an encouragement to the exportation of goods, and as a check against the importation; for the exporter can afford to sell the goods cheaper in proportion to the premium which he receives for his bill, while the discount on bills from abroad operates as a tax or duty on importation. Thus exchange has always, in ordinary times, a natural tendency to restore an equilibrium.

MEASURES, WEIGHTS, AND MONEYS.

For the purpose of easy reference amongst the extensive details contained under this head, the measures and weights of Great Britain, and the tables appertaining to them, are first enumerated and explained, and those of other countries are afterwards given according to the alphabetical order of the several places. The English equivalents are uniformly in relation to Imperial measures, and avoirdupois weights are always to be understood, unless otherwise stated. Troy grains and avoirdupois grains are identical in value, though the English grain has generally the former denomination, being originally derived from the standard troy pound. The names of places given at the top of the pages always refer to the contiguous matter immediately underneath them, this arrangement being considered the clearest for rapid reference.

GREAT BRITAIN.

The act, 5 Geo. IV. c. 74, for establishing uniformity of weights and measures, came into operation on the 1st of January, 1826. The measures of capacity are the only ones which it changed. The old wine gallon contained 231 cubic inches; the corn gallon, 268·8; and the old ale gallon, 282. These were altered to the uniform imperial gallon, containing 277·274 cubic inches. The imperial standards remain unaltered in the Weights and Measures Act, 1878, now in force.

Measures of Length.

3 barleycorns	make	1 inch.	
12 inches	,,	1 foot (12 inches).	
3 feet	,,	1 yard (36 inches).	
5½ yards	,,	1 rod, pole or perch (5½ yards or 16½ feet).	
4 poles or 100 links	,,	1 chain	(22 yards or 66 feet).
10 chains	,,	1 furlong	(220 yards or 660 feet).
8 furlongs	,,	1 mile	(1760 yards or 5280 feet).

A line is the $\frac{1}{12}$th part of an inch.
A nail is 2¼ inches (used in measuring cloth).
A palm is 3 inches.
A hand is 4 inches (used for measuring the height of horses).

A span is 9 inches.
A cubit is 1½ foot.
A military pace is 2½ feet.
An itinerary pace is 5 feet.
A Scotch ell is 37·06 inches }
An English ell is 45 inches } (used in measuring holland and other cloth).
A fathom is 2 yards or 6 feet (used in sounding depths).
A cable's length is 120 fathoms or 240 yards.
A league is 3 miles.
A degree of the equator is 69·1613 miles or 365172 feet.
A degree of the meridian is 69·046 miles or 364565 feet.

The old Scotch and Irish miles are $1\frac{1}{8}$ and $1\frac{3}{11}$ English.

Among ordinary mechanics, the inch is usually divided into eighths; but in scientific calculations it is mostly divided into decimals, or otherwise the foot is decimally divided.

Measures of Surface.

144 square inches	make	1 square foot.
9 square feet	,,	1 square yard.
30¼ square yards	,,	1 pole, rod or perch (30¼ square yards).
16 poles	,,	1 chain (484 square yards).
40 poles	,,	1 rood (1210 square yards).
4 roods, or 10 chains	,,	1 acre (4840 square yards).
640 acres	,,	1 square mile.

Measures of Capacity. 1. Dry Measure.

4 gills	make	1 pint	(34·659 cubic inches).
2 pints	,,	1 quart	(69·318 cubic inches).
4 quarts	,,	1 gallon	(277·274 cubic inches).
2 gallons	,,	1 peck	(2 gallons).
4 pecks	,,	1 bushel	(8 gallons).
4 bushels	,,	1 coomb	(4 bushels).
2 coombs	,,	1 quarter	(8 bushels).
5 quarters	,,	1 wey or load	(40 bushels).
2 weys	,,	1 last	(80 bushels or 10 quarters)

A pottle is 2 quarts or half a gallon.
A strike is 2 bushels.
A cubic foot is 1728 cubic inches.
A cubic yard is 27 cubic feet; which measure of earth is called a load.

2. *Wine and Spirit Measure.*

4 gills	make	1 pint	(34·659 cubic inches).
2 pints	,,	1 quart	(69·318 cubic inches).
4 quarts	,,	1 gallon	(277·274 cubic inches).
36 gallons	,,	1 tierce	(36 gallons).
1½ tierces	,,	1 hogshead	(54 gallons).
2 hogsheads	,,	{ 1 pipe, butt, or puncheon }	(108 gallons).

The larger quantities, such as hogsheads, puncheons, &c. are gauged, and charged according to the actual contents.

3. *Ale, Beer, and Porter Measure.*

4 gills [1]	make	1 pint.	
2 pints	,,	1 quart.	
4 quarts	,,	1 gallon	(277·274 cubic inches).
9 gallons	,,	1 firkin	(9 gallons).
2 firkins	,,	1 kilderkin	(18 gallons).
2 kilderkins	,,	1 barrel	(36 gallons).
3 kilderkins	,,	1 hogshead	(54 gallons).
2 hogsheads	,,	1 butt	(108 gallons).
2 butts	,,	1 tun	(216 gallons).

To reduce cubic inches to bushels.
Rule. Multiply by 5, and divide by 11091.
To reduce cubic inches to gallons.
Rule. Multiply by 40, and divide by 11091.

WEIGHTS.

Troy Weight.

24 grains	make	1 pennyweight	(24 grains).
20 pennyweights	,,	1 ounce	(480 grains).
12 ounces	,,	1 pound	(5760 grains).

By troy weight gold, silver, jewels, and precious stones are weighed. Diamonds and pearls are an exception; they

---

[1] In London the gill is commonly called a "quartern;" in the North of England the gill is termed a "noggin," and a half-pint is called a "gill."

are weighed by the carat, which contains 4 grains; but 5 diamond grains are only equal to 4 troy grains; the ounce troy containing 150 diamond carats.

The imperial standard pound troy, made in the year 1758, is that from which all other weights are obtained: $\frac{1}{12}$th of it is the troy ounce; $\frac{1}{20}$th of the ounce is a pennyweight; and $\frac{1}{24}$th of the pennyweight is a grain; so that 5760 grains is a troy pound, and 7000 such grains is a pound avoirdupois, the grain in each case being identical.

Apothecaries' Weight.

20 grains	make	1 scruple	(20 grains)	sign ℈.
3 scruples	,,	1 drachm	(60 grains)	sign ʒ.
8 drachms	,,	1 ounce	(480 grains)	sign ℥.
12 ounces	,,	1 pound	(5760 grains)	sign lb.

Apothecaries compound their medicines by these weights, but buy and sell by avoirdupois.

The *pound, ounce*, and *grain*, are the same as in *troy* weight.

Apothecaries' Fluid Measure.

60 minims (♏)	make	1 drachm (ƒ ʒ).	
8 drachms	,,	1 ounce (ƒ ℥).	
20 ounces	,,	1 pint.	
8 pints	,,	1 gallon.	

Avoirdupois Weight.

16 drams	make	1 ounce	(437½ grains).
16 ounces	,,	1 pound	(7000 grains).
14 pounds	,,	1 stone	(14 lbs.).
2 stone	,,	1 quarter	(28 lbs.).
4 quarters	,,	1 hundred (cwt.)	(112 lbs.).
20 cwt.	,,	1 **ton**	(2240 lbs.).

The new act, 1878, declares that "all articles sold by weight shall be by avoirdupois weight, except gold, silver, platinum, diamonds, and other precious metals or stones, and drugs when sold by retail; and that such excepted articles, and none others, may be sold by troy weight."

The stone formerly varied from 8 lb. to 16 lb. in different places; but by the act passed in 1834, the stone is to consist of 14 lb. avoirdupois, and the cwt. of 8 stone; and all contracts made by any other measure are null and void.

Hay and Straw.

36 pounds	make	1 truss of Straw.
56 pounds	,,	1 truss of Old Hay.
60 pounds	,,	1 truss of New Hay.
36 trusses	,,	1 load.
18 cwt.	,,	1 load of Old Hay.
19 cwt. 32 lbs.	,,	1 load of New Hay.
11 cwt. 64 lbs.	,,	1 load of Straw.
1 cubic yard of New Hay weighs		6 stone.
1 ———— Oldish Hay	,,	8 stone.
1 ———— Old Hay	,,	9 stone.

Hay is considered as new for three months, and is called old on the 1st of September.

To find the weight of Hay contained in a Stack.—Multiply the length of the stack by its breadth, and multiply the result by its height, all in feet; divide the product by 27, which will give the number of cubic yards; this multiply by 6, 8, or 9, according to the age of the hay, as above, and the product will be the weight in stones. In measuring the height deduct two-thirds of the distance in feet from the eaves to the top.

Coal.

14 pounds	make	1 stone.
28 pounds	,,	1 quarter cwt.
56 pounds	,,	1 half cwt.
1 sack of 112 pounds	,,	1 cwt.
1 double sack of 224 pounds	,,	2 cwt.
20 cwt. or 10 large sacks	,,	1 ton.
21 tons 4 cwt.	,,	1 barge or keel.
20 keels, or 424 tons	,	1 ship load.
140 cwt. or 7 tons	,,	1 room.

GREAT BRITAIN. 55

The Newcastle chaldron is a weight of 53 cwt.

By the 1st and 2nd of William IV., it is directed that all coals be sold by weight instead of measure; 10 sacks of 224 lbs. each to one ton.

To calculate the weight of Cattle.—Measure round the animal close behind the shoulder, then along the back, from the fore part of the shoulder-blade to the bone at the tail. Multiply the square of the girt by five times the length, both expressed in feet. Divide the product by 21, and you have the weight of the four quarters, in stones of 14 lbs. In very fat cattle, the weight is about a twentieth more than that ascertained in this manner, while very lean ones weigh about a twentieth less. The quarters are little more than half the weight of the living animal. The skin weighs about the eighteenth, and the tallow about the twelfth of the whole.

Miscellaneous Liquid Measures.

Hogshead of Claret	46	gallons.
Butt of Sherry	108	,,
Pipe of Port or Masden	115	,,
Pipe of Madeira or Cape	92	,,
Pipe of Teneriffe	100	,,
Pipe of Lisbon or Bucellas	117	,,
Butt of Tent, Malaga or Mountain	105	,,
Aum of Hock, Moselle, and other German wines	30	,,
Double aum of ditto	60	,,
Pipe of Marsala or Bronti	93	,,
Puncheon of Scotch Whisky	110 to 130	,,
Puncheon of Brandy	110 to 120	,,
Hogshead of Brandy	55 to 60	,,
Puncheon of Rum	90 to 100	,,

A hogshead is one-half ⎫
A quarter cask is one-fourth ⎬ of a pipe, butt or puncheon.
An octave is one-eighth ⎭

Money Table.

4 farthings make 1 penny (4 farthings).
12 pence „ 1 shilling (48 farthings).
20 shillings „ {1 pound or sovereign} (960 farthings).

Other coins in use:—

A half-sovereign is 10 shillings.
A crown is 5 shillings.
A half-crown is 2 shillings and 6 pence.
A florin is 2 shillings.
A sixpence or tester is 6 pence.
A fourpenny piece or groat is 4 pence.
A threepenny piece or bit is 3 pence.
A halfpenny is half a penny or 2 farthings.

Former coins now out of circulation:—

Moidore	27 shillings.
Jacobus	25 „
Carolus	23 „
Guinea	21 „
Mark	13 shillings and 4 pence.
Half-guinea	10 shillings and 6 pence.
Angel	10 shillings.
Seven shilling piece	7 shillings.
Noble	6 shillings and 8 pence.

Proposed Decimal Coinage.

(The mil = 0·24 penny = 0·96 farthing).
10 mils make 1 cent (2·4 pence).
10 cents „ 1 florin (24 pence).
10 florins „ 1 pound (240 pence).

The mil, cent, and their multiples are the only new coins required.

This simple and uniform system will soon be generally understood, and its advantages are obviously so great that it must eventually come into operation.

Scotland and Ireland. In all bill or money transactions relating to Scotland or Ireland, it is requisite to insert or mention the word *sterling*, to indicate that the established money values of England are intended.

ABYSSINIA (AFRICA).

Measures.—The principal measure of length is the Turkish pic, which contains 26·8 English inches or 0·6804 metre of France.

The measure for grain is the ardeb:—

> At Gondar, in the interior, the ardeb contains 10 madegas;
> „ Massowah, on the Red Sea, „ „ 24 „ ;
> and about 80 madegas make an English imperial bushel.

Weight.—The weights are the dirhem or drachm, the wakea or ounce, and the rottolo or pound:—

			ENGLISH VALUE.
10 drachms	make	1 wakea	400 grains.
12 wakeas	„	1 rottolo or liter	4800 grains or 10 troy ounces.
12 drachms	„	1 mocha	480 grains or 1 oz. troy.

Money.—Coins of other countries are in circulation, amongst which may be mentioned Venetian sequins, Spanish dollars and imperial or Austrian dollars. The last are called patakas or patacks:—

23 harfs	make 1 pataka or dollar.	4s. 2d.
2¼ patakas	„ 1 sequin.	9s. 4½d.

Payments of large amount are usually made in ingots of gold, weighed by the wakea or Abyssinian ounce, containing 400 English grains. The pataka is also a money of account, of fluctuating value, and about 12 patakas are reckoned as the price of the wakea.

> Aix-la-Chapelle; see Prussia.
> Aleppo (Syria); see Ottoman Asia.
> Alexandria; see Egypt.
> Algiers (Africa); see France.
> Alicante; see Spain.
> Altona; see Denmark.

America; see United States.
Amsterdam; see Holland.
Ancona; see Roman States.
Antwerp; see Belgium.

ARABIA.

Measures.—At Mocha the long measures are the guz (25 English inches) and the cobido or covid (19 inches). The baryd (4 farsakh) is 12 English miles.

Liquids.

		ENGLISH VALUE.
16 vakias make 1 noosfia		$\frac{1}{4}$ imperial gallon.
8 noosfias ,, 1 gudda		2 ,, gallons.

For dry measure, 40 mecmedas or kellas make the teman or tomand, which, in rice, weighs 168 lbs. avoirdupois.

Weights.

40 vakias	make 1 maund	3 lbs. avoirdupois.
10 maunds	,, 1 frazil	30 ,, ,,
15 frazils	,, 1 bahar	450 ,, ,,

Money.

80 caveers current make 1 piastre (3s. 5d. sterling).

Payments are however commonly made in Spanish dollars, valued at $1\frac{1}{5}$ piastre. The moneys coined in the country are commassees, which contain but little silver (only 7 carats), and pass at about 40 for the dollar, being used for small payments.

Archangel; see Russia.
Arragon; see Spain.
Athens; see Greece.
Augsburg; see Bavaria

AUSTRIA.

AUSTRIA: VIENNA.

Length.

			ENGLISH VALUE.
12 punkte	make	1 linie	0·0864 inch.
12 linien	,,	1 zoll	1·0371 ,,
12 zoll	,,	1 fuss	12·445 inches or 1·0371 foot.
6 fuss	,,	1 klafter	6·2226 feet or 2·0742 yards.
4000 klafter	,,	1 meile	8297 yards or 4·7142 miles.

The elle is 30·66 English inches or 2·555 feet.

Surface.—A joch, or day's work, supposed to be as much ground as can be ploughed with one team in a day, is 1600 Vienna square klafters or fathoms = 6884 square yards or 1·4223 acre, and it is divided into 3 metzen.

Liquid Capacity.

2 pfiff	make	1 seidel	0·0779 imperial gallons.
2 seidel	,,	1 kanne	0·1557 ,, ,,
2 kannen	,,	1 mass	0·3114 ,, ,,
10 mass	,,	1 viertel	3·1143 ,, ,,
4 viertel	,,	1 eimer	12·4572 ,, ,,
32 eimer	,,	1 fuder	398·6304 ,, ,,

Dry Capacity.

8 probmetzen	make	1 becher	0·0132 bushels.	
4 becher	,,	1 futtermassel	0·0529 ,,	
2 futtermassel	,,	1 muhlmassel	0·1057 ,,	
2 muhlmassel	,,	1 achtel	0·2115 ,,	
2 achtel	,,	1 viertel	0·4230 ,,	
4 viertel	,,	1 metze	1·6918 ,,	
30 metzen	,,	1 muth	50·7536 ,,	or 6·3442 quarter

Weight, Commercial.

4 pfennig	make	1 quentchen	67·5 grains or 0·0096 lb.
4 quentchen	,,	1 loth	270·2 ,, 0·0386 ,,
2 loth	,,	1 unze	540·4 ,, 0·0772 ,,
4 unzen	,,	1 vierding	2161·6 ,, 0·3088 ,,
2 vierding	,,	1 mark	4323·2 ,, 0·6176 ,,
2 mark	,,	1 pfund	8646·4 ,, 1·2352 ,,

AUSTRIA.

Weight, Apothecaries'.

			ENGLISH VALUE.
20 gran	make	1 scrupel	22·52 grains or 0·0469 oz. troy
3 scrupel	,,	1 drachme	67·55 ,, 0·1407 ,,
8 drachmen	,,	1 unze	540·4 ,, 1·1258 ,,
12 unzen	,,	1 pfund	6484·8 ,, 13·510 ,,

The mark (4333 grains) is the unit of gold and silver weight and the apothecaries' pound = $1\frac{1}{2}$ mark.

Money.

10 kreuzer piece	2·4 penny sterling.
100 new kreuzers make 1 gulden or florin	24 pence or two shillings.
2 gulden ,, 1 thaler or rixdollar	4 shillings.
The gold ducat	= 9s. 5d. sterling.
,, half-sovereign	= 13s. 11d. ,,

The standard of money is called 20 guldenfuss, as 20 gulden are coined from the Cologne mark of fine silver.

Since 1876 the only legal measures and weights are those of the metric system of France; see p. 75.

See also Bohemia and Venetian Lombardy.

BADEN (GERMANY).

Length.

10 punkte make 1 linie	0·118 inches.	
10 linien ,, 1 zoll	1·181 ,,	
10 zoll ,, 1 fuss	11·811 ,,	or 0·9842 feet.
10 fuss ,, 1 ruthe	118·110 ,,	9·8425 ,,

The ruthe is 3 French metres.

Surface.

100 square ruthen make 1 viertel	9688 square feet or 0·2224 acre.
4 viertel ,, 1 morgen	38752 ,, ,, 0·8896 ,,

Liquid Capacity.

10 glass make 1 mass	0·3501 gallons.
10 mass ,, 1 stütze	3·3014 ,,
10 stützen ,, 1 ohm	33·014 ,,
10 ohm ,, 1 fuder	330·140 ,,

The ohm is 15 French decalitres.

BADEN.

Dry Capacity.

			ENGLISH VALUE.		
10 becher	make	1 mässlein	0·0413 bushels.		
10 mässlein	,,	1 sester	0·4127	,,	
10 sester	,,	1 malter	4·1268	,,	or 0·5158 quarters.
10 malter	,,	1 zuber	41·2680	,,	5·1585 ,,

The malter is 15 French decalitres.

Weight.

10 ass	make	1 pfennig	7·7 grains.		
10 pfennig	,,	1 centass	77·2	,,	
10 centass	,,	1 zehnling	772	,,	or 0·1103 lb.
10 zehnling	,,	1 pfund	7720	,,	1·1029 ,,

The pfund is ½ French kilogramme.

The mark of Cologne (3609 grains troy) is used for weighing gold and silver.

For apothecaries' weight, see Nürnberg.

For the new imperial measures, weights, and moneys, now in use, see Germany, page 78, and France, page 75.

Barbadoes ; see West Indies.
Barcelona ; see Spain.
Basle, or Bâle ; see Switzerland.
Batavia ; see Java.

BAVARIA.

Length.

The Bavarian	foot	11·42 inches or 0·9517 feet.		
,, ,,	ell	32·796	,,	2·7330 ,,
,, Augsburg	foot	11·65	,,	0·9703 ,,
,, ,,	long ell	24·00	,,	2·0000 ,,
,, ,,	short ell	23·32	,,	1·9433 ,,
,, Nuremberg	foot	11·96	,,	0·9967 ,,
,, ,,	ell	26·00	,,	2·1667 ,,

Liquid Capacity.

The Bavarian	eimer		=	14·116 gallons.
,, Augsburg	mass		=	0·326 ,,
,, ,,	muid		=	15·080 ,,
,, Munich	eimer		=	8·122 ,,
,, Nuremberg	,,	visirmass		14·963 ,,
,, ,,	,,	schenkmass		13·064 ,,

BAVARIA.

Dry Capacity.

	ENGLISH VALUE.
The Bavarian scheffel	6·1172 bushels.
„ Augsburg „	12·087 „
„ „ „ (8 metzen)	5·650 „
„ Munich scheffel	9·976 „
„ Nuremberg malter	4·598 „

Weight.

The Bavarian pound	8642 grains or	1·2346 lb.
„ Augsburg mark	3643 „	0·5204 „
„ „ heavy pound	7580 „	1·0829 „
„ „ light „	7295 „	1·0421 „
„ Munich pound	8656 „	1·2366 „
„ Nuremberg mark	3670 „	0·5243 „
„ „ pound	7870 „	1·1243 „
„ „ old troy „	7360 „	1·0514 „
„ „ apothecaries' „	5520 „	0·7886 „

The Nuremberg apothecaries' pound is used for weighing medicines throughout Germany, and its subdivisions are the same as in England.—See Nürnberg.

Money (Austrian standard).

60 kreuzers make 1 florin	2s. sterling.
The rixdollar of 1800	4s. „
The florin of Nuremberg	20d. „

For the new imperial measures, weights, and moneys, now in use, see Germany, page 78.

BELGIUM.

The weights and measures are the same as those of France or Holland, though some of them are differently expressed, as aune for metre or ell, litron for litre or kannen, and livre for kilogramme or ponden.

Money.

100 centimes make { 1 franc   9·4d. sterling.
                         { 1 florin   20d. „

The value of Belgian money in francs is the same as that of France; and in florins, as well as the old Brabant money in schillings and grotes, it is the same as that of Holland. In the division of the florin, the stiver is 5 cents, so that 20 stivers make the florin; and its value is about $2\frac{1}{9}$ francs.

Bengal; see East Indies.
Bergen; see Sweden and Norway.
Berlin; see Prussia.
Bermudas; see West Indies.
Berne; see Switzerland.

BIRMAH (ASIA): RANGOON.

Length.

The paulgaut is 1 inch English.
The taim or cubit is 18 inches.
The saundaung or royal cubit is 22 inches.
The dha or bamboo is 7 royal cubits = 154 inches.
The dain or Birman league is 1000 dhas = 2·4306 miles.

Weight.

		ENGLISH VALUE.
100 ticals or 3 catties } make 1 vis		$3\frac{1}{4}$ lbs. avoirdupois.
150 vis	,, 1 candy	500 ,, ,,

The Birmans, like the Chinese, keep their accounts decimally, and have no coin. Silver bullion and lead are the currency of the country.

BOHEMIA: PRAGUE.

The Prague foot measures 11·88 English inches; and the ell, 23·2 inches.

For the existing weights, measures and money, see Austria.

Bologna; see Roman States.
Bombay; see East Indies.
Bonn; see Prussia.
Bordeaux; see France.
Boston; see United States.

BRAZIL (SOUTH AMERICA).

The old weights and measures are, with the exceptions noticed below, the same as those of Portugal. The metric system of France has lately been adopted.

The medida = ¾ English imperial gallon.
„ alqueire = 1·1004 imperial bushel.
           = 0·1378   „   quarter.
„ mark   = 7·3781 ounces troy.

The rate of exchange for government estimates is 27 pence to the milreis (paper currency). At this rate of exchange to reduce milreis to English pounds, divide by 10 and to the quotient at its ⅛th part.

For further information, see Portugal.

BREMEN (GERMANY).

Length.

ENGLISH VALUE.

10 linien make	1 zoll (inch)	0·95 inches.	
12 zoll „	1 fuss (foot)	11·38 „	or 0·9483 feet.
2 fuss „	1 elle (ell)	22·76 „	1·8967 „
8 ellen „	1 ruthe (rood)	15·174 feet	or 5·058 yards.
The klafter is 3 ellen		5·69 feet.	
The meile is 20,000 Rhenish feet		6865 yards or 3·9006 miles.	

Surveyors divide the fuss decimally.

Surface.

The morgen is 120 square ruthen (3070 square yards or 0·6343 acre).

Liquid Capacity.

4 mingel make	1 quartier	0·1772 gallons.
9 quartier „	1 viertel	1·5953 „
5 viertel „	1 anker	7·9763 „
4 anker „	1 ohm	31·9052 „
6 ohm „	1 fuder	191·4315 „
The stübchen (gallon) is 4 quartier		0·709 „
„ oxhoft (hogshead) is 6 anker		47·858 „

BREMEN.

Dry Capacity.

			ENGLISH VALUE.
4 spinte	make	1 viertel	0·5094 bushels.
4 viertel	,,	1 scheffel	2·0377 ,,
40 scheffel	,,	1 last	81·5088 ,, or 10·1886 quarters.

Weight.

4 ort	make	1 quentchen	60·1 grains or 0·1373 oz.
4 quentchen	,,	1 loth	240·3 ,, 0·5493 ,,
2 loth	,,	1 unze	480·6 ,, 1·0986 ,,
8 unzen	,,	1 mark	3845 ,, 0·5493 lb.
2 mark	,,	1 pfund	7690 ,, 1·0986 ,,

Gold and silver are weighed by the mark of Cologne (3609 grains troy). For apothecaries' weights, see Nürnberg.

Money.

5 schwaren make 1 grot	0·55d. sterling.
72 grot ,, 1 rixdollar } or thaler }	39·4d. ,,
48 grot (silver piece)	27d. ,,

For the new imperial measures, weights, and moneys, now in use, see Germany, page 78.

British Islands; see Great Britain.

BRITISH POSSESSIONS IN NORTH AMERICA.

Throughout the United Canadas, New Brunswick, Nova Scotia, Prince Edward's Island, Newfoundland and the territories of the Hudson's Bay Company, the weights and measures are those of Great Britain, but generally with the old measures of capacity in wine gallons and Winchester bushels, and therefore the same as in the United States.

The moneys of account are either in pounds, shillings and pence sterling, in the same denominations of money in a nominal currency, or in dollars and cents.

According to the Halifax currency, which prevails throughout these provinces, the Spanish or American dollar is valued at 5s. or 60d. currency, and what is called sterling is the valuation of the dollar at the former standard of 4s. 6d., and between this and the Halifax currency, the proportion is as 9 to 10, making 90l. nominal sterling (in dollars at 4s. 6d.) equal to 100l. in Halifax currency.

But if the sterling value of the dollar be estimated at 4s. 2d., the par of exchange should be 83l. 6s. 8d. nominal sterling for 100l. Halifax currency, the same being in the proportion of 5 to 6. This proportion exists in New Brunswick. In Canada and Nova Scotia it is 4 to 5. In Newfoundland the dollar passes for 5s., and its assumed value is 1s. 4d. sterling, making the par of exchange 86l. 13s. 4d. nominal sterling for 100l. currency, being in the proportion of 13 to 15; but as the dollar is here overrated as to its sterling value, bills on England usually bear a premium of about 5 per cent.

The decimal coinage of the United States has been recently adopted.

In Lower Canada, wheat is measured by the minot, an old French measure (1·0736 imperial bushel). Land is measured by the arpent, another old French measure (0·8449 acre).

BRUNSWICK (GERMANY).

Length.

	ENGLISH VALUE.
12 zoll make 1 schuh (shoe or foot)	11·23 inches or 0·9358 feet.
2 schuh ,, 1 elle	22·46 ,, 1·8717 ,,
The meile (34424 Rhineland feet) =	11816 yards or 6·7140 miles.

Surface.

The morgen (30720 square schuh) = 26904 square feet or 0·6176 acre.

Liquid Capacity.

2 nössel	make 1 quartier	0·205 gallons.	
4 quartier	,, 1 stübchen	0·82	,,
40 stübchen	,, 1 ohm	32·80	,,
The fuder (4 oxhoft) of wine = 240 stübchen		196·8 gallons.	
The fass (4 tonnen) of beer = 108	,,	88·56	,,

BRUNSWICK.

Dry Capacity.

			ENGLISH VALUE.
4 becher or löchers make	1 vierfass	0·2139 bushels.	
4 vierfass	,,	1 himt	0·8556 ,,
10 himt	,,	1 scheffel	8·5560 ,, or 1·0695 quarter.

Weight.

2 heller	make	1 pfennig	14·1 grains.	
4 pfennig	,,	1 quentchen	56·3 ,,	
4 quentchen	,,	1 loth	225·3 ,,	or 0·515 oz.
16 loth	,,	1 mark	3605 ,,	0·515 lb.
2 mark	,,	1 pfund	7210 ,,	1·030 ,,

The liespfund is 14 Brunswick pfund   14·42 lbs.
The centner ,, 114 ,, ,,   117·42 ,,
The schiffpfund (ship-pound) is 20 liespfund   288·40 ,,

Gold and silver are weighed by the mark of Cologne (3609 grains troy).

For apothecaries' weights, see Nürnberg.

For the new imperial measures, weights, and moneys, now in use, see Germany, page 78.

Brussels; see Belgium.
Cadiz; see Spain.
Cairo; see Egypt.
Calcutta; see East Indies.
Canada; see Great Britain, and British Possessions in North America.
Ceylon; see East Indies.

CANARY ISLANDS (IN ATLANTIC).

The measures, weights, and coins are from Spain, but several of them are somewhat variable and depreciated in value.

For long measure the pié is the Castilian foot (11·128 English inches).

The standard libra, or pound, is 1·0148 ℔. avoirdupois.

CANDIA (IN MEDITERRANEAN).

		ENGLISH VALUE.
The pic or ell	=	25·11 inches.
The mistate (for oil)	=	2·456 gallons.
The carga (for corn)	=	4·189 bushels.
The cantaro of 100 rottoli or 44 occas	=	116½ lbs.
40 paras make 1 piastre		2¼d. sterling.

Canton; see China.

CAPE OF GOOD HOPE (BRITISH COLONY).

For measures and weights, see Great Britain, page 44.

Money.—British currency is also used; and occasionally the former Dutch currency, according to which,

| 6 stivers | make 1 skilling | 2¼d. sterling. |
| 8 skillings | „ 1 rixdollar | 18d. „ |

Cape Verd Islands (in Atlantic); see Portugal.
Cassel (Germany); see Hesse Cassel.
Castile; see Spain.

CEYLON (INDIA): COLUMBO.

The parrah is 5·62 gallons; and the seer 1 quart.
The candy or bahar weighs 500 lbs. avoirdupois.

Measures of length and surface are the same as in England.

Since 1825, by order in council, the public accounts are kept in British money, with the following values of the coin in circulation:—

The rixdollar	1s. 6d. sterling.
The Spanish dollar	4s. 2d. „
The Sicca rupee	2s. „
The rupee of Madras and Bombay	1s. 10d. „

But the Sicca rupee has superseded the sterling currency in all commercial transactions.

Christiana; see Sweden and Norway.

CHINA.

Length.

			ENGLISH VALUE.
10 fêns	make 1 tsun		1·41 inches.
10 tsuns	,, 1 chik or covid		14·1 ,,
10 chiks	,, 1 cheüng or fathom		141 ,, or 11·75 feet.
10 cheüngs	,, 1 yin		117·5 feet.
Surveyers and engineers' chik			12·70 inches.
	Itinerary	,,	12·17 ,,
	Pekin	,,	13·12 ,,
	Canton (commercial)	,,	14·70 ,,
	Imperial	,,	12·612 ,,

The li or mile (1800 itinerary chiks) = 1826 English feet.

Capacity.

10 kops	make 1 shing tsong	0·12 gallons or 0·96 pints.
10 shings	,, 1 tau (12 catties)	1·2 ,, 9·6 ,,
10 taus	,, 1 hwŭh	12·0 ,, 96·0 ,,

The measures of dry capacity are nearly ⅓ greater than these.

Weight.

| 16 taels | make 1 catty or pound | 1⅓ avoirdupois lbs. |
| 100 catties | ,, 1 pecul or tan. | 133⅓ ,, ,, |

Therefore the tael or ounce (10 mace) = $\frac{1}{12}$ lb. or 583 grains.

Money.

10 cash ("le") make 1 candereen ("fun")	¾d. sterling.
10 candereens ,, 1 mace ("tsëen")	7 d. ,,
10 mace ,, 1 tael ("lëang")	70d. ,,

These moneys, excepting the cash, are imaginary, and are formed from weights of Sysee silver, under the same denominations; and the tael in the money and commercial weights are alike. The touch or fineness of Sysee silver is 0·980. The cash are casts of common metal with a square hole in the middle, through which they are strung like beads in various numbers. Silver ingots from ½ to 100 taels are used as money; but gold is considered as merchandise, and is sold in ingots, called shoes. If the metals be sysee or pure, 10 taels of silver are given for 1 of gold.

Coblentz; see Prussia.

COLOGNE (PRUSSIA).

In 1816 a uniform system of weights and measures was decreed for all the Prussian dominions, for which see Prussia. The Rhineland foot was adopted as the standard unit for measures of length, and the Cologne mark as the unit for weights. As the system of weights previously established at Cologne is still in use throughout Germany, it is here inserted for reference.

Weight.

			ENGLISH VALUE.		
4 pfennig	make	1 quentchen	56·4 grains.		
4 quentchen	,,	1 loth	225·56	,,	
2 loth	,,	1 unze	451·12	,,	
8 unzen	,,	1 mark	3609	,,	or 0·51557 lb.
2 mark	,,	1 pfund	7218	,,	1·03114 ,,

The standard copy of the Cologne mark in use at Hamburg has been ascertained to weigh 3608 English grains, being 1 grain lighter than the average of the Prussian standards.

For the new imperial measures, weights, and moneys, now in use, see Germany, page 78.

Constantinople; see Turkey.
Copenhagen; see Denmark.
Corsica (in Mediterranean); see France.

CRACOW (POLAND).

The Cracow foot = 14·032 inches or 1·1693 foot.

Cremona (Italy); see Venetian Lombardy.
Cuba; see West Indies.
Damascus; see Ottoman Asia.
Dantzic; see Prussia.
Demerara; see West Indies.

DENMARK.

Length.

			ENGLISH VALUE.
12 linien	make	1 tomme	1·03 inches.
12 tommen	,,	1 fod	12·357 ,, or 1·0298 feet.
2 fod	,,	1 aln	24·714 ,, 2·0595 ,,
The miil is 12000 aln			24,714 feet or 4·6807 miles.

Liquid Capacity.

		pott =	0·2126 gallons.
2 pott	make	1 kande	0·4252 ,,
2 kanden	,,	1 stübchen	0·8504 ,,
2 stübchen	,,	1 viertel	1·7008 ,,
4⅞ viertel = 39 pott	,,	1 anker	8·2914 ,,

Dry Capacity.

		pott =	0·02657 bushels.
18 pott	make	1 skieppe	0·47835 ,,
2 skieppen	,,	1 fjerding	0·9567 ,, or 0·1196 quarter
4 fjerding	,,	1 tonne	3·8268 ,, 0·47835 ,,
22 tonnen	,,	1 last	84·188 ,, 10·5235 ,,

The liquid and dry pott measures are identical in capacity.

Weight.

4 ort	make	1 quintin	60·3 grains or 0·1378 oz.
4 quintin	,,	1 lod	241·2 ,, 0·5514 ,,
2 lod	,,	1 unze	482·5 ,, 1·1029 ,,
8 unzen	,,	1 mark	3860 ,, 0·5514 lb.
2 mark	,,	1 pund	7720 ,, 1·1029 ,,
The lispund is 16 pund			17·646 lbs.
The skippund is 20 lispund			352·914 ,, or 3·151 cwt.

For gold and silver the pound is 7266 English grains, and is divided the same as the preceding.

The metric system of measures and weights is now being adopted.

DENMARK.

Money.

	ENGLISH VALUE.
16 skillinge make 1 mark	4·4d. sterling.
6 mark „ 1 { rigsbank or rix-banco } dollar	26·35d. „
The Danish specie dollar	51·4d. „

Since 1873 the monetary system has been—
100 öre make 1 kronor | 1s. 1·3d. sterling.

Dresden ; see Saxony.

EAST INDIES.

1. BENGAL : CALCUTTA.

Length.

3 jows or barleycorns make 1 unglee or finger		¾ inch.	
4 unglees	„ 1 moot or hand	3 inches.	
3 hands	„ 1 span	9 „	
2 spans	„ 1 haut or cubit	18 „	
2 cubits	„ 1 guz or yard	36 „	or 3 feet
2 yards	„ 1 fathom	6 feet.	
1000 fathoms	„ { 1 Bengal coss or mile }	6000 „	or 1·3/12 mile.

Liquids and grain are measured by weight.

Weight.

		FACTORY.	BAZAAR.
5 siccas	make 1 chittack	0·1167 lbs.	0·1283 lbs.
16 chittacks	„ 1 seer	1·8667 „	2·0533 „
40 seers	„ 1 maund	74·667 „	82·133 „
	30 maunds =	20 cwt.	22 cwt

These are the avoirdupois values of the Factory and Bazaar weights, the latter being 10 per cent. heavier than the former. The new Indian sicca weight or tola (180 grains), established in 1833, is ¼ grain heavier than the above, and it corresponds with the weight of the Company's silver rupee, and also with that of the gold mohur.

Money.

12 pie or pice make 1 anna	1·4d. sterling.
16 annas „ 1 Company's rupee	22·4d. „
The Sicca rupee =	23·3d. „
The gold mohur of 16 sicca rupees	32s. „

80 silver rupees, or 36 copper annas, weigh 80 new Indian tolas, or 1 seer, which proportions suggest a ready and available means of testing the correctness of the Bazaar weights.

2. MADRAS.

Length.—The covid for cloth measure is 18·6 inches; but the English yard is generally used.

Capacity.

			ENGLISH VALUE.
8 ollucks	make	1 measure or puddy	0·338 gallons.
8 measures	,,	1 marcal	2·704 ,,
5 marcals	,,	1 parah or chunam	13·52 ,,
80 parahs	,,	1 garce, weighing 8400 lbs.	135·2 bushels or 16·9 quarters.

Weight.

10 pagodas	make	1 pollam	0·078 lbs. avoirdupois.
8 pollams	,,	1 seer	0·625 ,, ,,
5 seers	,,	1 vis	3·125 ,, ,,
8 vis	,,	1 maund	25 lbs.
20 maunds	,,	1 candy	500 ,,

Money.—Same as Bengal.

3. BOMBAY.

Length.

16 tussoos make 1 hath	18 inches.
24 tussoos ,, 1 guz	27 ,,

Capacity.

16 adoulies make 1 parah	3·03 bushels.
8 parahs ,, 1 candy	24·24 ,,

Weight.

30 pice or 72 tanks make 1 seer	0·7 lbs.
40 seers ,, 1 maund	28 lbs. or ¼ cwt.
20 maunds ,, 1 candy	560 ,, 5 ,,

Money.

100 reas	make 1 quarter	5·6d. sterling.
4 quarters or 16 annas }	„ 1 rupee	22·4d. „

The gold mohur or 15 rupee piece is of the same weight and purity as the silver rupee, 15 alloy and 165 fine grains, and its value at the British mintage rate is 1l. 9s. 2d. The gold pagoda star is valued at 7s. 4d. sterling.

A lac is 100,000 and a crore is 10 millions of rupees.

At Singapore the Spanish dollar circulates at 2·18 rupees or 4s. 1d. sterling.

See also Ceylon.

EGYPT (AFRICA).

Length.—4 derahs make 1 gasab (2·832 English yards).

The principal measure for cloth and silk is the pic (26·8 English inches).

Ancient Measures.

The natural cubit (24 Egyptian fingers) is 17·71 inches.
 „ royal „ (28 „ „) 20·66 „

Surface.—The feddan al risach, or acre, is 400 square gasab = 3208 square yards or 0·6628 acre.

Capacity.—24 robs make 1 ardeb (4·9 imperial bushels or 0·6125 quarter).

Weight.

	ENGLISH VALUE.
144 drachmas or meticals make 1 rotl or pound	1·008 lbs.
100 rottoli „ 1 cantaro	100·8 „
The oke is 400 drachmas	2·8 „

Money.—40 paras make 1 piastre (2¼d. sterling).

Elsinore; see Denmark.
England; see Great Britain.
Falkland Islands; see Great Britain.
Finland; see Russia.
Flanders; see Belgium.
Florence; see Tuscany.

FRANCE.

1. METRICAL SYSTEM NOW IN USE.

Length.

		ENGLISH VALUE.
Millimètre	(1000th of a mètre)	0·03937 inches.
Centimètre	(100th of a mètre)	0·39371 ,,
Décimètre	(10th of a mètre)	3·93708 ,,
Mètre	(unit of length)	39·3708 ,, or 3·2809 feet.
Décamètre	(10 mètres)	32·809 feet. 10·9363 yards.
Hectomètre	(100 mètres)	328·09 ,, 109·3633 ,,
Kilomètre	(1000 mètres)	1093·63 yards or 0·62138 miles.
Myriamètre	(10,000 mètres)	10936·33 ,, 6·21382 ,,

Surface.

Centiare { (100th of an are or a square mètre) }	1·1960 square yards.	
Are { (square décamètre and unit of surface) }	119·6032 ,, ,, or 0·0247 acres.	
Decare (10 ares)	1196·033 ,, ,, 0·2474 ,,	
Hectare (100 ares)	11960·33 ,, ,, 2·4736 ,,	

Capacity.

Millitre { (1000th of a litre or cubic centimètre) }	0·06103 cubic inches.
Centilitre (100th of a litre)	0·61027 ,, ,,
Décilitre (10th of a litre)	6·10270 ,, ,,
Litre { (cubic decimètre and unit of capacity) }	61·02705 ,, ,, or 1·7608 pints.
Décalitre (10 litres)	610·2705 ,, ,, 2·2010 gallons.
Hectolitre (100 litres)	3·53166 cubic feet 22·0097 ,,
Kilolitre { (1000 litres or cubic mètre) }	35·31658 ,, ,, 220·0967 ,,
Myrialitre (10,000 litres)	353·1658 ,, ,, 2200·9667 ,,

Solid.

Décistère (10th of a stère)	3·5317 cubic feet.
Stère (cubic mètre)	35·3166 ,, ,,
Décastère (10 stères)	353·1758 ,, ,,

FRANCE.

Weight.

		ENGLISH VALUE.
Milligramme (1000th of a gramme)	0·0154 grains.
Centigramme (100th of a gramme)	0·1544 ,,
Décigramme (10th of a gramme)	1·5440 ,,
Gramme (unit of weight)	15·44 ,,
Décagramme (10 grammes)	154·4 ,,
Hectogramme (100 grammes)	1544 grains $\begin{cases} 3·2167 \text{ oz. troy, or} \\ 3·5291 \text{ oz. avoirdupois.} \end{cases}$
Kilogramme (1000 grammes)	32⅙ oz. troy or 2·2057 lbs. ,,
Myriagramme (10000 grammes)	321⅔ ,, 22·057 ,, ,,

The foregoing metric system is now adopted in Austria, Belgium, Greece, Holland, Germany, Italy, Norway, Portugal, Spain, Sweden, and Switzerland.

2. "SYSTEME USUEL."
(Formerly in use, but interdicted since 1840).

Length.

12 lignes make 1 pouce	1·094 inches.
12 pouces ,, 1 "pied usuel"	13·124 ,, or 1·0936 feet.
3 pieds ,, 1 mètre	39·371 ,, 3·2809 ,,
2 mètres ,, 1 toise	78·742 ,, 6·5618 ,,
The aune is $\begin{cases} 12 \text{ decimétres} \\ 1\frac{1}{5} \text{ mètre} \end{cases}$	47·245 ,, 3·9371 ,,

Weight.

72 grains make 1 gros	60·31 grains troy.
8 gros ,, 1 once	482·5 ,, = 1·0052 oz. troy.
8 onces ,, 1 mark	3860 ,, = 8·0417 ,,
2 marks ,, 1 livre = 500 grammes	7720 ,, $\begin{cases} = 1·3403 \text{ lb. troy,} \\ \text{or } 1·1029 \text{ ,, avoird.} \end{cases}$

3. ANCIENT SYSTEM.

Length.

12 lignes make 1 pouce or inch	1·066 inches.
12 pouces ,, 1 "pied de Roi" (0·3249 mètre)	12·79 ,,
6 pieds ,, 1 toise (1·9492 ,,)	6·395 feet.
The aune (1·1880 ,,)	46·85 inches.
"Lieue de poste" (2000 toises)	4263 yards or 2 4222 miles.

FRANCE. 77

Weight.

			ENGLISH VALUE.
72 grains make	1 gros	59·0 grains troy.	
8 gros ,,	1 once	472·2 ,, ,, =0·9837 oz. troy.	
8 onces ,,	1 mark	3777·5 ,, ,, =7·8698 ,,	
2 marks ,,	1 poids de marc	7555 ,, ,, {1·3116 lb. troy, or	
		{1·0793 ,, avoirdupois.	

Money.

10 décimes } make 1 franc	9·4d. sterling.
100 centimes }	
5 franc piece in silver	3s. 11d. ,,
10 franc piece in gold	7s. 11d. ,,
20 franc piece or Napoleon	15s. 10d. ,,

There are gold coins of 5, 10, 20, 50, and 100 francs, and silver coins of 1, 2, and 5 francs; also bronze coins of 1, 2, 5, and 10 céntemes.

FRANKFORT-ON-THE-MAINE (GERMANY).

Length.

The foot measures 11·27 English inches, and the ell 21·54 inches.

Liquid Capacity.

4 eich-mass } make 1 viertel	1·5784 gallons.
1½ neu-mass }	
20 viertel ,, 1 ohm	31·5674 ,,
6 ohm ,, 1 fuder	189·4044 ,,

The mass is also divided into 4 schoppen.

Dry Capacity.

4 schrot make 1 mässchen	0·01233 bushels.
4 mässchen ,, 1 gescheid	0·04932 ,,
4 gescheid ,, 1 sechter	0·1973 ,,
2 sechter ,, 1 metze	0·3946 ,,
2 metzen ,, 1 simmer	0·7892 ,,
4 simmer ,, 1 malter or achtel	3·1568 ,,

FRANKFORT.

Weight.

			ENGLISH VALUE.		
2 heller	make	1 pfennig	14·1 grains or	0·0322 oz.
4 pfennig	„	1 quentchen	56·4 „	0·1289 „
4 quentchen	„	1 loth	225·6 „	0·5157 „
2 loth	„	1 unze	451·2 „	1·0314 „
8 unzen	„	1 mark	3610 „	0·5157 lb.
2 mark	„	1 pfund	7220 „	1·0314 „

The heavy weight is 8 per cent. more than this table; and the zoll-centner is 110·24 lbs. avoirdupois.

Apothecaries' weight is the same as at Nürnberg.

Money.

The gold ducat is about 9s. 4d. sterling.

For the imperial measures, weights, and moneys, now in use, see Germany (below).

Frankfort-on-the-Oder; see Prussia.
Gallen, St. Gall; see Switzerland.
Geneva; see Switzerland.
Genoa; see Sardinia.

GERMANY.

See Austria, Baden, Bavaria, Bohemia, Bremen, Brunswick, Cologne, Frankfort, Hamburg, Hanover, Hesse, Lubeck, Prussia, and Saxony.

The measures and weights throughout Germany are now those of the metric system, the meter or stab (staff or ell) and the kilogramme being the principal units; the millimeter is also denominated strich (linie); see France, p. 75.

The present imperial monetary system for the whole of Germany is—

100 pfennige make 1 Reichs-mark | 1s. sterling.

GIBRALTAR.

The weights and measures are those of Great Britain.

Money.

10 decimas	make	1 real vellon	2·5d. sterling.
20 reals	„	1 dollar (Spanish)	4s. 2d. „
16 dollars	„	1 doubloon (Spanish)	66s. 8d. „

GIBRALTAR.

The Spanish hard dollar is a legal tender at this rate in all the British Colonies. By an order in council, issued in 1845, it was directed that Spanish, Mexican, and South American dollars shall be legal tenders at 4s. 2d. sterling; and that gold doubloons, or 16 dollar pieces, of the same countries, shall pass for 3l. 6s. 8d. Mercantile operations are carried on in dollars and cents.

Great Britain; see page 44.

GREECE.

Length.

	ENGLISH VALUE.
The short picha (used for silks)	25 inches.
„ long „ (woollens and linens)	27 „
Ancient Greek foot (16 Egyptian fingers)	11·81 „
„ Attic or Olympic foot	12·10 „
„ Pythic foot	9·75 „

The cubit is 1½ feet and the stadium 400 cubits or 600 feet.

Capacity.

The kila measures 0·9152 bushel or 0·1144 quarter.
The staro, of 3 bachels, measures 2·259 bushels.
The ancient keramion or metretes is 8·468 gallons.

Weight.

The pound is 6168 grains or 0·8811 lb. avoirdupois.
The cantaro, of 40 okes, is 112 lbs. or 1 cwt.

The French metric system is now uniformly established, only with a different nomenclature; see p. 75.

Money.

The silver drachma (=100 lepta or centimes)	8·4d. sterling.	
The 5 drachma silver piece	3s. 6d.	„
The 10 drachma „ „	7s. 0d.	„
The 20 drachma gold piece	14s. 2d.	„

GUINEA (AFRICA).

Length.—The jacktan = 12 English feet.

Weight.

		ENGLISH VALUE.
2 media-tabla make 1 aguirage		62 grains.
2 aguirages „ 1 piso or uzan		124 „
4 pisos „ 1 benda-offa		496 „
2 benda-offas „ 1 benda		992 „
The quinto is 3 media-tabla		93 „
„ seron „ 6 „ „		186 „
„ eggēba „ 10⅔ „ „		330 „

Money.—100 cents make 1 dollar (50*d*. sterling).

The decimal system was introduced in 1839, the previous money being that of Holland, of which 3 guilders were to represent 1 dollar currency.

HAMBURG (GERMANY).

8 achtel make 1 zoll	0·94 inches.
12 zoll „ 1 fuss	11·29 „ or 0·9408 feet.
2 fuss „ 1 elle	22·58 „ 1·8817 „ or 0·6272 yard.
The Rhenish foot	12·357 „ 1·0298 feet.

Engineers and surveyors use the Rhenish foot and inch, decimally divided.

The meile is 24000 Rhenish feet = 4·6807 English miles.

Surface.—The morgen is 117600 square fuss = 2·3895 acres.

Liquid Capacity.

2 össel make 1 quartier	55·2 cubic inches or 0·1992 gallons.
2 quartier „ 1 kanne	110·4 „ „ 0·3983 „
2 kannen „ 1 stübchen	220·9 „ „ 0·7967 „
2 stübchen „ 1 viertel	441·8 „ „ 1·5934 „
5 viertel „ 1 anker	2209 „ „ 7·9668 „
4 anker „ 1 ohm	8836 „ „ 31·867 „
6 ohm „ 1 fuder	53016 „ „ 191·292 ,
The elmer is 4 viertel	0·3734 gallons.
The tonne is 24 „	38·2404 „
The oxhoft is 30 „	47·8006 „

Dry Capacity.

			ENGLISH VALUE.			
4 spint.	make	1 himt	1606·5 cubic inches or		0·7242	bushels.
2 himt	,,	1 fass	3213	,, ,,	1·4485	,,
2 fass	,,	1 scheffel	6426	,, ,,	2·8969	,,
10 scheffel	,,	1 wispel	64260	,, ,,	28·9694	,,
3 wispel	,,	1 last	192780	,, ,,	86·9083	,,
					or 1·0864 lasts.	

Weight.

4 pfennig	make	1 quentchen	58·4 grains.		
4 quentchen	,,	1 loth	233·6	,,	
2 loth	,,	1 unze	467·2	,,	or 1·068 oz.
8 unzen	,,	1 mark	3738	,,	8·544 ,,
2 mark	,,	1 pfund	7476	,,	1·068 lbs.

The schiffpfund or 20 liespfund is 280 pfund.

Gold and silver are weighed by the Cologne mark; and medicine by Nürnberg apothecaries' weight.

Money.

12 pfennige	make 1 schilling		0·9d.	sterling.
16 schillinge	,, 1 mark		1s. 2¼d.	,,
3 mark	} 1 rixdollar }		3s. 7d.	,,
48 schillinge	,, (nominal)			
The mark banco (imaginary)			1s. 5¼d.	,,
The rixdollar specie			4s. 5·7d.	,,
The gold ducat			9s. 4d.	,,

The weight of the Cologne mark, according to the Hamburg standard, is 3608 English troy grains. This weight of *fine* silver is coined into 34 marks, and its value, at the rate of 5s. sterling per ounce standard, is 40s. 7d.; hence the value of the "mark current" as above. Banco is a nominal valuation of the Cologne mark of silver at 27¾ marks Banco, according to which the estimated value of the "mark banco" is 17½d. sterling.

For the new imperial measures, weights, and moneys, now in use, see Germany, p. 78.

HANOVER (GERMANY).
Length.

			ENGLISH VALUE.		
12 linien = 8 achtel }	make 1 zoll		0·95 inches.		
12 zoll	„	1 fuss	11·45	„	or 0·9542 feet.
2 fuss	„	1 elle	22·9	„	1·9083 „
8 ellen	„	1 ruthe	183·2	„	15·2667 „
The meile (25,400 fuss) =			24236 feet		4·5901 miles.

Surface.—The morgen of Calenberg is 30720 square fuss = 3108 square yards or 0·6438 acre.

Liquid Capacity.

2 nössel	make	1 quartier	0·214 gallons.	
2 quartier	„	1 kanne	0·428	„
2 kannen	„	1 stübchen	0·856	„
2 stübchen	„	1 viertel	1·7118	„
5 viertel	„	1 anker	8·559	„
4 anker	„	1 ohm	34·236	„
6 ohm	„	1 fuder	205·416	„
The eimer is	8 viertel		13·6944	„
The oxhoft is	30 viertel		51·354	„

Dry Capacity.

4 vierfass = 3 drittel }	make 1 himt		0·8556 bushels.		
6 himt	„	1 malter	5·1337	„	or 0·6417 quarters.
8 malter	„	1 wispel	41·07	„	5·1337 „
2 wispel	„	1 last	82·14	„	10·2675 „

Weight.

4 örtchen	make	1 quentchen	58·7 grains or 0·1341 oz.		
4 quentchen	„	1 loth	234·8	„	0·5366 „
2 loth	„	1 unze	469·5	„	1·0731 „
8 unzen	„	1 mark	3756	„	0·5366 lb.
2 mark	„	1 pfund	7512	„	1·0731 „

Gold and silver are weighed by the Cologne mark; and medicine by Nürnberg apothecaries' weight.

For the new imperial measures, weights, and moneys, now in use, see Germany, p. 78.

Havannah; see West Indies.
Hayti or Haiti (St. Domingo); see West Indies.

HESSE-CASSEL (GERMANY).
Length.

	ENGLISH VALUE.
12 linien make 1 zoll	0·9437 inches.
12 zoll ,, 1 (standard) fuss	11·324 ,, or 0·9437 foot.
The (surveyors') fuss	11·217 ,, 0·9347 ,,

Surface.—The acker (29400 square surveyors' fuss) = 2854 square yards or 0·5897 acre.

Liquid Capacity.

4 schoppen make 1 mass	0·44 gallons.
4 mass ,, 1 viertel	1·76 ,,
20 viertel ,, 1 ohm	35·2 ,,
6 ohm ,, 1 fuder	211·2 ,,

Dry Capacity.

4 mässchen make 1 metze	0·276 bushels.
4 metzen ,, 1 himt	1·105 ,,
2 himt ,, 1 scheffel	2·21 ,,

Weight.

4 quentchen make 1 loth	234 grains or 0·535 oz.
2 loth ,, 1 unze	468 ,, 1·07 ,,
16 unzen ,, 1 pfund	7490 ,, 1·07 lb.

Gold and silver by the Cologne mark; and apothecaries' weight the same as at Nürnberg.

For the new imperial measures, weights, and moneys, now in use, see Germany, p. 78.

HESSE-DARMSTADT (GERMANY).
Length.

10 linien make 1 zoll	0·984 inches.
10 zoll ,, 1 fuss	9·843 ,, (¼ French metre).
10 fuss ,, 1 klafter	98·427 ,, = 8·2022 feet or 2·7341 yards.

HESSE-DARMSTADT.

Surface.—The morgen is 40,000 square fuss or 400 square klafter = 2990 square yards or 0·6178 acre.

Liquid Capacity.

			ENGLISH VALUE.
4 schoppen	make	1 mass	0·44 gallons.
4 mass	„	1 viertel	1·76 „
20 viertel	„	1 ohm	35·2 „
6 ohm	„	1 fuder	211·2 „

Dry Capacity.

4 mässchen	make	1 gescheid	0·055 bushels.
4 gescheid	„	1 kümpf	0·22 „
4 kümpf	„	1 simmer	0·88 „
4 simmer	„	1 malter	3·52 „

Weight.

4 pfennig	make	1 quentchen	60·3 grains.		
4 quentchen	„	1 loth	241·1 „	or 0·5511 oz.
32 lothe	„	1 pfund	7716 „	1·1023 lb.

Previously to 1821 the weights and measures were those of Frankfort.

For the new imperial measures, weights, and moneys, now in use, see Germany, p. 78.

Hindoostan (Asia); see East Indies.

HOLLAND.

With the exception of the old nomenclature, the weights and measures of Holland, since 1817, have been according to the metric system of France.

Length.

The streep	is the	millimètre	0·03937 inches.	
„ duim	„	centimètre	0·39371 „	
„ palm	„	decimètre	3·93708 „	
„ elle	„	mètre	39·3708 „	or 3·2809 feet.
„ roede	„	décamètre	32 809 feet or	10·9363 yards.
„ mijl	„	kilomètre	1093·63 yards or	0·6214 mile.

HOLLAND.   85

Liquid Capacity.

The vingerhoed	is the	centilitre	0·61027 cubic inches.
,, maatje	,,	decilitre	6·10270 ,, ,,
,, kan	,,	litre	61·02705 ,, ,, or 1·7608 pints.
,, vat	,,	hectolitre	3·53166 cubic feet or 22·0097 gallons.

Dry Capacity.

The maatje	is the	decilitre	6·10270 cubic inches.
,, kop	,,	litre	61·02705 ,, ,,
,, schepel	,,	decalitre	610·2705 ,, ,, or 0·27512 bushels.
,, mudde } or zak }	,,	hectolitre	3·53166 cubic feet or 2·7512 ,,
,, last	is	30 mudde	82·536 bushels or 10·317 quarters } or 1·0317 last imperial. }

Weight.

The korrel	is the	decigramme	1·544 grains.
,, wigtje	,,	gramme	15·44 ,,
,, lood	,,	decagramme	154·4 ,,
,, ons	,,	hectogramme	1544 ,, or 3·5291 oz.
,, pond	,,	kilogramme	15440 ,, 2·2057 lbs.

Apothecaries' weight is similar to that of Great Britain.

Money.

		ENGLISH VALUE.
5 cents make 1 stiver		1d. sterling.
20 stivers } ,, 1 florin or }		
100 cents } guilder }		1s. 8d. ,,
The rixdollar } is { 50 stivers }		
(nominal) } { 2½ florins }		4s. 2d. ,,
12 groot make { 1 schilling }		
{ Flemish }		6d. ,,
The ducatoon		5s. 5d. ,,
The gold ducat		9s. 5d. ,,
,, ,, 10 florin piece		16s. 6¼d. ,,
,, ,, ryder		25s. 1d. ,,

Holstein; see Denmark.
Hungary; see Austria.

IONIAN ISLANDS (IN MEDITERRANEAN).

Since 1817, the weights and measures have been those of Great Britain with Italian designations.

The libbra sottile is the pound troy; the libbra grossa the pound avoirdupois; the centinajo is 100 libbre.

The dicotoli is the English pint; the chilo the bushel; the barile is 16 gallons; the stadio is the chain (22 yards), &c.

Money.—104 oboli make 1 Spanish dollar (50d. sterling).

Ireland; see Great Britain.

Jamaica; see West Indies.

ITALY.

The measures and weights now adopted throughout Italy are those of the metric system of France; see p. 75.

Money.

	ENGLISH VALUE.
100 centimes make 1 lira.	9·125d. sterling.

For former measures, weights, and moneys, see Modena, Naples, Parma, Roman States, Sardinia, Sicily, Tuscany, and Venetian Lombardy.

JAPAN (ASIA).

The long measure of Japan is the inc ($6\frac{1}{4}$ English feet).

Money.

10 rin	make	1 sen	$\frac{1}{2}d$. sterling.	
100 sen	„	1 yen	4s. 2d.	„

ISLAND OF JAVA: BATAVIA.

Length.—The ell is $27\frac{3}{4}$ English inches; the foot is the Rhineland foot (12·357 inches); the ikje (8 ells) is $83\frac{1}{4}$ inches.

Capacity.—The kanne is about $\frac{2}{3}$ of an English gallon.

Weight.—The weights are those of China.

Money.—The old florin of Batavia is valued at $19\frac{3}{4}d$. sterling; the rixdollar $87\frac{1}{2}d$.; and the new gulden or florin of the Netherlands is valued at 20d. sterling. See Holland.

Kiel; see Germany.

Konigsberg; see Prussia.

Leghorn; see Tuscany.
Leipzic; see Saxony.
Lille; see France.
Lisbon; see Portugal.
Lombardo-Veneto; see Venetian Lombardy.
London; see Great Britain, p. 50.

LÜBECK (GERMANY).

Length.

			ENGLISH VALUE.		
12 punkte make 1 linie			0·16 inches.		
6 linien	„	1 zoll	0·95	„	
12 zoll	„	1 fuss	11·45	„	or 0·9542 feet.
2 fuss	„	1 ello	22·90	„	1·9084 „

Liquid Capacity.

2 ort	make	1 planke	0·0996 gallons or 0·7968 pints.		
2 planken	„	1 quartier	0·1992	„	1·5936 „
2 quartier	„	1 kanne	0·3985	„	3·188 „
2 kannen	„	1 stübchen	0·797	„	
2 stübchen	„	1 viertel	1·594	„	
5 viertel	„	1 anker	7·97	„	
4 anker	„	1 ohm	31·88	„	
6 ohm	„	1 fuder	191·28	„	
The eimer is 4 viertel			6·376	„	
The oxhoft is 30	„		47·82	„	

Dry Capacity (Wheat, &c.).

4 fass	make	1 scheffel	0·92 bushels.		
4 scheffel	„	1 tonne	3·68	„	
3 tonnen	„	1 drömt	11·04	„	or 1·38 quarters.
8 drömt	„	1 last	88·32	„	11·04 „

The corresponding measures for oats are ⅛th larger.

Weight.

4 pfennig	make	1 quentchen	58·4 grains or 0·1336 oz.		
4 quentchen	„	1 loth	233·7	„	0·5343 „
2 loth	„	1 unze	467·5	„	1·0686 „
8 unzen	„	1 mark	3740	„	0·5343 lb.
2 mark	„	1 pfund	7480	„	1·0686 „

LUBECK.

Gold and silver are weighed by the Cologne mark (3609 grains); and apothecaries' weight is the same as Nürnberg.

Money (Hamburg standard).

12 pfennige make 1 schilling	1·1d. sterling.
16 schillinge „ 1 mark	17·6d. „

For the new imperial measures, weights, and moneys, now in use, see Germany, p. 78.

Lucca (Italy); see Tuscany.
Lucerne; see Switzerland.
Lyons; see France.
Madeira (in Atlantic); see Portugal.
Madras; see East Indies.
Madrid; see Spain.
Malaga; see Spain.

MALTA (IN MEDITERRANEAN).

Length.—The foot is 11·17 English inches; the palmo is 10·3 inches; and the canna is 8 palmi or 82·4 inches.

Capacity.

	ENGLISH VALUE.
The caffiso of oil is	4·580 gallons imperial.
„ barile of wine is	9·160 „ „
„ salma of corn is	7·969 bushels.

Weight.

The rottolo of 30 ounces is	1¾ lbs. avoirdupois.
„ cantaro of 100 rottoli is	175 „ „
64 rottoli make	1 cwt.

Money.

20 grani make	1 taro	1·65d. sterling
12 tari „	1 scudo	20d. „
2½ scudo } = 30 tari } „	1 pezza, } Sicilian dollar }	50d. „

As dollars and doubloons here form the principal circulating medium, it was ordered, in 1845, that the Spanish or South American dollar should pass for 4s. 2d. or 30 tari,

MALTA.

and the Sicilian dollar for 4s. or 28 tari and 16 grani. But the pezza or dollar of Sicily is usually valued at 50d. as above stated.

Marseilles ; see France.

MAURITIUS (IN INDIAN SEA).

For weights and measures, see Great Britain ; and also France ("ancient system").

Money.—100 cents make 1 current dollar (50d. sterling).

The Spanish dollar is valued at 4s. 4d. sterling, and is divided into halves, quarters, eighths, and sixteenths.

In 1843 an order in council established the pound sterling as the money for public accounts, and gave to the coins circulating in the colony the following values :—

Dollars of Spain, Mexico, and South America } 4s. 2d. = 1·01⅛ Mauritius dollars.
East India Company's rupee   1s. 10d. = 0·45¾   ,,   ,,
    ,,   ,,   ,,   gold mohur 29s. 2d. = 7·29½   ,,   ,,
5 francs . . . .   3s. 10d. = 0·96²¹⁄₂₉   ,,   ,,
Napoleon of 20 francs . .   15s. 10d. = 3·95¾   ,,   ,,

To obviate fractions, the population freely pass rupees at two to the Mauritius dollar ; which, therefore, represents only 3s. 8d. instead of 4s. Thus British coins are practically excluded, excepting in government transactions, and rupees have become the principal currency.

Subsequent to 1875 the metric system of France has been introduced into the Mauritius and its dependencies.

Primary standards of the metre, litre, and kilogram have recently been issued for that government by the Board of Trade.

MECKLENBURG-SCHWERIN } (GERMANY).
MECKLENBURG-STRELITZ

The weights and measures were the same as Hamburg, with the exception of measures of capacity which were those of Lubeck. For metric system now in use, see Germany.

Memel ; see Prussia.

MEXICO (NORTH AMERICA).

The weights and measures of Mexico are those of Spain, with some local variations difficult to ascertain and enumerate. See Spain.

Accounts are kept in pesos or dollars of 8 reals, the real being usually valued at 6 pence sterling.

Milan (Italy); see Venetian Lombardy.
Mocha; see Arabia.

MODENA (ITALY).

Length.

The piede, or foot, is 20·592 English inches.
6 piedi make 1 cavezzo (123·55 inches = 10·296 feet
= 3·432 yards).
The braccio (2¼ Genoa palmi) = 24·52 inches.

Surface.—The biolca, of 72 tavole (288 square cavezzi) = 3392 square yards or 0·7009 acre.

Liquid Capacity.

	ENGLISH VALUE.
3 boccali make 1 fiasco	0·4584 gallons.
20 fiasci ,, 1 barile	9·1680 ,,

Dry Capacity.

| 2 staja make 1 sacco | 3·876 bushels or 0·4845 quarter. |

Weight.

| 16 ferlini make 1 oncia | 411 grains or 0·9394 oz. |
| 12 oncie ,, 1 libbra or lira | 4932 ,, 0·7046 lb |

For more recent information see Italy, p. 86.

Montpelier; see France.

MOROCCO (AFRICA).

The cubit or canna	21 inches.
The pic	26 ,,
The commercial pound	1·190 lb.
The market pound	1·785 ,,

MOROCCO. 91

The principal coins in circulation are Spanish dollars and doubloons.

Moscow; see Russia.
Munich; see Bavaria.
Munster; see Prussia.
Nantes; see France.

NAPLES, THE TWO SICILIES (ITALY).

Length.

ENGLISH VALUE.

5 minuti make 1 oncia	0·865 inches.	
12 oncie „ 1 palmo	10·382 „ or 0·8652 feet.	
8 palmi „ 1 canna	83·055 „ 6·9213 „	
The pertica or passo is 7½ palmi	77·865 „ 6·4887 „	
The miglio is 7000 palmi	6056 feet or 1·1470 mile.	

Surface.

The mòggio is 900 square passi or 50625 square palmi } | 37898 square feet or 0·8700 acre.

Liquid Capacity (Wine, Spirits, &c.).

60 caratti make 1 barile	9·174 gallons.
12 barili „ 1 botto	110·088 „
2 botti „ 1 carro	220·176 „

Liquid Capacity (Oil).

6 misurelle make 1 quarto	0·1392 gallons.
16 quarti „ 1 stájo	2·228 „
16 staja „ 1 salma	35·647 „

The stájo is also divided into 20 pignate.

Dry Capacity.

24 misure make 1 tomolo	1·407 bushels.
36 tomoli „ 1 carro	50·660 „ or 6·332 quarters.

NAPLES.

Weight (Troy and Apothecaries').

20 accini make 1 trapeso or scrupolo	13¾ grains.
3 scrupoli ,, 1 dramma	41¼ ,,
10 dramme ,, 1 oncia	412½ ,, or 0·8594 oz. troy
12 oncie ,, 1 libbra	4950 ,, 0·8594 lb. ,,
The rotolo grosso is 33⅓ oncie	1·9643 lb. avoirdupois.
The rotolo piccolo is 18 oncie	1·0607 ,, ,,

Money.

10 grani make 1 carlino	4d. sterling.
2 carlini ,, 1 taro	8d. ,,
5 tari } ,, 1 ducat 100 grani }	39¾d. ,,
The gold oncetta (1818)	10s. 3d. ,,

For more recent information, see Italy, p. 86.

Netherlands; see Holland.

Neuchâtel or Neufchâtel; see Switzerland.

New South Wales; see Great Britain.

New York; see United States.

Newfoundland; see Great Britain.

Nice (Italy); see Sardinia.

North America; see United States.

Norway; see Sweden and Norway.

Nova Scotia; see British Possessions in North America.

NÜRNBERG, OR NUREMBERG (BAVARIA).

The apothecaries' weight of Nürnberg is used for medicines throughout Germany. Its pfund is ¾ of the old Nürnberg money pound, or 1¼ old Nürnberg mark, and is subdivided thus:—

20 gran make 1 scrupel	19·2 grains.
3 scrupel ,, 1 drachme	57·5 ,,
8 drachmen ,, 1 unze	460 ,, or 0·9583 oz. troy.
12 unzen ,, 1 pfund	5520 ,, 11½ ,, ,,

For further particulars of the weights, measures, and money of Nürnberg, see Bavaria.

Odessa; see Russia.

Oporto; see Portugal.

Ostend; see Belgium.

OTTOMAN ASIA: ALEPPO, SMYRNA, &c.

The Turkish pic (26·8 inches) is used for measures of length, and the oke for weight. At Damascus the pic is 23 inches.

Weight.

6 okes	make 1 batman	16·974 lbs.
7½ batman	„ 1 cantaro	127·305 „

Ottoman Empire; see Turkey.
Padua (Italy); see Venetian Lombardy.
Palermo (Sicily); see Sicily.
Paris; see France.

PARMA (ITALY).

Length.

ENGLISH VALUE.

12 atomi	make 1 punto	0·15 inches.		
12 punti	„ 1 oncia	1·78 „		
12 oncie	„ 1 braccio di legno	21·34 „	or	1·7783 feet.
6 braccia	„ 1 pertica	128·04 „		10·6700 „
The pié or foot	=	22·428 „		1·8690 „

Surface.—The biolca, of 6 stari (288 square pertica) = 3643 square yards or 0·7527 acre.

Liquid Capacity.—See Milan.

Dry Capacity.

8 quartaròli	make 1 mina	0·7066 bushels.
2 mine	„ 1 stájo	1·4132 „

Weight.

24 grani	make 1 denaro	17·5 grains.	
24 denari	„ 1 oncia	419·8 „	
12 oncie	„ 1 libbra	5038 „	or 0·7197 lb.
25 libbre	„ 1 rubbio	17·99 lbs.	

For gold and silver weight, see Venetian Lombardy.

Money.

The 5 lire silver piece (1815)	3s. 11d. sterling.
„ 20 „ „ „	15s. 10d. „
The gold sequin (zecchino)	9s. 5d. „

For more recent information see Italy, p. 86.

PERSIA (ASIA).

Length.

The royal guerze, or monkelser, is 37¼ English inches.
The common guerze is ⅔ the royal, or 25 ,, ,,
The arish is 38·27 English inches.

Capacity.

			ENGLISH VALUE.
4 sextarios make	1 chenicas		80·26 cubic inches.
2 chenicas ,,	1 capichas		160·52 ,, ,,
25 capichas ,,	1 artaba		4013 ,, ,, or 1·009 bushels.

Weight.

2 mascais make	1 dirhem	0·0423 lbs. or 0·0211 lbs.
50 dirhems ,,	1 rattel	2·1136 ,, 1·0568 ,,
6 rattels ,,	1 batman	12·6816 ,, 6·3408 ,,

Both sorts of weights are used with these divisions.
The batman of Shirez is 12·6816 lbs. avoirdupois.
,, ,, Tauris ,, 6·3408 ,, ,,
The dirhem used for weighing gold and silver = 150 grains troy.

Money.

5 dinars simple make	1 kasbequis	0·216d.	sterling.
2 kasbequis ,,	1 dinars-bisti	0·432d.	,,
5 dinars-bisti ,,	1 shatree	2·16d.	,,
2 shatrees ,,	1 mamoodi	4·32d.	,,
2 mamoodi ,,	1 abassi	8·64d.	,,
50 abassi ,,	1 toman	1l. 16s.	,,
The silver rupee		1s. 11d.	,,
,, gold ,,		1l. 9s. 2d.	,,

The above dinar and toman are imaginary moneys of account, not represented by coins. The Persian gold toman is worth about 11s. sterling; and the silver rupee about 18d.

Petersburg; see Russia.
Philadelphia; see United States.
Piedmont; see Sardinia.
Pondicherry (Asia); see France.

PORTUGAL.

Length.

			ENGLISH VALUE.		
12 pontos	make	1 linha or line	0·090 inches.		
12 linhas	„	{ 1 pollegada, thumb, or inch }	1·082 „		
8 pollegadas	„	1 palmo or span	8·656 „	or 0·7214 feet.	
5 palmos	„	1 vara or yard	43·28 „	3·6067 „	
2 varas	„	{ 1 braça or fathom }	86·56 „	7·214 „	

The grao (of barley, in width) is 2 linhas	0·1803 inches	
„ dedo or finger „ 8 linhas	0·7213 „	
„ covada or cubit „ 3 palmos	25·968 „	or 2·164 feet

The commercial covada measures 27 inches.
The Portuguese mile is 1·2786 miles English.
To reduce Portuguese palmos to English feet. Take $\frac{5}{7}$ths and increase the same by its $\frac{1}{100}$th part.

Surface.—The square palmo = 74·926 square inches or 0·5203 square feet.

The geira is 4840 square vara =: 62959 square feet or 1·4453 acres.

Liquid Capacity.

			LISBON.	OPORTO.	
4 quartilhos make	1 canada		0·3034	0·46	⎫
6 canadas	„	{ 1 pote, cantaro, or alqueire }	1·8202	2·76	⎬ gallons.
2 potes	„	1 almüde	3·6405	5·52	⎭

The almüde of { Lisbon = 3·6405, Oporto = 5·5200 } imperial gallons.

PORTUGAL.

Dry Capacity.

			LISBON.	OPORTO.	
8 outavas	make	1 alqueire	0·3720	0·4696	⎫
4 alqueires	,,	1 fanga	1·4878	1·8782	⎬ bushels.
15 fangas	,,	1 moio	22·317	28·173	⎭

ENGLISH VALUE.

The fanga of $\left.\begin{array}{l}\text{Lisbon} = 1\cdot4878 \\ \text{Oporto} = 1\cdot8782\end{array}\right\}$ imperial bushel.

Timber is measured in cubic polegadas; masonry in cubic palmas; earthwork in cubic braças.

A cubic palmo = 648·56 cubic inches or 0·37532 cubic foot.

Weight.

24 graos	make	1 scropulo	18·45 grains.	
3 scropulos	,,	1 outava	55·34 ,,	
8 outavas	,,	1 onça	442·69 ,,	
16 onça	,,	1 arratel or pound	7083 ,,	or 1·01186 lb.
32 arratels	,,	1 arroba	32·3795 lbs.	
4 arrobas	,,	1 quintal	129·518 ,,	

Gold and silver are weighed by the marco of 8 onças; and medicines are weighed by a libra of 12 onças = ¾ arratel.

In Portugal the measures and weights are now those of the metric system; see France, p. 75.

Money.

1 vintem	=	20 reas	1·1 pence sterling.	
1 testoon	=	100 ,,	5·6 ,,	,,
1 patàca	=	320 ,,	17·9 ,,	,,
1 crusàdo	=	400 ,,	22·4 ,,	,,
1 sello or new crusàdo	=	480 ,,	26·9 ,,	,,
Silver crown, 1 milrea	=	1000 ,,	56·0 ,,	,,
,, Brazillian dollar	=	1920 ,,	107·5 ,,	,,
Gold crown, 5 milreas	=	5000 ,,	1*l*. 3*s*. 4*d*.	,,
Moeda of 4000 gold	=	9000 ,,	2*l*. 2*s*. 0*d*.	,,
,, 6400 ,,	=	16000 ,,	3*l*. 14*s*. 8*d*.	,,
Milrea of Madeira			50*d*.	,,
,, ,, Azores			41·6*d*.	,,

The rei is an imaginary coin of reckoning, and the milrea = 1000 reis is usually written 1&000; also 1000 milreis, or one million of reis, is called a conto of reis, and written 1:000&000.

PORTUGAL. 97

In 1834, foreign moneys were ordered to be received as a legal tender at the rate of 4120 reis for an English sovereign, and 870 reis for a Spanish or Mexican dollar.

Prague; see Bohemia.
Prince Edward's Island; see British Possessions in North America.
Presburg; see Vienna.

PRUSSIA.
Length.

			ENGLISH VALUE.
12 scrupel	make	1 linie	0·086 inches.
12 linien	,,	1 zoll	1·03 ,,
12 zoll	,,	1 fuss / Rhein-fuss	12·357 ,, or 1·0298 feet.
12 Rhein-fuss	,,	1 ruthe	12·357 feet or 4·119 yards.
2000 ruthen	,,	1 post-meile	8238 yards or 4·6807 miles.
The elle is 25½ zoll or 2⅛ Rhein-fuss			26·258 inches = 2·1882 feet or 0·7294 yard.

Surface.

The morgen is 180 square ruthen | 3054 square yards or 0·6310 acre.

Liquid Capacity.

2 össel	make	1 quartier	70 cubic inches or 0·252 gallons.
30 quartier	,,	1 anker	2096 ,, ,, 7·559 ,,
2 anker	,,	1 eimer	4192 ,, ,, 15·118 ,,
2 eimer	,,	1 ohm	8384 ,, ,, 30·237 ,,
6 ohm	,,	1 fuder	50304 ,, ,, 181·4 ,,

Dry Capacity.

4 mässchen	make	1 metze	209·6 cubic inches or 0·0945 bushels.
4 metzen	,,	1 viertel	838·5 ,, ,, 0·3780 ,,
4 viertel	,,	1 scheffel	3354 ,, ,, 1·5121 ,,
12 scheffel	,,	1 malter	18·145 bushels or 2·2681 quarters.
6 malter	,,	1 last	108·870 ,, 1·3609 last.
The wispel is 18 scheffel			27·2175 ,, 3·4022 quarters.

Weight.

4 quentchen	make	1 loth	225·6 grains or 0·5156 oz.
2 loth	,,	1 unze	451·1 ,, 1·0311 ,,
8 unzen	,,	1 mark (Cologne)	3609 ,, 0·5156 lbs.
2 mark	,,	1 pfund	7218 ,, 1·0311 ,,

PRUSSIA.

The pfund is $\frac{1}{66}$ of a Rhineland cubic foot of distilled water, weighed, and reduced to a vacuum, at the temperature of 15° Réaumur or 65$\frac{3}{4}$° Fahrenheit.

			ENGLISH VALUE.
The liespfund	is	16½ pfund	17·014 lbs.
The centner = 5 stein	„	110 „	113·426 „
The schiffpfund = 3 centner	„	330 „	310·277 „

For weighing Gold and Silver.

288 grains make 1 Cologne mark = 3609 English grains.

Apothecaries' Weight.

20 gran	make	1 scrupel	18·8 grains or 0·039 oz. troy.
3 scrupel	„	1 drachme	56·4 „ 0·117 „ „
8 drachmes	„	1 unze	451·1 „ 0·940 „ „
12 unzen	„	1 pfund	5413·5 „ 11·278 „ „

Money.

| 12 pfennige | make | 1 groschen | 1¼d. sterling. |
| 30 silver groschen | „ | 1 rix dollar or thaler | 2s. 10¾d. „ |

The dollar used formerly to be divided into 24 good groschen.

The Cologne mark weight of *fine* silver is coined into 14 Prussian dollars, from which the above sterling value is calculated.

The gold ducat is estimated at 9s. 3d., and the Frederick at 16s. 4d. sterling.

For the new imperial measures, weights, and moneys, now in use, see Germany, p. 78.

Quebec ; see Great Britain.
Rangoon ; see Birmah.
Revel ; see Russia.
Riga ; see Russia.
Rio de Janeiro ; see Brazil.
Rochelle ; see France.

ROMAN STATES (ITALY).

1. ROME.

Length (Commercial).

		ENGLISH VALUE.
The pié or foot measures		11·592 inches or 0·966 feet.
The palmo	,,	8·796 ,, 0·733 ,,
The canne	,,	78·4 ,, 6·533 ,,
The braccio	,,	30·732 ,, 2·561 ,,

For Cloth, &c.

The palmo measures	8·347 inches.
The fathom of 3 palmi ,,	25·041 ,,
The fathom of 4 palmi ,,	33·388 ,,

Length (Architects, &c.).

10 decimi	make	1 oncia	0·73 inches.	
12 oncie	,,	1 palmo	8·79 ,,	or 0·7325 feet.
10 palmi	,,	1 canna	87·9 ,,	7·3250 ,,
The pié	is	16 oncie	11·72 ,,	0·9767 ,,
The catèna } 10 stajol o }	,,	57½ palmi	42·119 feet.	
The ancient foot	=		11·62 inches.	

Liquid Capacity.

4 quartucci	make	1 foglietta	0·1003 gallons.
4 fogliette	,,	1 boccale	0·4012 ,,
32 boccali	,,	1 barile	12·84 ,,
16 barili	,,	1 botte	205·44 ,,

Dry Capacity.

4 quartucci	make	1 scorzo	0·3682 bushels.
1¾ scorzi	,,	1 starello	0·5063 ,,
4 starelli	,,	1 quarta	2·0251 ,,
4 quarte	,,	1 rubbio	8·1004 ,,
The stairo	is	½ quarta	0·6750 ,,

Weight.

24 grani	make	1 denaro	18·2 grains.	
24 denari	,,	1 oncia	436·2 ,,	or 0·9970 oz.
12 oncie	,,	1 libbra	5234 ,,	,, 0·7477 lb.
The ancient libbra			4966	,, 0·7094 ,,

ROMAN STATES.

Money.

	ENGLISH VALUE.
10 bajocchi make 1 paolo	5d. sterling.
10 paoli ,, 1 Roman scudo (silver crown)	50½d. ,,
The 10 scudi gold piece =	42s. 8d. ,,
The gold sequin =	9s. 2d. ,,
,, ,, pistole =	13s. 6d. ,,

For more recent information see Italy, p. 86.

2. BOLOGNA.

Liquid Capacity.

4 fogliette	make 1 boccale	0·288 gallons.
15 boccali	,, 1 quarterone	4·32 ,,
4 quarteroni	,, 1 corba	17·28 ,,

Dry Capacity.

4 quarticini	make 1 quarterone	0·27 bushels.
4 quarteroni	,, 1 stájo	1·08 ,,
2 staja	,, 1 corba	2·16 ,,

The capacity of the corba is the same in both liquid and dry measure.

Weight.

4 grani	make 1 carato	2·9 grains.	
10 carati	,, 1 ferlino	29·1 ,,	
2 ferlini	,, 1 ottavo	58·2 ,,	
8 ottavi	,, 1 oncia	465·5 ,,	or 1·064 oz.
12 oncie	,, 1 libbra	5586 ,,	0·798 lb.

For more recent information see Italy.

3. ANCONA.

The foot	measures	15·384 inches or 1·282 feet.
,, braccio or ell	,,	25⅓ ,, 2⅛ ,,
,, soma	,,	18·9 gallons.
,, rubbio	,,	7·73 bushels.
,, commercial libbra weighs		5094 grains or 0·7277 lb.

For more recent information see Italy, p. 86.

Rotterdam ; see Holland.
Rouen ; see France.

RUSSIA.

Length.

			ENGLISH VALUE.
16 verschoks	make	1 archine	28 inches.
3 archines	,,	1 sachine	7 feet.
500 sachines	,,	1 verst or werst	3500 ,, or 0·6629 miles.
The Lithuania meile is 28530 Rhein-fuss			9793 yards or 5·5641 ,,

Liquid Capacity.

100 tscharkeys	make	1 vedro	750 cubic inches or 2·7049 gallons.
3 vedros	,,	1 anker	2250 ,, ,, 8·1147 ,,
40 vedros	,,	1 sarokowaja	324·588 ,,

Dry Capacity.

2 garnetz	make	1 tschetwerka	400 cubic inches or 0·1803 oushels.
4 tschetwerkas	,,	1 tschetwerik	1600 ,, ,, 0·7213 ,,
2 tschetweriks	,,	1 pajak	3200 ,, ,, 1·4426 ,,
2 pajaks	,,	1 osmin	6400 ,, ,, 2·8852 ,,
2 osmins	,,	tschetwert	12800 ,, ,, 5·7704 ,,

Weight.

96 doli	make	1 zolotnic	65·8 grains or 0·1504 oz.
3 zolotnics	,,	1 loth	197·4 ,, 0·4513 ,,
8 zolotnics	,,	1 lana	526·5 ,, 1·2035 ,,
12 lanas (32 loths)	,,	1 funt or pound	6318·5 ,, 0·90264 lbs.
40 funts	,,	1 pud	36·1056 lbs.
10 puds	,,	1 berkowitz	361·056 ,,
3 berkowits	,,	1 packen	1083·168 ,,

Money.

100 copecs make 1 silver ruble (37½d. sterling).
The gold ducat   (1796)   9s. 5d. sterling.
,,   ,,   imperial (1801)   32s. 2d.   ,,
The paper ruble, 2s. 6d. to 2s. 9d. sterling.

St. Domingo (Hayti); see West Indies.
St. Gall, or St. Gallen; see Switzerland.
St. Petersburg; see Russia.

SARDINIA (ITALY).

1. GENOA.

Length.

			ENGLISH VALUE.	
12 atomi	make	1 punto	0·140 inches.	
12 punti	,,	1 oncia	1·686 ,,	
12 oncie	,,	1 piede liprando	20·228 ,,	or 1·6857 foot.
8 oncie	,,	1 piede manuale	14·712 ,,	1·2260 ,,
5½ oncie	,,	1 palmo	9·808 ,,	0·8173 ,,
2½ palmi	,,	1 braccio	22·885 ,,	1·9071 ,,
9 palmi nearly	,,	1 canna	87·60 ,,	7·30 ,,

Liquid Capacity.

90 amole or 50 pinte	} make 1 barile	16·337 gallons.
2 barili	,, 1 mezzaruòla	32·674 ,,

Dry Capacity.

12 gombette	make	1 quarto	0·415 bushels.
8 quarti	,,	1 mina	3·321 ,,

Weight.

24 grani	make	1 denaro	17 grains.	
24 denari	,,	1 oncia	407·7 ,,	or 0·8491 oz. troy.
12 oncie	,,	1 libbra	4892 ,,	0·8494 lb. ,,
18 oncie	,,	1 rottolo	7338 ,,	1·0483 ,, avoirdupois.

Money.—100 centesimi make 1 lira nuova (9·12*d.* sterling). Also see Italy, p. 86.

2. PIEDMONT : TURIN.

Length.

12 atomi	make 1 punto	0·140 inches.
12 punti	,, 1 oncia	1·686 ,,
12 oncie	,, 1 piede liprando	20·228 ,,
6 piede liprando	,, 1 trabucco	10·114 feet.
2 trabucci	,, 1 pertica	20·228 ,,
The raso or ell is 14 oncie		23 60 inches.
The miglio is 4333⅓ piedi liprando		7305 feet or 1·3835 miles.

SARDINIA. 103

Surface.—The giornata is 100 square pertica or 14400 square piedi liprando = 40917 square feet or 0·9393 acre.

Liquid Capacity.

			ENGLISH VALUE.
2 quartini	make	1 boccale	0·1722 gallons.
2 boccali	„	1 pinta	0·3444 „
6 pinte	„	1 rubbio	2·0664 „
6 rubbi	„	1 brenta	12·3984 „
10 brente	„	1 carro	123·984 „

Dry Capacity.

20 cucchiari	make	1 copello	0·066 bushels.
4 copelli	„	1 quartière	0·264 „
2 quartieri	„	1 mina	0·527 „
2 mine	„	1 stájo	1·054 „
3 staja	„	1 sacco	3·162 „

Weight.

24 granotini	make	1 grano	0·8 grains.		
24 grani	„	1 denaro	19·8 „		
3 denari	„	1 ottavo	59·3 „		
8 ottavi	„	1 oncia	474·4 „	or 0·9883 oz troy.	
12 oncie	„	1 libbra	5693 „	0·9883 lb. „
The marco	is	8 oncie	3795 „	0·6589 „ „

For gold and silver the carato is 4 grani.
The apothecaries' pound is 1¼ marco.
For more recent information see Italy, p. 86.

Savoy; see Sardinia.

SAXONY (GERMANY): DRESDEN AND LEIPZIC.

Length.

10 linien	make	1 zoll	0·929 inches.	
12 zoll	„	1 fuss	11·148 „	or 0·9290 foot.
2 fuss	„	1 elle	1·858 feet	0·6193 yards.
8 ellen	„	1 ruthe	14·864 „	4·9547 „
The post-meile is 24000 fuss			7432 yards or 4·2227 miles.	
Leipzic (architects') fuss			11·13 inches or 0·9275 foot.	

SAXONY.

Liquid Capacity.

			ENGLISH VALUE.
4 quartier	make	1 nössel	0·1325 gallons.
2 nössel	,,	1 kanne	0·2650 ,,
63 kannen	,,	1 eimer	16·6942 ,,
2 eimer	,,	1 ohm	33·3883 ,,
6 ohm	,,	1 fuder	200·330 ,,
The anker is		54 kannen	14·309 ,,
,, oxhoft ,,		3 eimer	50·083 ,,
,, fass ,,		5 ,,	83·471 ,,

The Dresden liquid measures are ⅙th less.

Dry Capacity.

4 mässchen	make	1 metze	0·1786 bushels.		
4 metzen	,,	1 viertel	0·7146 ,,		
4 viertel	,,	1 scheffel	2·8583 ,,	or 0·3573 quarters.	
12 scheffel	,,	1 malter	34·3 ,,	4·2875 ,,
2 malter	,,	1 wispel	68·6 ,,	8·5750 ,,

Weight.

15 gran	make	1 pfennig	14·1 grains.	
4 pfennig	,,	1 quentlein	56·4 ,,	or 0·1289 oz.
4 quentlein	,,	1 loth	225·5 ,,	0·5154 ,,
2 loth	,,	1 unze	451 ,,	1·0309 ,,
8 unzen	,,	1 mark	3608 ,,	0·5154 lb.
2 mark	,,	1 pfund	7216 ,,	1·0309 ,,

Money (same standard as Prussia).

12 pfennige	make	1 neu-groschen	1⅙d. sterling.
30 neu-groschen	,,	1 dollar or thaler	34¾d. ,,
	The gold ducat		9s. 5d. ,,

For the new imperial measures, weights, and moneys, now in use, see Germany, p. 78.

Seville; see Spain.

SIAM (ASIA).

Length

12 nins make 1 kub	10 inches
2 kubs ,, 1 sok	20 ,,
2 soks ,, 1 ken	40 ,,
2 kens ,, 1 wa	80 ,,
The roënong of 2000 vouhs =	12612 feet or 2·3886 miles.

SIAM. 105

Liquid Capacity.

100 nins make 1 thanan or cocoa-nut-shell | 57·87 cubic inches.

Dry Capacity.

ENGLISH VALUE.

2000 nin make 1 thangsat or bucket | 0·52178 bushel = 1157·41 cubic inches.

Weight.

4 bats or ticals make 1 tael | 936·25 grains or 0·13375 lbs.
20 taels make 1 chang or catty | 2·675 lbs.
50 changs or catties make 1 hap or pecul | 133·75 ,,

Money.—The coins are gold ticals, which pass for 10 silver ticals, each of the latter being worth about 26*d.*

Siberia; see Russia.

SICILY.

Length.

The palmo is 9·53 inches.
The canna (8 palmi) is 76·25 inches.
Archimedes foot ,, 8·76 inches.

Capacity.

The salma of Messina is 19·226 gallons.
,, ,, Syracuse is 16·823 ,,
,, ,, grossa (dry measure) is 9·472 bushels.
,, ,, generale ,, ,, 7·630 ,,

Weight.

The libbra of 12 oncie is 0·7 lb.
,, rottolo grosso of 33 oncie ,, 1·925 lb.
,, ,, sottile of 30 oncie ,, 1·75 ,,

Money.

20 grani make 1 taro | 4·1*d.* sterling.
12 tari ,, 1 scudo | 4*s.* 1·4*d.* ,,
2½ scudi = 30 tari } ,, 1 oncia | 10*s.* 3½*d.* ,,

For more recent information see Italy, p. 86.

Smyrna (Asia); see Turkey.

SPAIN: MADRID AND CASTILE.

Length.

			ENGLISH VALUE.	
12 puntos	make	1 linea	0·077 inches.	
12 lineas	,,	1 pulgada	0·927 ,,	
6 pulgadas	,,	1 sesma	5·564 ,,	
2 sesmas	,,	1 pie (foot)	11·128 ,,	or 0·9273 feet.
3 pie	,,	1 vara	33·384 ,,	2·782 ,.
4 varas	,,	1 estadal	133·536 ,,	11·128 ,,
The dedo	is	9 lineas	0·6955 ,,	
The palmo	,,	12 dedos	8·346 ,,	
The legua	,,	8000 vara	22256 feet or 4·2152 miles.	

Liquid Capacity (Wine, &c.).

4 copas	make	1 quartillo	0·1105 gallons.
4 quartillos	,,	1 azumbre	0·4422 ,,
8 azumbres	,,	1 arroba or cantaro	3·5380 ,,

The arrōba for oil contains 2·78 English gallons and is divided into 4 quartillos or 100 panillas. The standards of the arrōba are 84 libras of water and 25 libras of oil.

Dry Capacity.

4 ochavillos	make	1 racion	0·0081 bushels.	
4 raciones	,,	1 quartillo	0·0323 ,,	
2 quartillos	,,	1 medio	0·0646 ,,	
2 medios	,,	1 almude	0·1292 ,,	
12 almudes	,,	1 fanega	1·5503 ,,	
12 fanegas	,,	1 cahiz	18·6034 ,,	or 2·3254 quarters.

Weight.

12 granos	make	1 tomin	0·2 grains.	
3 tomines	,,	1 adarme	27·7 ,,	
2 adarmes	,,	1 ochava or dracma	55·5 ,,	
8 ochavas	,,	1 onza	443·8 ,,	or 0·0634 lb.
8 onzas	,,	1 marco	3550·5 ,,	0·5072 ,,
2 marcos	,,	1 libra	7101 ,,	1·0144 ,,

SPAIN.

Money.

20 reals vellon or ⎫ make ⎧ 1 duro, piastre, ⎫ (50d. sterling).
10½ reals of plate ⎭   ⎩ or hard dollar   ⎭

The reals are each of them divided into 34 maravedises.

The legal money of Spain is founded on the reals vellon which are now more commonly divided into 10 decimas or 100 centenas.

The Spanish dollar is by law also divided into 100 cents.

 The gold pistole is estimated at 15s. 11d. sterling.
  ,, ,, doubloon ,, ,, 65s. 10d. ,,
 The silver peseta is one-fifth of the piastre.

The metric system of measures, weights, and decimal system of moneys is now adopted. See France, p. 75.

 Stockholm; see Sweden and Norway.
 Strasbourg; see Germany.

SWEDEN AND NORWAY.

Length.

			ENGLISH VALUE.
12 linies	make	1 tum	0·9742 inches.
12 tums	,,	1 fot	11·6904 ,, or 0·9742 foot.
2 fots	,,	1 aln	23·3808 ,, 1·9484 ,,
3 alns	,,	1 famn	5·8452 feet.
The stång is		8 aln	15·5872 ,,
The mil	,,	6000 famn	11630 yards or 6·6423 miles.

The fot is now decimally divided.

Liquid Capacity.

4 jungfrus	make	1 qwarter	0·5756 pints.
4 qwarters	,,	1 stop	2·3024 ,,
2 stops	,,	1 kanna	4·6048 ,, or 0·5756 gallon.
48 kannas	,,	1 tunna	27·6288 gallons.

Dry Capacity.

4 orts	make	1 qwarter	0·0090 bushels.
4 qwarters	,,	1 stop	0·0360 ,,
2 stops	,,	1 kanna	0·0720 ,,
7 kannas	,,	1 fjerding	0·5038 ,,
4 fjerdings	,,	1 spann	2·0150 ,,
2 spanns	,,	1 tunna of ⎫ 32 kappe ⎭	4·0300 ,, or 0·50375 quarter.

Weight (Commercial).

The smallest denomination of weight in Sweden is the as = 0·7418 English grain.

		ENGLISH VALUE.
4 qwintin make 1 lod		205·1 grains.
2 lods „ 1 untz		410·2 „
16 untzs „ 1 skälpund = 8848 as		6563·2 „ or 0·9376 lb.
The lispund is 20 skälpunds		18·752 lbs.
The skeppund is 400 skälpunds		375·040 „

Weight (Gold and Silver).

4 qwintin make 1 lod	203·2 grains or 0·4234 oz. troy.
2 lods „ 1 untz	406·5 „ 0·8469 „ „
8 untzs „ 1 mark = 4384 as	3252 „ 6·7750 „ „

Apothecaries' Weight.

20 grains make 1 scrupel	19·1 grains or 0·0398 oz. troy.
3 scrupels „ 1 drachma	57·3 „ 0·1194 „ „
8 drachmas „ 1 untz	458·4 „ 0·9550 „ „
12 untzs „ 1 skalpund = 7416 as	5501 „ 11·4604 „ „

Money.

12 runstycken make 1 skilling	1·1d. sterling.
48 skillings „ 1 riksdaler (silver specie)	53d. „
The gold ducat	9s. 2d. „

The riksdaler banco is only ⅔ the above value.

In Norway the riksdaler is divided into 120 skillings.

The measures, weights, and currency of Sweden and Norway are since 1875 established on the decimal system. The monetary basis is now the following:—

Money.

| 100 öre make 1 riksdaler | 1s. 1·3d. sterling. |

The measures and weights, now in use, are those of the metric system of France. See p. 75.

SWITZERLAND.

1. BERNE.

Length.

			ENGLISH VALUE.		
12 secundes make	1 linie		0·08 inches.		
12 linien	"	1 zoll	0·96	"	
12 zoll	"	1 fuss	11·54	"	or 0·9617 feet.
10 fuss	"	1 ruthe	115·40	"	9·6167 "
The ell		=	21·40	"	1·7833 "
The Swiss meile is 26666⅔ fuss		=	8548 yards or 4·8568 miles.		

Surface.—The juchart or feld acker, is 400 square ruthen = 4110 square yards or 0·8492 acre.

Liquid Capacity.

2 bechers make	1 vierteli	0·0919 gallons.	
4 vierteli	" 1 mass	0·3676	"
25 mass	" 1 eimer	9·19	"
4 eimer	" 1 saum	36·76	"
6 saum	" 1 landfass	220·56	"

Dry Capacity.

2 sechszehnerli make	1 achterli	0·0482 bushels.	
2 achterli	" 1 immi	0·0964	"
2 immi	" 1 mässli	0·1927	"
2 mässli	" 1 mäss	0·3854	"
12 mäss	" 1 mütt	4·6250	"

Weight (Commercial).

4 pfennig make	1 quent	63 grains.		
4 quent	" 1 loth	251·9	"	
2 loth	" 1 unze	503·7	"	or 1·1514 oz.
16 unzen	" 1 pfund	8060	"	1·1514 lb.

Weight (Gold and Silver).

4 pfennig make	1 quent	59·5 grains.			
4 quent	" 1 loth	238·1	"	or 0·4961 oz. troy.	
2 loth	" 1 unze	476·3	"	0·9923 "	"
8 unzen	" 1 marc	3810·3	"	7·9381 "	"
2 marc	" 1 pfund	7620·6	"	15·8762 "	"

SWITZERLAND.

Apothecaries' Weight.

			ENGLISH VALUE.
20 gran	make 1 scrupel	19·05 grains.	
3 scrupel	,, 1 drachma	57·15 ,,	
8 drachmen	,, 1 unze	457·2 ,,	or 0·9525 oz. troy.
12 unzen	,, 1 pfund	5486·4 ,,	11·430 ,, ,,

Money.

The silver franc (1803)	1s. 2d. sterling.
,, gold ducat	9s. 2d. ,,
,, ,, pistole	18s. 10d. ,,

Since 1878 the legal system of measures, weights, and moneys have been uniformly the same as those of France. See p. 75.

2. LUCERNE.

Length.—The elle is 2 schuh (or Rhein-fuss) = 24·714 inches, or 2·0595 feet.

Liquid Capacity.

10 primas make 1 schoppen	0·0951 gallons.
4 schoppen ,, 1 mass	0·3803 ,,
100 mass ,, 1 saum	38·035 ,,
The ohm is 30 mass	11·410 ,,

Dry Capacity.

10 primas make 1 becher	0·0598 bushels.
16 primas ,, 1 immi	0·0956 ,,
10 immi ,, 1 viertel	0·9561 ,,
4 viertel ,, 1 mütt	3·8245 ,,
4 mütt ,, 1 malter	15·2980 ,,

Weight.

| 4 quentchen make 1 loth | 226 grains or 0·5167 oz. |
| 36 loth ,, 1 pfund | 8138 ,, 1·1626 lb. |

The French metric system of measures and weights has been adopted since 1878; see p. 75.

3. ZURICH.

Length.

12 linien make 1 zoll	0·984 inches.	
12 zoll ,, 1 fuss	11·81 ,,	or 0·9842 foot.
(Builders) ,,	11·86 ,,	0·9883 ,,
The ell =	23·64 ,,	1·9700 ,,

SWITZERLAND. 111

Liquid Capacity.

			ENGLISH VALUE.
2 stotzen	make 1 quärtli		0·1807 gallons.
2 quärtli	„ 1 mass		0·3614 „
15 mass	„ 1 viertel		5·421 „
4 viertel	„ 1 eimer		21·684 „
	The kopt is 2 mass		0·7228 „

The land-mass measures are ⅕th greater.

Dry Capacity.

2¼ immi	make 1 mässli	0·0355 bushels.
4 mässli	„ 1 vierling	0·1422 „
4 vierling	„ 1 viertel	0·5688 „
4 viertel	„ 1 mütt	2·275 „
4 mütt	„ 1 malter	9·10 „

Weight.

12 quenten make 1 loth	226 grains or 0·5167 oz.
2 loth „ 1 unze	452 „ 1·0334 „
18 unzen „ 1 pfund	8138 „ 1·1626 lb.

Money.

| The silver crown (1781) | 3s. 8d. sterling. |
| „ gold ducat | 6s. 4d. „ |

Since 1873 the measures, weights, and moneys have been the same as those of France; see page 75.

4. GENEVA.
Length.

| The foot | = | 23·028 inches or 1·9190 feet. |
| The ell | = | 45·04 „ 3·7533 „ |

Liquid Capacity.

2 pots	make 1 quarteron	0·415 gallons.
24 quarterons	„ 1 setter	9·954 „
12 setters	„ 1 char	119·448 „

Dry Capacity.—The coupe (or sack) = 2·135 bushels.

Weight.

24 grains	make 1 dernier	19·7 grains.
24 dernier	„ 1 unze	472·2 „ or 1·0794 oz.
18 unzen	„ 1 pfund	8500 „ 1·2143 lb.

Money (same as France).—100 cents make 1 lira nuova (franc) (9¼d. sterling).

Since 1873 the measures and weights have been those of the metric system of France; see p. 75.

5. BASLE OR BÂLE.
Length.

		ENGLISH VALUE.	
The foot	=	11·74 inches or	0·9783 feet.
The aune (large ell)		46·4 ,,	3·8667 ,,
The brasse (small ell)		21·4 ,,	1·7833 ,,

Liquid Capacity.

32 pott make 1 ohm	11 gallons.
3 ohm ,, 1 saum	33 ,,

Dry Capacity.

2 becher make 1 köpflein	⅛ bushel.
32 köpflein ,, 1 coupe or sack	3⅝ bushels.

Weight.

72 grains make 1 gros	59·4 grains.	
8 gros ,, 1 unze	475 ,, or	1·0857 oz.
16 unzen ,, 1 pfund	7600 ,,	1·0857 lb.

Money.

100 raps make 1 Swiss frank	13½d. sterling.
The silver crown	3s. 6d. ,,

Since 1873 the measures, weights, and moneys have been the same as those of France; see p. 75.

6. ST. GALLEN.
Length.

The ell for cloth	=	24·2 inches.
,, ,, silks	=	31·6 ,,

Liquid Capacity.—The cimer (containing 32 mass)=11 gallons.

Dry Capacity.—The mütt (containing 4 viertel) = 2·1 bushels.

Weight.

The heavy pound	=	9016 grains or	1·288 lb.
The light ,,	=	7175 ,,	1·025 ,,

SWITZERLAND.

Money.—60 kreuzers make 1 florin or guilder (20¼d. sterling).

Since 1873 the measures, weights, and moneys have been the same as those of France; see p. 75.

7. NEUFCHATEL.

		ENGLISH VALUE.
The foot	=	11·81 inches.
The ell	=	43·80 ,,

The French metric system is now adopted throughout Switzerland.

Toulon; see France.

Trieste; see Austria.

TRIPOLI (BARBARY).

Length.

The Turkish dreah or pik = 3 palmi, is 26·42 English inches.
The arbi dreah or lesser pic is 19·03 ,, ,,

Liquid Capacity.—The barile = 24 bozze, is 8956 cubic inches or 14·267 imperial gallons.

Dry Capacity.

2 nufs-orbah make 1 orbah	409·4 cubic in. or 0·1846 bushels.
4 orbahs ,, 1 temen	1637·7 ,, ,, 0·7383 ,,
4 temen ,, 1 ueba	6551 ,, ,, 2·9533 ,,

Weight.

16 kharoubas make 1 dram	48 grains.
10 drams ,, 1 okie (oz.)	480 ,, or 1 oz. troy.
16 okies ,, 1 rottol	7680 ,, 1⅓ lb. ,,
100 rottols ,, 1 cantar	109⅔ lb. avoird. or 133¼ lbs. troy.

The metrical (for gold and silver) is 73·6 grains.

Money.

| 40 paras make 1 piastre | 2½d. |
| 20 piastres ,, 1 mahhub | 4s. 2d. |

Turin; see Sardinia.

TURKEY.

Length.—The pic or pike is 26·8 inches.

Liquid Capacity.

The almud is 319·4 cubic inches or 1·152 imperial gallon.
The almud of oil should weigh 8 okes or 22⅔ lbs. avoirdupois.

TURKEY.

Dry Capacity.

The killow contains 2023 cubic inches or 0·912 bushel.
The fortin is 4 killows = 8092 cubic inches or 3·648 bushels.

The killow of rice is supposed to weigh 10 okes.

Weight.

		ENGLISH VALUE.
100 drams	make 1 chequee	4950 grains or 0·7072 lbs.
4 chequees	,, 1 oke	19800 ,, 2·8286 ,,
45 okes	,, 1 kintal or cantaro	127·3 lbs.
The rotolo of 18 drams		8910 grains or 1·2729 lbs.

Money.—40 paras make 1 piastre (2¼d. sterling).
Piastre of Selim (1801) (13d. ,,).
,, - - - (1818) (9d. ,,).

TUSCANY (ITALY): FLORENCE AND LEGHORN.

Length.

12 denari	make 1 soldo	1·15 inches.
10 soldi	,, 1 palmo	11·49 ,, or 0·9375 feet.
2 palmi	,, 1 braccio	22·98 ,, 1·9150 ,,
4 braccia	,, 1 canna	7·66 feet or 2·5533 yards.
5 braccia	,, 1 canna architects and surveyors	9·575 ,, 3·1916 ,,
The miglio is 2833⅓ braccia		5426 ,, 1·0277 mile.

Liquid Capacity (Wine, &c.).

2 quartucci	make 1 mezzetta	0·1254 gallons.
2 mezzette	,, 1 boccale	0·2508 ,,
2 boccali	,, 1 fiasco	0·5016 ,,
20 fiasci	,, 1 barile weighing 133½ libbre	10·032 ,,

Liquid Capacity (Oil).

2 boccali	make 1 fiasco	0·5016 gallons.
16 fiasci	,, 1 barile weighing 120 libbre	8·026 ,,
2 barili	make 1 soma	16·052 ,,

The barile of spirits also weighs 120 libbre.

TUSCANY.

Dry Capacity.

			ENGLISH VALUE.	
2 bussoli	make	1 quartuccio	0·0105 bushels.	
2 quartucci	,,	1 mezzetta	0·0210 ,,	
2 mezzette	,,	1 metadella	0·0419 ,,	
4 metadelle	,,	1 quarto	0·1676 ,,	
2 quarti	,,	1 mina	0·3352 ,,	
2 mine	,,	1 stájo	0·6704 ,,	
3 staja	,,	1 sacco	2·0112 ,,	
8 sacci	,,	1 mòggio	16·0896 ,,	or 2·0112 quarters.

Weight.

24 grani	make	1 denaro	18·2 grains or 0·0026 lb.
3 denari	,,	1 dramma	54·6 ,, 0·0078 ,,
8 dramme	,,	1 oncia	436·7 ,, 0·0624 ,,
12 oncie	,,	1 libbra	5240·2 ,, 0·7486 ,,

Money.

100 centesimi make 1 lira (7·82d. sterling).
The gold sequin is estimated at 9s. 5d. sterling.
,, ,, rosini ,, 17s. ,,
,, ,, rusponi ,, 28s. 5d. ,,

For more recent information see Italy, p. 86.

UNITED STATES (NORTH AMERICA).

The weights and measures of the United States are precisely the same as those of Great Britain, with the exception of the measures of capacity, for which the old standards are still retained. Thus the unit of DRY CAPACITY is the old Winchester bushel = 2150·42 cubic inches = 7·75556 imperial gallons = 0·96944 [1] imperial bushel.

Also the unit of LIQUID CAPACITY is the old Wine gallon = 231 cubic inches = 0·83311 [1] imperial gallon.

Money.

	ENGLISH VALUE.
100 cents make 1 dollar	50d. sterling.
The gold eagle, 10 dollar piece	2l. 3s. 6d. ,,

---

[1] $0 \cdot 96944 = 1 - \frac{1}{36}\left(1 + \frac{1}{10}\right)$, and $0 \cdot 83311 = 1 - \frac{1}{6}\left(1 + \frac{1}{1000}(1 + \frac{1}{3})\right)$
which expressions may expedite reductions to English imperial measures.

In mercantile transactions between the United States and Great Britain, the dollar is valued at a fixed par of 4s. 6d. sterling, making 444·44 dollars, equal to 100l. sterling; and the variation of exchange is made by a corresponding per centage, premium or discount, on the sterling amounts. But by an Act of Congress, passed July 27, 1842, the Custom-house valuation is fixed at the rate of 4·84 dollars to the pound sterling, thereby making the value of the dollar about 4s. 1·6d. This compared with the par of 4s. 6d. per dollar is equal to a premium of nearly 9 per cent.—See British Possessions in North America.

Valencia; see Spain.

VENETIAN LOMBARDY: MILAN AND VENICE.

Length.

		ENGLISH VALUE.
12 atomi make 1 punto		0·16 inches.
12 punti „ 1 oncia		1·95 „
12 oncie „ 1 braccio		23·42 „
Milan foot		15·62 „
Venice „		13·69 „

New Decimal Metric System introduced in 1803.

10 atomi make 1 dito		0·394 inches.
10 diti „ 1 palmo		3·937 „
10 palmi „ 1 metro or braccio		39·3708 „ or 3·2809 feet.
1000 metri „ 1 miglio		1093·63 yards or 0·6214 mile.

Surface.—100 square palmi make 1 tornatura, identical with the French *are*, 119·6 square yards or nearly $\frac{1}{40}$ acre.

Capacity.

10 coppi make 1 pinta		0·2201 gallon or 0·0275 bushel.
10 pinte „ 1 mina		0·2751 bushels.
10 mine „ 1 soma		2·7512 „ or 0·3439 quarter

Weight.

10 grani make 1 denaro		0·0353 oz.
10 denari „ 1 grosso		0·3527 „
10 grossi „ 1 oncia		3·5274 „
10 oncie „ 1 libbra metrica		2·2046 lbs.

VENETIAN LOMBARDY.

For Gold and Silver.

			ENGLISH VALUE.		
24 grani	make	1 denaro	18·9 grains.		
24 denari	,,	1 oncia	453·4 ,,	or 0·9445 oz. troy.	
8 oncie	,,	1 marco	3627 ,,	7·5562 ,,	,,

Money.

5 centesimi	make	1 soldo Austriaca	(0·4d. sterling).
20 soldi Austriachi	,,	1 lira Austriaca	(8·1d. ,,).

The value of the lira is the same as the 20 kreuzer piece, or ⅓rd of the Austrian florin; and therefore the 3 lire piece is of the same value as the florin.

The silver ducat is estimated at 3s. 2d. sterling.
,, gold sequin (zecchino) ,, 9s. 5d. ,,
,, ,, pistole ,, 15s. 7d. ,,

For more recent information see Italy, p. 86.
Venice; see Venetian Lombardy.
Verona; see Venetian Lombardy.
Vienna; see Austria.
Warsaw (Poland); see Russia.

WEST INDIES.

The weights and measures are generally those of Great Britain.

Money.—100 cents make 1 dollar (50d. sterling). Pounds, shillings, and pence sterling are also used in accounts, the dollar being reckoned at the government par of 4s. 2d., as above, and the 16 dollar piece (Spanish onza), known as the doubloon, at 64s. sterling.

In most of the West India islands, a fixed valuation was established between the currency and nominal sterling, and the variations of exchange were effected by a per centage on the actual sterling, as at present in the Halifax currency. The legislature, in Jamaica, declared, in 1838, that 166l. 13s. 4d., Jamaica currency, should represent 100l. sterling, the same being in the proportion of 5 to 3, and the

Spanish coinage was regulated accordingly. The late currencies have, however, been abolished in all the islands.

The British currency was established in Bermuda in 1842, and all existing contracts were directed to be settled at the rate of 1⅔l. currency per pound sterling, which corresponds with the rate of currency at Jamaica.

WÜRTEMBURG (GERMANY).

Length.

		ENGLISH VALUE.
10 punkte make 1 linie		0·1126 inches.
10 linien „ 1 zoll		1·126 „
10 zolle „ 1 fuss		11·26 „ or 0·9383 foot.
The klafter (6 fuss)	=	5·63 feet.
The ruthe (10 fuss)	=	9·3833 „
The (Stuttgard) elle	=	24·18 inches or 2·015 feet.

Surface.—The morgen (38400 square fuss)=3757 square yards or 0·7763 acre.

Liquid Capacity.

4 schoppen make 1 mass	0·4043 gallons.
10 mass „ 1 immi	4·0432 „
16 immi „ 1 eimer	64·692 „
6 eimer „ 1 fuder	388·15 „

Dry Capacity.

4 viertelein make 1 ecklein	0·0192 bushels.
2 ecklein „ 1 mässlein	0·0384 „
2 mässlein „ 1 achtel	0·0767 „
2 achtel „ 1 viertel	0·1534 „
4 viertel „ 1 simri	0·6136 „
8 simri „ 1 scheffel	4·9090 „

For apothecaries' weight, see Nürnberg.

For the new imperial measures, weights, and moneys, now in use, see Germany, p. 78.

Zurich; see Switzerland.

GENERAL TABLES.

In reducing foreign measures to the corresponding values in other denominations, architects, engineers, and practical men generally, of different countries, often experience considerable perplexity, in consequence of the necessity of frequent and varied references, and the tediousness of arithmetical calculation. The following Tables are designed to simplify and expedite all such reductions. Table I. contains a list of the principal linear measures of the various countries, states, and cities throughout Europe, arranged in an alphabetical order. The names of the places occupy the first column, and the columns of figures on the right exhibit, to four places of decimals, the value of an unit of each respective measure, when estimated in English inches, English feet, Florence braccia, French metres, French pieds, Napolitan palmi, Rhineland feet, Roman palmi, Venice feet, and Vienna feet. Thus, in the line opposite the "Ancona foot," we read that it is equivalent to 15·384 English inches, 1·2820 English feet, 0·6695 parts of a Florence braccio, 0·3908 parts of a metre, 1·2029 French feet, 1·4818 Napolitan palmi, or 1·2449 Rhineland feet, &c. Table II. shows the comparisons of square or superficial measures, on precisely the same plan as Table I.

The following examples will practically exemplify their use.

Example 1. Reduce 326 Bergamo linear feet to the corresponding measure in Rhineland feet.

In Table I., opposite to Bergamo foot, and under Rhineland feet, we see that a Bergamo foot is equal to 1·3896 Rhineland feet. Therefore, multiplying 1·3896 by 326, we get 453·0 for the required measure in Rhineland feet.

Example 2. Reduce 218·54 Frankfort feet into English feet.

Referring to Table I., opposite Frankfort foot, and in the second column under English feet, we observe that a Frankfort foot is equal to 0·9392 parts of an English foot. Therefore, multiplying 218·54 by 0·9392, the result is 205·25 English feet.

Example 3. Reduce 215·36 Malta square feet into Venice square feet.

In Table II., opposite Malta foot and under Venice feet, we take out 0·6656 for the parts of a Venice foot which are measured by one Malta foot. Consequently by multiplying 215·36 by 0·6656, the answer is 143·34 Venice superficial feet.

Example 4. Reduce 562·18 Palermo square palmi to English feet.

Referring to Table II., as before, we find a Palermo square palmo is equal to 0·6308 parts of an English square foot. Hence multiply 562·18 by 0·6308, and the required result is found to be 354·62 English square feet.

TABLE I.

CONVERSION OF STANDARD LINEAR MEASURES.

Linear Measure.		English Inches.	Feet.	Florence Braccia.	French Mètres.	Pieds.
Aix-la-Chapelle	foot	11·410	0·9508	0·4966	0·2898	0·8921
Amsterdam	foot	11·150	0·9292	0·4853	0·2832	0·8719
Ancona	foot	15·384	1·2820	0·6695	0·3908	1·2029
Anspach	foot	11·720	0·9767	0·5101	0·2977	0·9164
Antwerp	foot	11·240	0·9367	0·4892	0·2855	0·8789
Aquileia	foot	13·530	1·1275	0·5888	0·3437	1·0579
Augsburg	foot	11·650	0·9708	0·5070	0·2959	0·9109
Austria	foot	12·445	1·0371	0·5416	0·3161	0·9731
Baden	foot	11·811	0·9842	0·5140	0·3000	0·9235
Basle	foot	11·740	0·9783	0·5109	0·2982	0·9179
Bavaria	foot	11·420	0·9517	0·4970	0·2901	0·8930
Bergamo	foot	17·172	1·4310	0·7473	0·4362	1·3427
Berlin	foot	12·190	1·0158	0·5305	0·3096	0·9531
Berne	foot	11·540	0·9617	0·5022	0·2931	0·9024
Bohemia	foot	11·670	0·9725	0·5079	0·2964	0·9125
Bologna	foot	14·928	1·2440	0·6497	0·3792	1·1672
Bremen	foot	11·380	0·9483	0·4952	0·2890	0·8898
Brescia	foot	18·720	1·5600	0·8147	0·4755	1·4637
———	braccio	25·104	2·0920	1·0925	0·6376	1·9629
Breslau	foot	11·190	0·9325	0·4870	0·2842	0·8750
Brunswick	foot	11·230	0·9358	0·4887	0·2852	0·8781
Brussels	foot	11·450	0·9542	0·4983	0·2908	0·8953
Cagliari	palmo	7·970	0·6642	0·3469	0·2024	0·6232
Calenberg	foot	11·500	0·9583	0·5005	0·2921	0·8992
Carrara	palmo	9·808	0·8173	0·4268	0·2491	0·7669
Chamberry	foot	13·284	1·1070	0·5781	0·3374	1·0387
China	math. foot	13·120	1·0933	0·5710	0·3332	1·0258
———	imp. foot	12·612	1·0510	0·5489	0·3203	0·9861
Clêves	foot	11·660	0·9717	0·5075	0·2962	0·9117
Cologne	foot	10·830	0·9025	0·4713	0·2751	0·8468

TABLE I.

CONVERSION OF STANDARD LINEAR MEASURES.

Linear Measure.		Napolitan Palmi.	Rhineland Feet.	Roman Palmi.	Venice Feet.	Vienna Feet.
Aix-la-Chapelle	foot	1·0990	0·9233	1·2971	0·8334	0·9168
Amsterdam	foot	1·0740	0·9023	1·2676	0·8145	0·8960
Ancona	foot	1·4818	1·2449	1·7489	1·1237	1·2361
Anspach	foot	1·1289	0·9484	1·3324	0·8561	0·9418
Antwerp	foot	1·0827	0·9096	1·2779	0·8210	0·9032
Aquileia	foot	1·3032	1·0949	1·5382	0·9883	1·0872
Augsburg	foot	1·1221	0·9427	1·3244	0·8509	0·9361
Austria	foot	1·1987	1·0071	1·4148	0·9090	1·0000
Baden	foot	1·1376	0·9557	1·3426	0·8627	0·9490
Basle	foot	1·1308	0·9500	1·3346	0·8575	0·9433
Bavaria	foot	1·1000	0·9242	1·2983	0·8342	0·9177
Bergamo	foot	1·6540	1·3896	1·9522	1·2543	1·3798
Berlin	foot	1·1741	0·9864	1·3858	0·8904	0·9795
Berne	foot	1·1116	0·9339	1·3120	0·8429	0·9273
Bohemia	foot	1·1241	0·9443	1·3267	0·8524	0·9377
Bologna	foot	1·4379	1·2080	1·6971	1·0904	1·1995
Bremen	foot	1·0961	0·9209	1·2937	0·8312	0·9144
Brescia	foot	1·8031	1·5148	2·1281	1·3674	1·5042
——	braccio	2·4180	2·0314	2·8539	1·8337	2·0172
Breslau	foot	1·0778	0·9055	1·2721	0·8174	0·8991
Brunswick	foot	1·0816	0·9087	1·2766	0·8202	0·9023
Brussels	foot	1·1029	0·9266	1·3017	0·8364	0·9201
Cagliari	palmo	0·7677	0·6450	0·9061	0·5822	0·6404
Calenberg	foot	1·1076	0·9306	1·3073	0·8400	0·9240
Carrara	palmo	0·9447	0·7936	1·1150	0·7164	0·7881
Chamberry	foot	1·2794	1·0750	1·5102	0·9703	1·0674
China	math. foot	1·2637	1·0616	1·4915	0·9583	1·0542
——	imp. foot	1·2148	1·0206	1·4338	0·9212	1·0134
Clêves	foot	1·1231	0·9436	1·3256	0·8517	0·9369
Cologne	foot	1·0432	0·8764	1·2312	0·7911	0·8702

122 CONVERSION OF STANDARD

Linear Measure.		English Inches.	Feet.	Florence Braccia.	French Mètres.	Pieds.
Constantinople	pic	26·800	2·2333	1·1663	0·6807	2·0955
Copenhagen	foot	12·357	1·0298	0·5378	0·3139	0·9663
Cracow	foot	14·032	1·1693	0·6107	0·3564	1·0972
Dantzic	foot	11·290	0·9408	0·4913	0·2868	0·8828
Denmark	foot	12·357	1·0298	0·5378	0·3139	0·9663
Dordrecht	foot	14·160	1·1800	0·6162	0·3597	1·1072
Dresden	foot	11·150	0·9292	0·4853	0·2832	0·8719
Embden	foot	11·660	0·9717	0·5075	0·2962	0·9117
England	foot	12·000	1·0000	0·5222	0·3048	0·9383
Farrari	foot	15·804	1·3170	0·6878	0·4014	1·2357
Florence	foot	11·940	0·9950	0·5196	0·3033	0·9336
———	braccio	22·978	1·9148	1·0000	0·5836	1·7966
France	foot	12·790	1·0658	0·5566	0·3249	1·0000
———	mètre	39·371	3·2809	1·7134	1·0000	3·0784
Frankfort	foot	11·270	0·9392	0·4905	0·2863	0·8813
Geneva	foot	23·028	1·9190	1·0022	0·5849	1·8006
Genoa	palmo	9·808	0·8173	0·4268	0·2491	0·7669
——	canna	87·600	7·3000	3·8123	2·2250	6·8495
Gottingen	foot	11·450	0·9542	0·4983	0·2908	0·8953
Gotha	foot	11·320	0·9433	0·4926	0·2875	0·8851
Greece	foot	11·810	0·9842	0·5140	0·3000	0·9235
Groningen	foot	11·490	0·9575	0·5000	0·2918	0·8984
Hamburg	foot	11·290	0·9408	0·4913	0·2868	0·8828
Hanover	foot	11·450	0·9542	0·4983	0·2908	0·8953
Harlem	foot	11·250	0·9375	0·4896	0·2857	0·8797
Heidelberg	foot	10·960	0·9133	0·4770	0·2784	0·8569
Hildesheim	foot	11·050	0·9208	0·4809	0·2807	0·8640
Inspruck	foot	12·500	1·0417	0·5440	0·3175	0·9774
Königsberg	foot	12·110	1·0092	0·5270	0·3076	0·9469
Leghorn	foot	11·904	0·9920	0·5181	0·3024	0·9308
Leipsic	foot	11·130	0·9275	0·4844	0·2827	0·8703
Leyden	foot	12·340	1·0283	0·5370	0·3134	0·9649
Liege	foot	11·320	0·9433	0·4926	0·2875	0·8851
Lindau	com. foot	11·400	0·9500	0·4961	0·2896	0·8914
———	long foot	12·400	1·0333	0·5396	0·3149	0·9695

LINEAR MEASURES.

Linear Measure.		Napolitan Palmi.	Rhineland Feet.	Roman Palmi.	Venice Feet.	Vienna Feet.
Constantinople	pic	2·5814	2·1687	3·0467	1·9575	2·1534
Copenhagen	foot	1·1903	1·0000	1·4049	0·9026	0·9930
Cracow	foot	1·3515	1·1355	1·5952	1·0249	1·1275
Dantzic	foot	1·0874	0·9136	1·2835	0·8246	0·9072
Denmark	foot	1·1903	1·0000	1·4049	0·9026	0·9930
Dordrecht	foot	1·3639	1·1458	1·6098	1·0343	1·1378
Dresden	foot	1·0740	0·9023	1·2676	0·8145	0·8960
Embden	foot	1·1231	0·9436	1·3256	0·8517	0·9369
England	foot	1·1558	0·9710	1·3642	0·8765	0·9642
Farrari	foot	1·5223	1·2789	1·7967	1·1544	1·2699
Florence	foot	1·1501	0·9662	1·3574	0·8721	0·9594
———	braccio	2·2132	1·8593	2·6122	1·6783	1·8463
France	foot	1·2319	1·0350	1·4540	0·9342	1·0277
———	mètre	3·7922	3·1859	4·4758	2·8757	3·1635
Frankfort	foot	1·0856	0·9120	1·2813	0·8232	0·9056
Geneva	foot	2·2181	1·8635	2·6179	1·6821	1·8504
Genoa	palmo	0·9447	0·7936	1·1150	0·7164	0·7881
———	canna	8·4376	7·0886	9·9586	6·3985	7·0388
Gottingen	foot	1·1029	0·9266	1·3017	0·8364	0·9201
Gotha	foot	1·0903	0·9160	1·2869	0·8268	0·9096
Greece	foot	1·1376	0·9557	1·3426	0·8627	0·9490
Groningen	foot	1·1067	0·9298	1·3062	0·8393	0·9233
Hamburg	foot	1·0874	0·9136	1·2835	0·8246	0·9072
Hanover	foot	1·1029	0·9266	1·3017	0·8364	0·9201
Harlem	foot	1·0836	0·9104	1·2789	0·8217	0·9040
Heidelberg	foot	1·0556	0·8868	1·2459	0·8005	0·8806
Hildesheim	foot	1·0643	0·8941	1·2562	0·8071	0·8879
Inspruck	foot	1·2040	1·0115	1·4211	0·9131	1·0044
Königsberg	foot	1·1665	0·9800	1·3768	0·8846	0·9731
Leghorn	foot	1·1466	0·9633	1·3533	0·8695	0·9565
Leipsic	foot	1·0720	0·9006	1·2653	0·8130	0·8943
Leyden	foot	1·1886	0·9985	1·4028	0·9013	0·9915
Liege	foot	1·0903	0·9160	1·2869	0·8268	0·9096
Lindau	com. foot	1·0980	0·9225	1·2960	0·8327	0·9160
———	long foot	1·1943	1·0034	1·4096	0·9057	0·9963

CONVERSION OF STANDARD

Linear Measure.		English Inches.	Feet.	Florence Braccia.	French Mètres.	Pieds.
Lisbon	archit. foot	13·331	1·1109	0·5802	0·3386	1·0424
———	com. foot	12·960	1·0800	0·5640	0·3292	1·0134
Lombardy	arch. foot	15·611	1·3009	0·6794	0·3965	1·2206
Lorraine	foot	11·300	0·9417	0·4918	0·2870	0·8836
Lübeck	foot	11·450	0·9542	0·4983	0·2908	0·8953
Lucca	braccio	23·496	1·9580	1·0225	0·5968	1·8372
Luneburg	foot	11·450	0·9542	0·4983	0·2908	0·8953
Macedonia	foot	13·918	1·1598	0·6057	0·3535	1·0882
Magdeburg	foot	11·160	0·9300	0·4857	0·2835	0·8726
Malta	foot	11·170	0·9308	0·4861	0·2837	0·8734
Manheim	foot	11·410	0·9508	0·4966	0·2898	0·8921
Mantua	braccio	25·104	2·0920	1·0925	0·6376	1·9629
———	brasso	18·252	1·5210	0·7943	0·4636	1·4272
Maestricht	foot	11·050	0·9208	0·4809	0·2807	0·8640
Mentz	foot	11·850	0·9875	0·5157	0·3010	0·9266
Middleburg	foot	11·810	0·9842	0·5140	0·3000	0·9235
Milan	foot	15·620	1·3017	0·6798	0·3968	1·2214
———	dec. foot	10·260	0·8550	0·4465	0·2606	0·8023
———	braccio	23·420	1·9517	1·0192	0·5949	1·8313
———	metro-braccio	39·371	3·2809	1·7134	1·0000	3·0784
Modena	foot	20·592	1·7160	0·8962	0·5230	1·6101
Monaco	foot	9·250	0·7708	0·4025	0·2349	0·7232
Moscow	foot	13·170	1·0975	0·5732	0·3345	1·0298
Munich	foot	11·490	0·9575	0·5000	0·2918	0·8984
Naples	palmo	10·382	0·8652	0·4518	0·2637	0·8118
Naples	canna	83·055	6·9213	3·6146	2·1096	6·4942
Neufchâtel	foot	11·810	0·9842	0·5140	0·3000	0·9235
Normandy	foot	11·720	0·9767	0·5101	0·2977	0·9164
Nuremberg	foot	11·960	0·9967	0·5205	0·3038	0·9352
Oldenburg	foot	11·650	0·9708	0·5070	0·2959	0·9109
Osnaburg	foot	11·000	0·9167	0·4787	0·2794	0·8601
Padua	foot	13·930	1·1608	0·6062	0·3538	1·0892
Palæste	foot	12·138	1·0115	0·5283	0·3083	0·9491
Palermo	palmo	9·530	0·7942	0·4148	0·2421	0·7452
Parma	foot	22·428	1·8690	0·9761	0·5697	1·7537

LINEAR MEASURES.

Linear Measure.		Napolitan Palmi.	Rhineland Feet.	Roman Palmi.	Venice Feet.	Vienna Feet.
Lisbon	archit. foot	1·2840	1·0787	1·5155	0·9737	1·0712
———	com. foot	1·2483	1·0487	1·4733	0·9466	1·0414
Lombardy	arch. foot	1·5036	1·2632	1·7747	1·1402	1·2544
Lorraine	foot	1·0885	0·9144	1·2847	0·8254	0·9080
Lübeck	foot	1·1029	0·9266	1·3017	0·8364	0·9201
Lucca	braccio	2·2631	1·9013	2·6711	1·7162	1·8880
Luneburg	foot	1·1029	0·9266	1·3017	0·8364	0·9201
Macedonia	foot	1·3405	1·1262	1·5822	1·0166	1·1183
Magdeburg	foot	1·0749	0·9031	1·2687	0·8152	0·8967
Malta	foot	1·0759	0·9039	1·2698	0·8159	0·8975
Manheim	foot	1·0990	0·9233	1·2971	0·8334	0·9168
Mantua	braccio	2·4180	2·0314	2·8539	1·8337	2·0172
———	brasso	1·7580	1·4770	2·0750	1·3332	1·4666
Maestricht	foot	1·0643	0·8941	1·2562	0·8071	0·8879
Mentz	foot	1·1414	0·9589	1·3472	0·8656	0·9522
Middleburg	foot	1·1376	0·9557	1·3426	0·8627	0·9490
Milan	foot	1·5046	1·2640	1·7758	1·1410	1·2551
———	dec. foot	0·9883	0·8303	1·1664	0·7494	0·8244
———	braccio	2·2558	1·8952	2·6625	1·7107	1·8819
———	metro-braccio	3·7922	3·1859	4·4758	2·8757	3·1635
Modena	foot	1·9834	1·6663	2·3410	1·5041	1·6546
Monaco	foot	0·8909	0·7485	1·0515	0·6756	0·7432
Moscow	foot	1·2685	1·0657	1·4972	0·9620	1·0582
Munich	foot	1·1067	0·9298	1·3062	0·8393	0·9233
Naples	palmo	1·0000	0·8402	1·1803	0·7584	0·8343
Naples	canna	8·0000	6·7210	9·4421	6·0667	6·6738
Neufchâtel	foot	1·1376	0·9557	1·3426	0·8627	0·9490
Normandy	foot	1·1289	0·9484	1·3324	0·8561	0·9418
Nuremberg	foot	1·1520	0·9678	1·3597	0·8736	0·9610
Oldenburg	foot	1·1221	0·9427	1·3244	0·8509	0·9361
Osnaburg	foot	1·0596	0·8902	1·2506	0·8035	0·8839
Padua	foot	1·3417	1·1272	1·5836	1·0175	1·1193
Palæste	foot	1·1692	0·9822	1·3799	0·8866	0·9753
Palermo	palmo	0·9180	0·7712	1·0835	0·6961	0·7658
Parma	foot	2·1603	1·8149	2·5497	1·6382	1·8021

Linear Measure.		English Inches.	Feet.	Florence Braccia.	French Mètres.	Pieds.
Parma	braccio	21·340	1·7783	0·9287	0·5420	1·6686
Pavia	foot	18·480	1·5400	0·8042	0·4694	1·4450
——	braccio	18·300	1·5250	0·7964	0·4648	1·4309
Persia	arish	38·270	3·1892	1·6655	0·9721	2·9924
Phileterian	foot	13·937	1·1614	0·6065	0·3540	1·0897
Piacenza	foot	22·428	1·8690	0·9761	0·5697	1·7537
Piedmont liprando ft.		20·228	1·6857	0·8803	0·5138	1·5817
—— common ft.		13·484	1·1237	0·5868	0·3425	1·0544
Poland	foot	14·032	1·1693	0·6107	0·3564	1·0972
Pomerania	foot	11·500	0·9583	0·5005	0·2921	0·8992
Portugal archit. foot		13·331	1·1109	0·5802	0·3386	1·0424
Prague	foot	11·880	0·9900	0·5170	0·3018	0·9289
Prussia	foot	12·357	1·0298	0·5378	0·3139	0·9663
Pythian	foot	9·749	0·8124	0·4243	0·2476	0·7623
Ratsburg	foot	11·450	0·9542	0·4983	0·2908	0·8953
Revel	foot	10·530	0·8775	0·4583	0·2675	0·8234
Reggio	braccio	20·850	1·7375	0·9074	0·5296	1·6303
Rhineland	foot	12·357	1·0298	0·5378	0·3139	0·9663
Riga	foot	10·790	0·8992	0·4696	0·2741	0·8437
Rimini	braccio	21·390	1·7825	0·9309	0·5433	1·6725
Rome	common foot	11·592	0·9660	0·5045	0·2944	0·9064
——.	archit. foot	11·720	0·9767	0·5101	0·2977	0·9164
——.	palmo	8·796	0·7330	0·3828	0·2234	0·6878
——.	braccio	30·732	2·5610	1·3375	0·7806	2·4030
——. palmo d'archit.		8·790	0·7325	0·3825	0·2233	0·6873
Rome canna d'archit.		87·900	7·3250	3·8254	2·2326	6·8731
Rostock	foot	11·380	0·9483	0·4952	0·2890	0·8898
Rotterdam	foot	12·357	1·0298	0·5378	0·3139	0·9663
Russia	foot	13·750	1·1458	0·5984	0·3492	1·0751
Sardinia	palmo	9·808	0·8173	0·4268	0·2491	0·7669
Sicily	palmo	9·530	0·7942	0·4148	0·2421	0·7452
—— Archimedes' ft.		8·760	0·7300	0·3812	0·2225	0·6850
Sienna	foot	14·868	1·2390	0·6471	0·3776	1·1625
Spain	foot	11·130	0·9275	0·4844	0·2827	0·8703
Stade	foot	11·450	0·9542	0·4983	0·2908	0·8953

LINEAR MEASURES.

Linear Measure.		Napolitan Palmi.	Rhineland Feet.	Roman Palmi.	Venice Feet.	Vienna Feet.
Parma	braccio	2·0555	1·7268	2·4260	1·5587	1·7147
Pavia	foot	1·7800	1·4954	2·1009	1·3498	1·4849
——	braccio	1·7627	1·4808	2·0804	1·3367	1·4704
Persia	arish	3·6862	3·0968	4·3507	2·7954	3·0751
Philcterian	foot	1·3424	1·1278	1·5844	1·0180	1·1199
Piacenza	foot	2·1603	1·8149	2·5497	1·6382	1·8021
Piedmont liprando ft.		1·9484	1·6369	2·2996	1·4775	1·6254
—— common ft.		1·2988	1·0912	1·5330	0·9849	1·0835
Poland	foot	1·3515	1·1355	1·5952	1·0249	1·1275
Pomerania	foot	1·1076	0·9306	1·3073	0·8400	0·9240
Portugal archit. foot		1·2840	1·0787	1·5155	0·9737	1·0712
Prague	foot	1·1443	0·9613	1·3506	0·8678	0·9546
Prussia	foot	1·1903	1·0000	1·4049	0·9026	0·9930
Pythian	foot	0·9390	0·7889	1·1083	0·7121	0·7833
Ratsburg	foot	1·1029	0·9266	1·3017	0·8364	0·9201
Revel	foot	1·0143	0·8521	1·1971	0·7691	0·8461
Reggio	braccio	2·0083	1·6872	2·3703	1·5229	1·6753
Rhineland	foot	1·1903	1·0000	1·4049	0·9026	0·9930
Riga	foot	1·0393	0·8732	1·2267	0·7882	0 8670
Rimini	braccio	2·0603	1·7309	2·4317	1·5624	1·7187
Rome	common foot	1·1166	0·9380	1·3178	0·8467	0·9315
——	archit. foot	1·1289	0·9484	1·3324	0·8561	0·9418
——	palmo	0·8472	0·7118	1·0000	0·6425	0 7068
——	braccio	2·9601	2·4869	3·4937	2·2448	2·4694
—— palmo d'archit.		0·8467	0·7113	0·9993	0·6421	0·7063
Rome canna d'archit.		8·4666	7·1130	9·9929	6·4205	7·0630
Rostock	foot	1·0961	0·9209	1·2937	0·8312	0·9144
Rotterdam	foot	1·1903	1·0000	1·4049	0·9026	0·9930
Russia	foot	1·3244	1·1126	1·5631	1·0043	1·1048
Sardinia	palmo	0·9447	0·7936	1·1150	0·7164	0·7881
Sicily	palmo	0·9180	0·7712	1·0835	0·6961	0·7658
—— Archimedes' ft.		0·8438	0·7089	0·9959	0·6399	0·7039
Sienna	foot	1·4322	1·2031	1·6902	1·0860	1·1947
Spain	foot	1·0720	0·9006	1·2653	0·8130	0·8943
Stade	foot	1·1029	0·9266	1·3017	0·8364	0·9201

Linear Measure.		English Inches.	Feet.	Florence Braccia.	French. Mètres.	Pieds.
Stettin	old foot	11·120	0·9267	0·4840	0·2825	0·8695
Strasburg	foot	11·390	0·9492	0·4957	0·2893	0·8906
Stuttgard	foot	11·260	0·9383	0·4900	0·2860	0·8804
Sweden	foot	11·690	0·9742	0·5088	0·2969	0·9141
Trent	foot	14·412	1·2010	0·6272	0·3661	1·1269
Turin	liprando foot	20·228	1·6857	0·8803	0·5138	1·5817
——	common foot	13·484	1·1237	0·5868	0·3425	1·0544
——	ras	23·496	1·9580	1·0225	0·5968	1·8372
Turkey	pic	26·800	2·2333	1·1663	0·6807	2·0955
Ulm	foot	11·390	0·9492	0·4957	0·2893	0·8906
Utrecht	foot	10·740	0·8950	0·4674	0·2728	0·8398
Venice	foot	13·691	1·1409	0·5958	0·3477	1·0705
Verona	foot	13·404	1·1170	0·5833	0·3405	1·0481
Vicenza	foot	13·632	1·1360	0·5933	0·3463	1·0659
Vienna	foot	12·445	1·0371	0·5416	0·3161	0·9731
Warsaw	foot	11·725	0·9771	0·5103	0·2978	0·9168
——	Cracow foot	14·032	1·1693	0·6107	0·3564	1·0972
Wismar	foot	11·580	0·9650	0·5040	0·2941	0·9055
Würtemberg	foot	11·260	0·9383	0·4900	0·2860	0·8804
Zell	foot	11·450	0·9542	0·4983	0·2908	0·8953
Ziriczee	foot	12·210	1·0175	0·5314	0·3101	0·9547
Zurich	foot	11·810	0·9842	0·5140	0·3000	0·9235

LINEAR MEASURES.

Linear Measure.		Napolitan Palmi.	Rhineland Feet.	Roman Palmi.	Venice Feet.	Vienna Feet.
Stettin	old foot	1·0711	0·8999	1·2642	0·8123	0·8936
Strasburg	foot	1·0971	0·9217	1·2949	0·8320	0·9153
Stuttgard	foot	1·0845	0·9111	1·2800	0·8224	0·9047
Sweden	foot	1·1260	0·9460	1·3290	0·8539	0·9394
Trent	foot	1·3882	1·1662	1·6384	1·0527	1·1580
Turin	liprando foot	1·9484	1·6369	2·2996	1·4775	1·6254
——	common foot	1·2988	1·0912	1·5330	0·9849	1·0835
——	ras	2·2631	1·9013	2·6711	1·7162	1·8880
Turkey	pic	2·5814	2·1687	3·0467	1·9575	2·1534
Ulm	foot	1·0971	0·9217	1·2949	0·8320	0·9153
Utrecht	foot	1·0345	0·8691	1·2210	0·7845	0·8630
Venice	foot	1·3187	1·1079	1·5564	1·0000	1·1001
Verona	foot	1·2911	1·0847	1·5238	0·9791	1·0770
Vicenza	foot	1·3130	1·1031	1·5497	0·9957	1·0954
Vienna	foot	1·1987	1·0071	1·4148	0·9090	1·0000
Warsaw	foot	1·1294	0·9488	1·3330	0·8564	0·9421
——	Cracow foot	1·3515	1·1355	1·5952	1·0249	1·1275
Wismar	foot	1·1154	0·9371	1·3165	0·8458	0·9305
Würtemberg	foot	1·0845	0·9111	1·2800	0·8224	0·9047
Zell	foot	1·1029	0·9266	1·3017	0·8364	0·9201
Ziriczee	foot	1·1761	0·9880	1·3881	0·8918	0·9811
Zurich	foot	1·1376	0·9557	1·3426	0·8627	0·9490

TABLE II.

CONVERSION OF STANDARD SQUARE MEASURES.

Square Measure.		English Square Inches.	Square Feet.	Florence Square Braccia.	French Square Mètres.	Square Pieds.
Aix-la-Chapelle	foot	130·19	0·9040	0·2466	0·0840	0·7959
Amsterdam	foot	124·32	0·8634	0·2355	0·0802	0·7602
Ancona	foot	236·67	1·6435	0·4482	0·1527	1·4470
Anspach	foot	137·36	0·9539	0·2602	0·0886	0·8398
Antwerp	foot	126·34	0·8774	0·2393	0·0815	0·7725
Aquileia	foot	183·06	1·2713	0·3467	0·1181	1·1192
Augsburg	foot	135·72	0·9425	0·2570	0·0876	0·8297
Austria	foot	154·88	1·0756	0·2933	0·0999	0·9469
Baden	foot	139·48	0·9686	0·2642	0·0900	0·8528
Basle	foot	137·83	0·9571	0·2610	0·0889	0·8426
Bavaria	foot	130·42	0·9057	0·2470	0·0841	0·7974
Bergamo	foot	294·88	2·0478	0·5585	0·1902	1·8029
Berlin	foot	148·60	1·0319	0·2814	0·0959	0·9084
Berne	foot	133·17	0·9249	0·2522	0·0859	0·8143
Bohemia	foot	136·19	0·9458	0·2579	0·0879	0·8326
Bologna	foot	222·85	1·5475	0·4221	0·1438	1·3624
Bremen	foot	129·50	0·8993	0·2453	0·0835	0·7917
Brescia	foot	350·44	2·4335	0·6637	0·2261	2·1425
——	braccio	630·21	4·3764	1·1936	0·4066	3·8530
Breslau	foot	125·22	0·8696	0·2372	0·0808	0·7656
Brunswick	foot	126·11	0·8757	0·2388	0·0814	0·7710
Brussels	foot	131·10	0·9105	0·2483	0·0846	0·8016
Cagliari	palmo	63·52	0·4412	0·1203	0·0410	0·3884
Calenberg	foot	132·25	0·9183	0·2505	0·0853	0·8085
Carrara	palmo	96·20	0·6680	0·1822	0·0621	0·5881
Chamberry	foot	176·46	1·2255	0·3342	0·1138	1·0789
China	math. foot	172·13	1·1953	0·3260	0·1110	1·0523
——	imp. foot	159·06	1·1046	0·3013	0·1026	0·9725
Clêves	foot	135·96	0·9442	0·2575	0·0877	0·8313
Cologne	foot	117·29	0·8145	0·2221	0·0757	0·7171

TABLE II.

CONVERSION OF STANDARD SQUARE MEASURES.

Square Measure.		Napolitan Square Palmi.	Rhineland Square Feet.	Roman Square Palmi.	Venice Square Feet.	Vienna Square Feet.
Aix-la-Chapelle	foot	1·2078	0·8524	1·6824	0·6945	0·8405
Amsterdam	foot	1·1535	0·8141	1·6069	0·6633	0·8027
Ancona	foot	2·1957	1·5497	3·0587	1·2627	1·5281
Anspach	foot	1·2744	0·8995	1·7753	0·7329	0·8869
Antwerp	foot	1·1722	0·8273	1·6329	0·6741	0·8158
Aquileia	foot	1·6984	1·1987	2·3660	0·9767	1·1820
Augsburg	foot	1·2591	0·8887	1·7540	0·7241	0·8762
Austria	foot	1·4369	1·0142	2·0017	0·8263	1·0000
Baden	foot	1·2941	0·9134	1·8027	0·7442	0·9006
Basle	foot	1·2786	0·9024	1·7811	0·7353	0·8898
Bavaria	foot	1·2100	0·8540	1·6856	0·6959	0·8421
Bergamo	foot	2·7358	1·9309	3·8110	1·5733	1·9039
Berlin	foot	1·3785	0·9730	1·9203	0·7928	0·9594
Berne	foot	1·2356	0·8721	1·7212	0·7106	0·8599
Bohemia	foot	1·2635	0·8918	1·7601	0·7266	0·8793
Bologna	foot	2·0675	1·4592	2·8800	1·1889	1·4388
Bremen	foot	1·2014	0·8480	1·6736	0·6909	0·8361
Brescia	foot	3·2512	2·2947	4·5290	1·8696	2·2626
——	braccio	5·8468	4·1267	8·1448	3·3623	4·0689
Breslau	foot	1·1617	0·8199	1·6183	0·6681	0·8085
Brunswick	foot	1·1699	0·8257	1·6297	0·6728	0·8142
Brussels	foot	1·2164	0·8585	1·6945	0·6995	0·8465
Cagliari	palmo	0·5894	0·4160	0·8210	0·3389	0·4102
Calenberg	foot	1·2269	0·8659	1·7091	0·7055	0·8538
Carrara	palmo	0·8924	0·6299	1·2431	0·5132	0·6210
Chamberry	foot	1·6372	1·1555	2·2807	0·9415	1·1394
China	math. foot	1·5969	1·1271	2·2245	0·9183	1·1113
——	imp. foot	1·4757	1·0415	2·0557	0·8486	1·0270
Clêves	foot	1·2614	0·8903	1·7572	0·7254	0·8778
Cologne	foot	1·0882	0·7680	1·5159	0·6258	0·7573

CONVERSION OF STANDARD

Square Measure.		English Square Inches.	Square Feet.	Florence Square Braccia.	French Square Mètres.	Square Pieds.
Constantinople	pic	718·24	4·9877	1·3603	0·4634	4·3912
Copenhagen	foot	152·70	1·0605	0·2892	0·0985	0·9336
Cracow	foot	196·90	1·3673	0·3729	0·1270	1·2038
Dantzic	foot	127·46	0·8851	0·2414	0·0822	0·7793
Denmark	foot	152·70	1·0605	0·2892	0·0985	0·9336
Dordrecht	foot	200·51	1·3924	0·3798	0·1294	1·2259
Dresden	foot	124·32	0·8634	0·2355	0·0802	0·7602
Embden	foot	135·96	0·9442	0·2575	0·0877	0·8313
England	foot	144·00	1·0000	0·2727	0·0929	0·8804
Farrari	foot	249·77	1·7345	0·4731	0·1611	1·5271
Florence	foot	142·56	0·9900	0·2700	0·0920	0·8716
———	braccio	527·99	3·6664	1·0000	0·3406	3·2279
France	foot	163·58	1·1360	0·3098	0·1055	1·0000
———	mètre	1550·08	10·7640	2·9358	1·0000	9·4768
Frankfort	foot	127·01	0·8821	0·2406	0·0819	0·7766
Geneva	foot	530·29	3·6826	1·0044	0·3421	3·2422
Genoa	palmo	96·20	0·6680	0·1822	0·0621	0·5881
———	canna	7673·76	53·2900	14·5340	4·9506	46·9160
Gottingen	foot	131·10	0·9105	0·2483	0·0846	0·8016
Gotha	foot	128·14	0·8898	0·2427	0·0827	0·7834
Greece	foot	139·48	0·9686	0·2642	0·0900	0·8528
Groningen	foot	132·02	0·9168	0·2500	0·0852	0·8072
Hamburg	foot	127·46	0·8851	0·2414	0·0822	0·7793
Hanover	foot	131·10	0·9105	0·2483	0·0846	0·8016
Harlem	foot	126·56	0·8789	0·2397	0·0817	0·7738
Heidelberg	foot	120·12	0·8341	0·2275	0·0775	0·7343
Hildesheim	foot	122·10	0·8479	0·2312	0·0788	0·7465
Inspruck	foot	156·25	1·0851	0·2960	0·1008	0·9553
Königsberg	foot	146·65	1·0185	0·2778	0·0946	0·8967
Leghorn	foot	141·71	0·9841	0·2684	0·0914	0·8664
Leipsic	foot	123·88	0·8602	0·2346	0·0799	0·7574
Leyden	foot	152·28	1·0574	0·2884	0·0982	0·9309
Liege	foot	128·14	0·8898	0·2427	0·0827	0·7834
Lindau com.	foot	129·96	0·9025	0·2461	0·0838	0·7945
——— long	foot	153·76	1·0677	0·2912	0·0992	0·9400

SQUARE MEASURES. 183

Square Measure.		Napolitan Square Palmi.	Rhineland Square Feet.	Roman Square Palmi.	Venice Square Feet.	Vienna Square Feet.
Constantinople	pic	6·6635	4·7031	9·2824	3·8320	4·6372
Copenhagen	foot	1·4168	1·0000	1·9736	0·8147	0·9860
Cracow	foot	1·8267	1·2893	2·5446	1·0505	1·2712
Dantzic	foot	1·1825	0·8346	1·6173	0·6800	0·8229
Denmark	foot	1·4168	1·0000	1·9736	0·8147	0·9860
Dordrecht	foot	1·8602	1·3129	2·5913	1·0697	1·2946
Dresden	foot	1·1535	0·8141	1·6069	0·6633	0·8027
Embden	foot	1·2614	0·8903	1·7572	0·7254	0·8778
England	foot	1·3360	0·9429	1·8611	0·7683	0·9297
Farrari	foot	2·3173	1·6355	3·2280	1·3326	1·6126
Florence	foot	1·3226	0·9335	1·8425	0·7606	0·9204
———	braccio	4·8983	3·4572	6·8234	2·8168	3·4088
France	foot	1·5176	1·0711	2·1141	0·8727	1·0561
———	mètre	14·3810	10·1500	20·0330	8·2699	10·0080
Frankfort	foot	1·1785	0·8318	1·6416	0·6777	0·8201
Geneva	foot	4·9199	3·4725	6·8536	2·8293	3·4239
Genoa	palmo	0·8924	0·6299	1·2431	0·5132	0·6210
———	canna	71·1930	50·2480	99·1745	40·9410	49·5450
Gottingen	foot	1·2164	0·8585	1·6945	0·6995	0·8465
Gotha	foot	1·1888	0·8390	1·6560	0·6836	0·8273
Greece	foot	1·2941	0·9134	1·8027	0·7442	0·9006
Groningen	foot	1·2248	0·8645	1·7062	0·7044	0·8524
Hamburg	foot	1·1825	0·8346	1·6473	0·6800	0·8229
Hanover	foot	1·2164	0·8585	1·6945	0·6995	0·8465
Harlem	foot	1·1742	0·8287	1·6357	0·6752	0·8171
Heidelberg	foot	1·1143	0·7865	1·5523	0·6408	0·7755
Hildesheim	foot	1·1328	0·7995	1·5780	0·6514	0·7883
Inspruck	foot	1·4487	1·0232	2·0195	0·8337	1·0089
Königsberg	foot	1·3607	0·9604	1·8955	0·7825	0·9469
Leghorn	foot	1·3147	0·9279	1·8314	0·7560	0·9149
Leipsic	foot	1·1493	0·8111	1·6010	0·6609	0·7998
Leyden	foot	1·4127	0·9971	1·9679	0·8124	0·9831
Liege	foot	1·1888	0·8390	1·6560	0·6836	0·8273
Lindau	com. foot	1·2057	0·8510	1·6796	0·6934	0·8391
———	long foot	1·4265	1·0068	1·9871	0·8203	0·9927

CONVERSION OF STANDARD

Square Measure.		English Square Inches.	English Square Feet.	Florence Square Braccia.	French Square Mètres.	French Square Pieds.
Lisbon archit.	foot	177·72	1·2341	0·3366	0·1147	1·0865
—— com.	foot	167·96	1·1664	0·3181	0·1084	1·0269
Lombardy arch.ft.		243·70	1·6923	0·4616	0·1572	1·4899
Lorraine	foot	127·69	0·8868	0·2419	0·0824	0·7807
Lübeck	foot	131·10	0·9105	0·2483	0·0846	0·8016
Lucca	braccio	552·06	3·8337	1·0456	0:3562	3·3752
Luneburg	foot	131·10	0·9105	0·2483	0·0846	0·8016
Macedonia	foot	193·71	1·3451	0·3669	0·1250	1·1842
Magdeburg	foot	124·55	0·8649	0·2359	0·0803	0·7614
Malta	foot	124·77	0·8664	0·2363	0·0805	0·7628
Manheim	foot	130·19	0·9040	0·2466	0·0840	0·7959
Mantua	braccio	630·21	4·3764	1·1936	0·4066	3·8530
——	brasso	333·14	2·3134	0·6310	0·2149	2·0368
Maestricht	foot	122·10	0·8479	0·2312	0·0788	0·7465
Mentz	foot	140·42	0·9752	0·2660	0·0906	0·8585
Middleburg	foot	139·48	0·9686	0·2642	0·0900	0·8528
Milan	foot	243·98	1·6944	0·4621	0·1574	1·4918
—— dec.	foot.	105·27	0·7310	0·1994	0·0679	0·6436
——	braccio	548·50	3·8091	1·0389	0·3539	3·3535
—— met.-braccio		1550·08	10·7640	2·9358	1·0000	9·4768
Modena	foot	424·03	2·9447	0·8031	0·2736	2·5925
Monaco	foot	85·56	0·5941	0·1620	0·0552	0·5231
Moscow	foot	173·45	1·2045	0·3285	0·1119	1·0604
Munich	foot	132·02	0·9168	0·2500	0·0852	0·8072
Naples	palmo	107·79	0·7486	0·2042	0·0695	0·6591
——	canna	6898·30	47·9050	13·0653	4·4504	42·1754
Neufchâtel	foot	139·48	0·9686	0·2642	0·0900	0·8528
Normandy	foot	137·36	0·9539	0·2602	0·0886	0·8398
Nuremberg	foot	143·04	0·9934	0·2709	0·0923	0·8746
Oldenburg	foot	135·72	0·9425	0·2570	0·0876	0·8297
Osnaburg	foot	121·00	0·8403	0·2292	0·0781	0·7398
Padua	foot	194·04	1·3475	0·3675	0·1252	1·1853
Palæste	foot	147·33	1·0232	0·2790	0·0951	0·9008
Palermo	palmo	90·82	0·6308	0·1720	0·0586	0·5553
Parma	foot	503·02	3·4932	0·9527	0·3245	3·0754

SQUARE MEASURES.

Square Measure.		Napolitan Square Palmi.	Rhineland Square Feet.	Roman Square Palmi.	Venice Square Feet.	Vienna Square Feet.
Lisbon archit.	foot	1·6488	1·1637	2·2968	0·9482	1·1474
—— com.	foot	1·5583	1·0998	2·1705	0·8961	1·0844
Lombardy arch.	ft.	2·2609	1·5957	3·1495	1·3002	1·5734
Lorraine	foot	1·1847	0·8362	1·6504	0·6813	0·8245
Lübeck	foot	1·2164	0·8585	1·6945	0·6995	0·8465
Lucca	braccio	5·1218	3·6149	7·1348	2·9454	3·5643
Luneburg	foot	1·2164	0·8585	1·6945	0·6995	0·8465
Macedonia	foot	1·7970	1·2684	2·5033	1·0334	1·2506
Magdeburg	foot	1·1555	0·8155	1·6096	0·6645	0·8041
Malta	foot	1·1575	0·8170	1·6124	0·6656	0·8055
Manheim	foot	1·2078	0·8524	1·6824	0·6945	0·8405
Mantua	braccio	5·8468	4·1267	8·1448	3·3623	4·0689
——	brasso	3·0907	2·1814	4·3055	1·7774	2·1509
Maestricht	foot	1·1328	0·7995	1·5780	0·6514	0·7883
Mentz	foot	1·3028	0·9195	1·8148	0·7492	0·9067
Middleburg	foot	1·2941	0·9134	1·8027	0·7442	0·9006
Milan	foot	2·2637	1·5977	3·1534	1·3018	1·5754
—— dec.	foot.	0·9767	0·6893	1·3605	0·5616	0·6797
——	braccio	5·0889	3·5917	7·0890	2·9264	3·5414
—— met.-	braccio	14·3810	10·1500	20·0330	8·2699	10·0080
Modena	foot	3·9340	2·7766	5·4802	2·2624	2·7378
Monaco.	foot	0·7937	0·5620	1·1057	0·4565	0·5524
Moscow	foot	1·6092	1·1357	2·2416	0·9254	1·1199
Munich	foot	1·2248	0·8645	1·7062	0·7044	0·8524
Naples	palmo	1·0000	0·7059	1·3932	0·5751	0·6960
—— ——	canna	64·0000	45·1710	89·1540	36·8040	44·5390
Neufchâtel	foot	1·2941	0·9134	1·8027	0·7442	0·9006
Normandy	foot	1·2744	0·8995	1·7753	0·7329	0·8869
Nuremberg	foot	1·3272	0·9367	1·8488	0·7632	0·9236
Oldenburg	foot	1·2591	0·8887	1·7540	0·7241	0·8762
Osnaburg	foot	1·1227	0·7924	1·5639	0·6456	0·7813
Padua	foot	1·8002	1·2706	2·5077	1·0352	1·2528
Palæste	foot	1·3669	0·9648	1·9041	0·7861	0·9513
Palermo	palmo	0·8427	0·5948	1·1739	0·4846	0·5864
Parma	foot	4·6668	3·2938	6·5010	2·6837	3·2477

136   CONVERSION OF STANDARD

Square Measure.	English Square Inches.	Square Feet.	Florence Square Braccia.	French Square Mètres.	Square Pieds.
Parma braccio	455·40	3·1624	0·8625	0·2938	2·7842
Pavia foot	341·51	2·3716	0·6468	0·2203	2·0880
—— braccio	334·89	2·3256	0·6343	0·2161	2·0475
Persia arish	1464·59	10·1710	2·7740	0·9449	8·9545
Phileterian foot	194·24	1·3488	0·3679	0·1253	1·1875
Piacenza foot	503·02	3·4932	0·9527	0·3245	3·0754
Piedmont lipr. ft.	409·17	2·8416	0·7750	0·2640	2·5017
—————— com. ft.	181·82	1·2627	0·3444	0·1173	1·1117
Poland foot	196·90	1·3673	0·3729	0·1270	1·2038
Pomerania foot	132·25	0·9183	0·2505	0·0853	0·8085
Portugal arch. ft.	177·72	1·2341	0·3366	0·1147	1·0865
Prague foot	141·13	0·9801	0·2673	0·0911	0·8629
Prussia foot	152·70	1·0605	0·2892	0·0985	0·9336
Pythian foot	95·04	0·6600	0·1800	0·0613	0·5811
Ratsburg foot	131·10	0·9105	0·2483	0·0846	0·8016
Revel foot	110·88	0·7700	0·2100	0·0715	0·6779
Reggio braccio	434·72	3·0188	0·8233	0·2805	2·6578
Rhineland foot	152·70	1·0605	0·2892	0·0985	0·9336
Riga foot	116·42	0·8086	0·2205	0·0751	0·7119
Rimini braccio	457·53	3·1773	0·8666	0·2952	2·7973
Rome com. foot	134·37	0·9332	0·2545	0·0867	0·8216
—— archit. foot	137·36	0·9539	0·2602	0·0886	0·8398
—— palmo	77·37	0·5373	0·1465	0·0499	0·4730
—— braccio	944·46	6·5587	1·7888	0·6093	5·7743
—— palmo d'arch.	77·26	0·5366	0·1463	0·0498	0·4724
Rome canna d'arc.	7726·41	53·6560	14·6340	4·9847	47·2390
Rostock foot	129·50	0·8993	0·2453	0·0835	0·7917
Rotterdam foot	152·70	1·0605	0·2892	0·0985	0·9336
Russia foot	189·06	1·3129	0·3581	0·1220	1·1558
Sardinia palmo	96·20	0·6680	0·1822	0·0621	0·5881
Sicily palmo	90·82	0·6308	0·1720	0·0586	0·5553
——Archimedes' ft.	76·74	0·5329	0·1453	0·0495	0·4692
Sienna foot	221·06	1·5351	0·4187	0·1426	1·3515
Spain foot	123·88	0·8602	0·2346	0·0799	0·7574
Stade foot	131·10	0·9105	0·2483	0·0846	0·8016

SQUARE MEASURES.

Square Measure.	Napolitan Square Palmi.	Rhineland Square Feet.	Roman Square Palmi.	Venice Square Feet.	Vienna Square Feet.
Parma braccio	4·2249	2·9819	5·8855	2·4296	2·9402
Pavia foot	3·1684	2·2362	4·4137	1·8220	2·2050
—— braccio	3·1070	2·1929	4·3281	1·7867	2·1622
Persia arish	13·5880	9·5905	18·9290	7·8141	9·4563
Phileterian foot	1·8020	1·2719	2·5103	1·0363	1·2541
Piacenza foot	4·6668	3·2938	6·5010	2·6837	3·2477
Piedmont lipr. ft.	3·7963	2·6794	5·2883	2·1831	2·6419
———— com. ft.	1·6869	1·1906	2·3500	0·9701	1·1740
Poland foot	1·8267	1·2893	2·5446	1·0505	1·2712
Pomerania foot	1·2269	0·8659	1·7091	0·7055	0·8538
Portugal arch. ft.	1·6488	1·1637	2·2968	0·9482	1·1474
Prague foot	1·3094	0·9242	1·8241	0·7530	0·9113
Prussia foot	1·4168	1·0000	1·9736	0·8147	0·9860
Pythian foot	0·8817	0·6223	1·2283	0·5071	0·6136
Ratsburg foot	1·2164	0·8585	1·6945	0·6995	0·8465
Revel foot	1·0287	0·7261	1·4330	0·5916	0·7159
Reggio braccio	4·0331	2·8466	5·6182	2·3193	2·8067
Rhineland foot	1·4168	1·0000	1·9736	0·8147	0·9860
Riga foot	1·0802	0·7624	1·5048	0·6212	0·7518
Rimini braccio	4·2448	2·9960	5·9132	2·4411	2·9541
Rome com. foot	1·2467	0·8799	1·7367	0·7169	0·8676
—— archit. foot	1·2744	0·8995	1·7753	0·7329	0·8869
—— palmo	0·7178	0·5066	1·0000	0·4128	0·4995
—— braccio	8·7623	6·1844	12·2060	5·0389	6·0979
—— palmo d'arch.	0·7168	0·5059	0·9986	0·4122	0·4989
Rome canna d'arc.	71·6840	50·5940	99·8570	41·2230	49·8860
Rostock foot	1·2014	0·8480	1·6736	0·6909	0·8361
Rotterdam foot	1·4168	1·0000	1·9736	0·8147	0·9860
Russia foot	1·7540	1·2379	2·4433	1·0086	1·2206
Sardinia palmo	0·8924	0·6299	1·2431	0·5132	0·6210
Sicily palmo	0·8427	0·5948	1·1739	0·4846	0·5864
—- Archimedes' ft.	0·7119	0·5025	0·9917	0·4094	0·4955
Sienna foot	2·0509	1·4475	2·8569	1·1794	1·4272
Spain foot	1·1493	0·8111	1·6010	0·6609	0·7998
Stade foot	1·2164	0·8585	1·6945	0·6995	0·8465

Square Measure.		English Square Inches.	English Square Feet.	Florence Square Braccia.	French Square Mètres.	French Square Pieds.
Stettin	old foot	123·65	0·8588	0·2342	0·0798	0·7561
Strasburg	foot	129·73	0·9010	0·2457	0·0837	0·7932
Stuttgard	foot	126·79	0·8804	0·2401	0·0818	0·7751
Sweden	foot	136·66	0·9491	0·2588	0·0882	0·8356
Trent	foot	207·71	1·4424	0·3934	0·1340	1·2699
Turin	liprando foot	409·17	2·8416	0·7750	0·2640	2·5017
———.	com. foot	181·82	1·2627	0·3444	0·1173	1·1117
———.	ras	552·06	3·8337	1·0456	0·3562	3·3752
Turkey	pic	718·24	4·9877	1·3603	0·4634	4·3912
Ulm	foot	129·73	0·9010	0·2457	0·0837	0·7932
Utrecht	foot	115·35	0·8010	0·2185	0·0744	0·7052
Venice	foot	187·44	1·3017	0·3550	0·1209	1·1460
Verona	foot	179·67	1·2477	0·3403	0·1159	1·0984
Vicenza	foot	185·83	1·2905	0·3520	0·1199	1·1362
Vienna	foot	154·88	1·0756	0·2933	0·0999	0·9469
Warsaw	foot	137·48	0·9547	0·2604	0·0887	0·8405
———	Cracow foot	196·90	1·3673	0·3729	0·1270	1·2038
Wismar	foot	134·10	0·9312	0·2540	0·0865	0·8199
Würtemberg	foot	126·79	0·8804	0·2401	0·0818	0·7751
Zell	foot	131·10	0·9105	0·2483	0·0846	0·8016
Ziriczee	foot	149·08	1·0353	0·2824	0·0962	0·9115
Zurich	foot	139·48	0·9686	0·2642	0·0900	0·8528

SQUARE MEASURES.

Square Measure.		Napolitan Square Palmi.	Rhineland Square Feet.	Roman Square Palmi.	Venice Square Feet.	Vienna Square Feet.
Stettin	old foot	1·1473	0·8098	1·5982	0·6598	0·7984
Strasburg	foot	1·2037	0·8496	1·6768	0·6922	0·8377
Stuttgard	foot	1·1762	0·8302	1·6385	0·6764	0·8185
Sweden	foot	1·2679	0·8949	1·7663	0·7292	0·8824
Trent	foot	1·9270	1·3601	2·6844	1·1082	1·3410
Turin	liprando foot	3·7963	2·6794	5·2883	2·1831	2·6419
———	com. foot	1·6869	1·1906	2·3500	0·9701	1·1740
———	ras	5·1218	3·6149	7·1348	2·9454	3·5643
Turkey	pic	6·6635	4·7031	9·2824	3·8320	4·6372
Ulm	foot	1·2037	0·8496	1·6768	0·6922	0·8377
Utrecht	foot	1·0701	0·7553	1·4907	0·6154	0·7447
Venice	foot	1·7390	1·2274	2·4225	1·0000	1·2102
Verona	foot	1·6669	1·1765	2·3220	0·9586	1·1600
Vicenza	foot	1·7241	1·2169	2·4017	0·9915	1·1998
Vienna	foot	1·4369	1·0142	2·0017	0·8263	1·0000
Warsaw	foot	1·2755	0·9002	1·7768	0·7335	0·8876
———	Cracow foot	1·8267	1·2893	2·5446	1·0505	1·2712
Wismar	foot	1·2441	0·8781	1·7331	0·7155	0·8658
Würtemberg	foot	1·1762	0·8302	1·6385	0·6764	0·8185
Zell	foot	1·2164	0·8585	1·6945	0·6995	0·8465
Ziriczee	foot	1·3831	0·9762	1·9267	0·7954	0·9625
Zurich	foot	1·2941	0·9134	1·8027	0·7442	0·9006

TABLE III.

ITINERARY OR ROAD MEASURES.

Distance.		English Yards.	English Miles.
Arabia	mile, 6000 Arabian feet	2146	1·2193
,,	baryd, of 4 farsakh	21120	12·0000
Austria	mile, post, 24,000 Vienna feet	8297	4·7142
,,	,, marine, 60 to the degree	2025	1·1508
Baden	,,	9721	5·5234
,,	,, post, 14,815 Baden feet	4860	2·7617
Bavaria	,, 25,046 Bavarian feet	8059	4·5792
,,	,, of Anspach	9443	5·3652
Belgium	,, old measure	2132	1·2111
,,	,, marine, 60 to the degree	2025	1·1508
,,	,, metrical (kilomètre)	1094	0·6214
,,	league, 20 to the degree	6076	3·4522
Berne, Switzerland, league, 18,000 Berne feet		5770	3·2784
Birmah	dain or league, 1000 dhas	4278	2·4306
Bohemia	league, 16 to the degree	7595	4·3154
,,	,, 12 ,, ,,	10126	5·7534
Brabant	,, 20 ,, ,,	6076	3·4522
Brazil	,, 18 ,, ,,	6751	3·8360
Bremen	mile, 20,000 Rhenish feet	6865	3·9006
Brunswick	,, 34,424 ,, ,,	11816	6·7140
China	li, or mile	609	0·3458
Dantzic	mile, 27,000 Dantzic feet	8467	4·8110
Denmark	,, 12,000 alns	8238	4·6807
,,	league, 14½ to the degree	8381	4·7618
East Indies :	Bengal coss or mile, 1000 fathoms	2000	1·1364
England	mile, statute	1760	1·0000
,,	,, geographical, 60 to the degree	2025	1·1508
,,	league ,, 20 ,, ,,	6076	3·4523
Flanders	,, 20,000 Rhein-fuss	6865	3·9006
France	mile, old measure	2132	1·2111
,,	,, marine, 60 to the degree	2025	1·1508
,,	,, metrical (kilomètre)	1094	0·6214
,,	league, post, 2000 ancient toises	4263	2·4222
,,	post (2 post leagues)	8527	4·8445

ITINERARY OR ROAD MEASURES. 141

Distance.		Yards.	English Miles.
France	league, common, 25 to the degree	4861	2·7617
,,	,, marine, 20 ,, ,,	6076	3·4521
,,	,, mean, 2450 toises .	5223	2·9674
Genoa	post, of 4000 French toises .	8527	4·8445
Germany	mile, geographical, 15 to the degree	8101	4·6030
,,	post (2 German miles) . . .	16203	9·2060
,,	mile long, 12 to the degree . .	10126	5·7534
,,	,, short	6859	3·8972
,,	,, metric, 1500 metres . .	1640	0·9320
Hamburg	,, 24,000 Rhenish feet . .	8238	4·6807
Hanover	,, old measure, 18,192 elles .	11572	6·5750
,,	,, (since 1818) 11,700 ,, .	7442	4·2287
,,	,, of 25,400 Calenberg feet .	8114	4·6102
Hebrew	ancient Eastern mile of 4000 cubits	2432	1·3820
Holland	mile, old measure, 19 to the degree	6396	3·6340
,,	,, marine, 20 ,, ,,	6076	3·4521
,,	,, legal (Netherlandic) . .	1094	0·6214
Holstein	,, 12,000 alns	8238	4·6807
Hungary	,, league, 13½ to the degree	9002	5·1145
India	Bengal coss or mile, 1000 fathoms	2000	1·1364
,,	league . . . . . 30 to the degree	4051	2·3015
,,	Carnatic league, 35 ,, ,,	3472	1·9727
Ireland	old mile, 320 poles of 7 yards	2240	1·2727
Italy	mile, 60 to the degree . . .	2025	1·1508
,,	post, of 8 Italian miles . . .	16203	9·2062
Lithuania	league, 12·44 to the degree .	9769	5·5503
Livonia	,, 17 to the degree . .	7148	4·0615
Lombardo-Veneto, metrical mile (kilomètre)		1094	0·6214
Lübeck	mile, marine	2028	1·1520
Mecklenburg	,, ...	8238	4·6807
,,	league, 12 to the degree . .	10126	5·7534
Naples	mile, of 7000 palmi	2018	1·1468
Netherlands	,, metrical (kilomètre) . .	1094	0·6214
Norway	,, ...	12182	6·9216
Oldenberg	,, 30,000 Oldenberg feet .	9708	5·5160
Persia	parasang, 20 to the degree .	6076	3·4522
Piedmont	post, of 4000 French toises .	8527	4·8445
Poland	league, long, 15 to the degree	8101	4·6028
,,	,, short, 20 ,, ,,	6076	3·4521
Portugal	mile . . 54 ,, ,,	2250	1·2787

ITINERARY OR ROAD MEASURES.

Distance.		English	
		Yards.	Miles.
Portugal	mile, marine, 60 to the degree	2025	1·1508
„	league (3 miles), 18 „ „	6751	3·8360
„	„ marine, 20 „ „	6076	3·4521
Prussia	mile, of 24,000 Rhineland feet	8238	4·6807
„	„ geographical, 15 to the degree	8101	4·6028
Rome	„ 74½ to the degree . .	1630	0·9261
„	„ metrical (kilomètre) . .	1094	0·6214
„	„ geographical, 60 to the degree	2025	1·1508
„	ancient millārium, 1000 Roman-passus or paces, or 5000 ancient feet	1614	0·9170
Russia	werst or verst, 500 sachines .	1167	0·6629
„	Lithuania mile, 28,530 Rhein-fuss	9793	5·5641
Sardinia	mile, 4333⅓ piede liprando .	2435	1·3834
Saxony	post mile, 24,000 fuss . . .	7432	4·2227
„	league, 12½ to the degree . .	9853	5·5985
Scotland	old mile, 1920 Scotch ells . .	1977	1·1230
Siam, Asia	roëneng, 2000 vouahs . . .	4204	2·3886
Silesia	mile	7086	4·0260
„	league, 1125 Silesian ells, being 17 to the degree . . .	7148	4·0615
Spain	mile	1522	0·8648
„	„ marine, 60 to the degree	2025	1·1508
„	league, common, 8000 varas .	7419	4·2152
„	„ legal, 5000 „ 26¼ to the degree	4637	2·6345
„	„ marine, 20 „ „	6076	3·4523
Suabia or Swabia:	mile, 12 „ „	10126	5·7534
Sweden	mile, 6000 Swedish fathoms .	11690	6·6423
Switzerland	mile, 26,666⅔ fuss	8548	4·8568
„	league, 13·3 to the degree . .	9137	5·1915
Tuscany	mile, 2833¼ bracci	1809	1·0277
Turkey	berri, 66½ to the degree . .	1827	1·0383
United States of North America		1760	1·0000
Weimar	mile	7443	4·2292
Westphalia	league, 10 to a degree . . .	12152	6·9016
Würtemberg	mile, 26,000 Stuttgard feet .	8132	4·6206

MEASURES OF TIME.

TIME, in the abstract, is truly measured by the space or distance described by a moving body or machine when the velocity of the same is sustained with perfect uniformity.

A SOLAR DAY is measured by the duration of a complete rotation of the earth round its axis with respect to the sun. The motion of the earth's rotation in space is uniform; but as it is here estimated with reference to the sun, it is affected by the movement of the earth in its orbit round the sun, the velocity of which is subject to a gradual acceleration and retardation, both on account of the ellipticity of the orbit and of the perturbations produced by the planets. To obviate this fluctuation, clocks are adjusted to an average or mean solar day, which is subdivided as follows:—

60 seconds make 1 minute
60 minutes „ 1 hour
24 hours „ 1 day.

In astronomical reckoning the day is supposed to commence at noon, and is counted throughout the twenty-four hours.

In civil reckoning the day commences at midnight, and is divided into two equal portions of twelve hours each, called morning and evening.

A WEEK is a period of seven days, and has been in use amongst eastern countries to the remotest periods of antiquity. The English names of these days are Sunday, Monday, Tuesday, Wednesday, Thursday, Friday, and Saturday.

A SOLAR YEAR, also called a tropical or civil year, is determined by a revolution of the earth in its orbit round the sun, estimated with respect to the sun and the equinox. In ordinary phraseology it is the time in which the sun moves from the vernal equinox to the vernal equinox again, and its average value, called a MEAN SOLAR YEAR, has been found by astronomers to be 365·24222 mean solar days, or 365 days 5 hours 48 min. 48 sec.

A CALENDAR MONTH is an interval varying from 28 to 31 days, and was probably first derived from the synodic revolution of the moon, or lunar month according to the periodical phases, the mean value of which period has been found to be 29·5305887 days, or 29 days 12 hours 44 min. 2·8 sec. The year is divided into twelve calendar months, each of which consists of an integral number of days, viz. :—

January	31 days
February	28 ,,
March	31 ,,
April	30 ,,
May	31 ,,
June	30 ,,
July	31 ,,
August	31 ,,
September	30 ,,
October	31 ,,
November	30 ,,
December	31 ,,
	365 days.

A BISSEXTILE YEAR, frequently called *leap-year*, consists of 366 days, an additional day being intercalated in the month of February, which is then made 29 days. This is occasionally required for the purpose of adjusting the calendar, so that the course of the seasons and the labours of agriculture, which depend on the situation of the sun, shall be correctly indicated. Before the time of Julius Cæsar, the Roman calendar was in great confusion, and, guided by Sosigenes, his astronomer, he adjusted it by making every fourth civil year into a bissextile of 366 days. The correction so made is called the *Julian correction*, and the length of a mean Julian year, or year of the Julian Calendar, is hence $365\frac{1}{4}$ or 365·25 days.

In the Ecclesiastical Calendar the intercalary day is placed between the 24th and 25th of February; in the Civil Calendar it is accounted the 29th.

THE GREGORIAN CALENDAR.

Independently of the gradual and progressive improvement in astronomical knowledge and astronomical data, the length of the mean Julian year was practically ascertained to be in excess of the actual mean solar or tropical year, which contains only 365·24222 days; and it was found that the vernal equinox, which, at the Council of Nice, in the year 325, was supposed to correspond to the 21st of March [1], after the lapse of about 1200 years, had retrograded to the 11th. To get rid of the accumulated error, and so restore the equinox to its supposed former place, Pope Gregory XIII., in 1582, directed ten days to be suppressed in the calendar, by calling the 5th of October the 15th for that year; and as the error of the Julian intercalation, according to the calculations of Aloysius Lilius, a celebrated astronomer and physician of Naples, was found to amount to about three days in 400 years, it was ordered that the intercalations on centenary years should thenceforward be omitted, excepting those which are multiples of 400. This important adjustment is usually called the *reformation of the calendar*, and it has since been adopted in almost all Christian countries, under the name of the *Gregorian Calendar*, or *New Style*, the Julian Calendar formerly in use being called the *Old Style*.

For the sake of distinctness we shall here state the Gregorian rule of intercalation.

1. For years that are not even centuries:

If the year, when divided by 4, leaves a remainder, such

---

[1] There is some slight inaccuracy in this; but it is of no consequence.

year is ordinary; if there be no remainder, the year is bissextile.

2. For years that are even centuries:
If the number of centuries, when divided by 4, leaves a remainder, the year is ordinary; if there be no remainder, it is bissextile.

Thus, 1857, 1858, 1859, 1861 are ordinary;
1856, 1860, 1864, 1868 are bissextile.

Also, 1900, 2100, 2200, 2300 are ordinary;
2000, 2400, 2800, 3200 are bissextile.

Hence every period of 400 years consists of
97 bissextile years or 35502 days,
303 ordinary „ 110595 „
_____
146097 days;

and therefore, taking the 400th part of this number, an average or mean Gregorian year is 365·24250 days.

Now the actual value of the mean solar or tropical year is 365·24222 days, so that the Gregorian rule causes the year to be only 0·00028 day in excess, which will amount to a day in about 3600 years[2]. This small error might be corrected by carrying the rule one step further and changing multiples of 4000 into ordinary years instead of bissextiles.

The Gregorian Calendar was immediately adopted in Denmark, France, Holland, Italy, Portugal, and Spain, as well as by Catholics in other countries. It was established in Germany and Switzerland in 1700, and was not adopted in Great Britain until the year 1752, no less than 170 years after its first formation.

The Act of Parliament passed in 1751, and is entitled "An Act for regulating the Commencement of the Year, and for correcting the Calendar now in use." The preamble recites,

---

[2] The Julian error (0·00778 day in excess) amounts to a day in 129 years.

that according to the legal supputation in England, the year began on the 25th of March; that this practice had produced various inconveniences, not only from its differing from the usage of neighbouring nations, but also from the legal computation in Scotland, and from the common usage throughout the whole kingdom; that the Julian Calendar then in use had been discovered to be erroneous, by means whereof the vernal or spring equinox, which at the time of the General Council of Nice, A.D. 325, happened on the 21st of March, now fell on the 9th or 10th of that month; that this error was still increasing; that a method of correcting the calendar had been received and established, and was generally practised by almost all other nations of Europe; and that it would be of general convenience to merchants and others corresponding with foreign nations if the like correction were received and established in his Majesty's dominions. It was therefore enacted,

1. That throughout all his Majesty's dominions in Europe, Asia, Africa, and America, the supputation according to which the year of our Lord began on the 25th of March shall not be used after the last day of December, 1751, and that the 1st day of January next following shall be reckoned as the 1st day of the year 1752, and so in all future years.

2. That from and after the 1st day of January, 1752, the several days of each month shall go on and be reckoned and numbered in the same order, and the feast of Easter and other moveable feasts thereon depending shall be ascertained according to the same method, as they now are, until the 2nd of September, 1752; that the natural day next immediately following the 2nd of September, 1752, shall be called and reckoned as the *fourteenth* day of September, omitting the eleven intermediate nominal days of the common calendar; that the days which follow next after the said 14th of September shall be reckoned in numerical order from that day; and all public and private proceedings whatsoever after the 1st of January, 1752, were ordered to be dated accordingly.

3. That the several years of our Lord 1800, 1900, 2100, 2200, 2300, or any other hundredth years of our Lord which shall happen in time to come (excepting only every fourth hundredth year of our Lord, whereof the year 2000 shall be the first), shall not be deemed Bissextile or Leap-years, but shall be considered as common years, consisting of 365 days only; and that the years of our Lord 2000, 2400, 2800, and every other fourth hundredth year of our Lord, from the year 2000 inclusive, and also all other years of our Lord, which by the present supputation are considered bissextile or leap-years, shall for the future be esteemed bissextile or leap-years, consisting of 366 days.

4. That whereas according to the rule then in use for calculating Easter-day, that feast was fixed to the first Sunday after the first full moon next after the 21st of March; and if the full moon happens on a Sunday, then Easter-day is the Sunday after, which rule had been adopted by the General Council of Nice, A.D. 325; but as the method of computing the full moons then used in the Church of England, and according to which the table to find Easter prefixed to the Book of Common Prayer was formed, had become considerably erroneous, it was enacted that the said method should be discontinued, and that from and after the 2nd of September, 1752, Easter-day, and the other moveable and other feasts, were henceforward to be reckoned according to the Calendar, Tables, and Rules annexed to the Act, and attached to the Books of Common Prayer.

DIFFERENCE OF STYLE.—From 1582 to 1700 the difference of style continued to be 10 days; but 1700 being bissextile in the Julian Calendar and ordinary in the Gregorian, the difference of the styles from 1700 to 1800 was 11 days. The Julian leap-year 1800 was also ordinary in the Gregorian Calendar, and therefore the difference during the present century is 12 days. After 1900 it will be 13 days, and will so continue till 2100, because the year 2000 will be a leap-year in both styles. Thus if c denote the number of completed centuries, the number of days' difference between the old and new styles, which has accumulated since the second century, will be correctly represented by the formula,

$$c - \binom{c}{4}_w - 2,$$

where w denotes the integral quotient of $\dfrac{c}{4}$, rejecting any fraction or remainder.

Hence the following table, the extension of which is evident without calculation :—

Date.	Difference.	Date.	Difference.
1500 to 1700	10 days	2900 to 3000	20 days
1700 „ 1800	11 „	3000 „ 3100	21 „
1800 „ 1900	12 „	3100 „ 3300	22 „
1900 „ 2100	13 „	3300 „ 3400	23 „
2100 „ 2200	14 „	3400 „ 3500	24 „
2200 „ 2300	15 „	3500 „ 3700	25 „
2300 „ 2500	16 „	3700 „ 3800	26 „
2500 „ 2600	17 „	3800 „ 3900	27 „
2600 „ 2700	18 „	3900 „ 4100	28 „
2700 „ 2900	19 „	&c.	&c.

The difference requires to be added to the day of the month according to the old style to deduce the same day in the new style, and *vice versâ*. Thus 1872, June 10, old style, is the same day as 1872, June 22, new style.

DOMINICAL LETTER.—The dominical or Sunday letter is one of the first seven letters of the alphabet, and is used for the purpose of determining the day of the week corresponding to any given date. In the Ecclesiastical Calendar the first letter A is placed opposite to the first day of the year or January 1, the second letter B is placed opposite the second day or January 2, and so on through the seven letters; after which they are in like manner repeated over and over again to the end of the year. Then the letter which falls opposite the first Sunday in the year will also fall opposite every following Sunday throughout the year, because the number of letters is the same as the number of days in the week. In ordinary years the letter so indicated is the dominical letter. But in bissextile or leap-years an interruption takes place in the order of the letters on account of the intercalary day, and it is made as a matter of convenience in chronological tabulations. As the intercalary day falls on the 24th of February, the 24th and 25th days are denoted by the same letter, so that after the 24th day of February the dominical letter goes back one place. In the Civil Calendar, however, according to which calculations are generally made, the intercalary day is supposed to be added at the end of February, so that the change of letter takes place on entering March.

As an ordinary year contains 365 days or 52 weeks and 1 day over, and a bissextile year contains 52 weeks and 2 days over, it is evident from the foregoing account that for a series of consecutive years the dominical letters stand in a retrograde order, and go back one letter after every ordinary year and two letters after a bissextile year, the first change in the latter case occurring at the intercalary day, and the second at the end of the year. Thus a bissextile or leap-year has always two dominical letters, one to be used before and the other after the intercalary day.

For any proposed year Y of the Gregorian Calendar, at any near or remote period of time, let c denote the number of completed centuries and y the year of the current century, so that $Y = 100\,c + y$; then the number of bissextile years, from the year 1 of the calendar up to the year Y inclusive, will be $\left(\dfrac{Y}{4}\right)_w - c + \left(\dfrac{c}{4}\right)_w$, and the dominical letter may be determined in the following manner:—

From the year 0 to the year $Y = 100\,c + y$ the total number of intervening years is Y, and, if ordinary and bissextile years are separately enumerated, the numbers are, according to what precedes—

Ordinary years $= Y - \left\{ \left(\dfrac{Y}{4}\right)_w - c + \left(\dfrac{c}{4}\right)_w \right\}$

Embolismic years $= \left(\dfrac{Y}{4}\right)_w - c + \left(\dfrac{c}{4}\right)_w$

In passing from January 1 of the year 0 to January 1 of the proposed year Y, the advancement of the day of the week, consisting of one day for every ordinary year, and two days for every intervening bissextile year, is obviously—

$= Y + \left\{ \left(\dfrac{Y}{4}\right)_w - c + \left(\dfrac{c}{4}\right)_w \right\}$ rejecting sevens

$= (100c + y) + \left\{ \left(\dfrac{100c+y}{4}\right)_w - c + \left(\dfrac{c}{4}\right)_w \right\}$,, ,,

$= 124c + y + \left(\dfrac{y}{4}\right)_w + \left(\dfrac{c}{4}\right)_w$,, ,,

But as $\left(\dfrac{c}{4}\right)_w$ denotes the quotient and $\left(\dfrac{c}{4}\right)_r$ the remainder, when the number c is divided by 4, we have the relation

$$4\left(\dfrac{c}{4}\right)_w = c - \left(\dfrac{c}{4}\right)_r$$

Therefore, $\left(\dfrac{c}{4}\right)_w = 2c - 2\left(\dfrac{c}{4}\right)_r$, rejecting sevens,

Similarly, $\left(\dfrac{y}{4}\right)_w = 2y - 2\left(\dfrac{y}{4}\right)_r$,, ,,

Hence, by substitution, the advancement of the day of the week, or the number by which the dominical letter has retrograded, is

$L_0 - L = 126c + 3y - 2\left(\dfrac{c}{4}\right)_r - 2\left(\dfrac{y}{4}\right)_r$, rejecting sevens,

$= 3y - 2\left(\dfrac{c}{4}\right)_r - 2\left(\dfrac{y}{4}\right)_r$,, ,,

The value of the sought number L is therefore

$$L = 2\left(\frac{c}{4}\right)_r + 2\left(\frac{y}{4}\right)_r - 3y + L_0 \text{ rejecting sevens}$$

$$= 2\left(\frac{c}{4}\right)_r + 2\left(\frac{y}{4}\right)_r + 4y + L_0 \quad \text{,,} \quad \text{,,}$$

$$= 2\left(\frac{c}{4}\right)_r + 2\left(\frac{y}{4}\right)_r + 4\left(\frac{y}{7}\right)_r + L_0 \quad \text{,,} \quad \text{,,}$$

It will be observed that sevens are rejected in this investigation because they do not alter the day of the week, and that sevens may be added to negative numbers for a like reason.

The letter L denotes the number that indicates the ordinal position in which the dominical letter stands in the alphabet. Similarly L_0 denotes the corresponding ordinal number for the year 0. Thus $L_0 - L$ determines how much the dominical number or dominical letter has retrograded through the entire period, that is, after sevens have been freely rejected from the result.

The number L_0 refers to a given year, and is, therefore, a constant. It will be requisite to ascertain its value in order to finally adapt the foregoing formula to direct calculation. Before proceeding to do so it may be well to observe by way of explanation of the object attained by the preceding deductions, that the terms of the expression have been severally put in the form of remainders, so as to simplify the computation as much as possible by the exclusive use of very small numbers.

By testing the last expression with any given year for which L is known, it is found that the constant $L_0 = 1$. Therefore the dominical letter for any year, $Y = 100c + y$, in the Gregorian Calendar may always be found from the simple and general formula,

$$L = 2\left(\frac{c}{4}\right)_r + 2\left(\frac{y}{4}\right)_r + 4\left(\frac{y}{7}\right)_r + 1 \text{ (rejecting sevens)};$$

MEASURES OF TIME. 153

where the small letter *r* is placed to indicate that it is the *remainder* of each division that enters into the calculation. The resulting number *L* may be called the *dominical number*, as it will indicate the numerical order of the required letter.

Thus if *L* be 1 | 2 | 3 | 4 | 5 | 6 | 7 |
The letter will be A | B | C | D | E | F | G |

If the proposed year be bissextile, the letter so calculated will be the second letter of the year, or that which applies after the intercalary day in February.

The preceding formula may be put down in the following rule:—

Rule.—Divide the number of centuries by 4; the years of the current century by 4, and the same by 7 : put down the three remainders; multiply them respectively by 2, 2, 4; add together the three products with an additional unit, and the sum after rejecting sevens, if necessary, will be the dominical number, or the ordinal number in which the dominical letter stands in the alphabet.

Example.—Required the dominical letter for the year 1942. The centuries are here 19, and the years of the current century 42; the three remainders are therefore 3, 2, 0; the three products are 6, 4, 0; which added together with an additional unit give 11; therefore rejecting 7, the ordinal number of the required letter is 4; it is therefore D, the fourth letter of the alphabet.

The dominical letter or letters of any proposed year may be obtained, by inspection, from the following table, to which an auxiliary table is added, showing the means by which the dominical letter is made to indicate the day of the week answering to any given date.

PERPETUAL TABLE OF DOMINICAL LETTERS.

Year of the Current Century (y).				Completed Centuries (c).			
				$\left(\frac{c}{4}\right)_r = 1$ 1700 2100 &c.	$\left(\frac{c}{4}\right)_r = 2$ 1800 2200 &c.	$\left(\frac{c}{4}\right)_r = 3$ 1900 2300 &c.	$\left(\frac{c}{4}\right)_r = 0$ 2000 2400 &c.
			0	C	E	G	BA
1	29	57	85	B	D	F	G
2	30	58	86	A	C	E	F
3	31	59	87	G	B	D	E
4	32	60	88	FE	AG	CB	DC
5	33	61	89	D	F	A	B
6	34	62	90	C	E	G	A
7	35	63	91	B	D	F	G
8	36	64	92	AG	CB	ED	FE
9	37	65	93	F	A	C	D
10	38	66	94	E	G	B	C
11	39	67	95	D	F	A	B
12	40	68	96	CB	ED	GF	AG
13	41	69	97	A	C	E	F
14	42	70	98	G	B	D	E
15	43	71	99	F	A	C	D
16	44	72		ED	GF	BA	CB
17	45	73		C	E	G	A
18	46	74		B	D	F	G
19	47	75		A	C	E	F
20	48	76		GF	BA	DC	ED
21	49	77		E	G	B	C
22	50	78		D	F	A	B
23	51	79		C	E	G	A
24	52	80		BA	DC	FE	GF
25	53	81		G	B	D	E
26	54	82		F	A	C	D
27	55	83		E	G	B	C
28	56	84		DC	FE	AG	BA

TABLE SHOWING THE DAY OF THE WEEK.

Month.	Dominical Letter.						
Jan. Oct.	A	B	C	D	E	F	G
Feb. Mar. Nov.	D	E	F	G	A	B	C
Apr. July	G	A	B	C	D	E	F
May	B	C	D	E	F	G	A
June	E	F	G	A	B	C	D
August	C	D	E	F	G	A	B
Sept. Dec.	F	G	A	B	C	D	E
1 8 15 22 29	Sun.	Sat.	Frid.	Thur.	Wed.	Tues.	Mon.
2 9 16 23 30	Mon.	Sun.	Sat.	Frid.	Thur.	Wed.	Tues.
3 10 17 24 31	Tues.	Mon.	Sun.	Sat.	Frid.	Thur.	Wed.
4 11 18 25	Wed.	Tues.	Mon.	Sun.	Sat.	Frid.	Thur.
5 12 19 26	Thur.	Wed.	Tues.	Mon.	Sun.	Sat.	Frid.
6 13 20 27	Frid.	Thur.	Wed.	Tues.	Mon.	Sun.	Sat.
7 14 21 28	Sat.	Frid.	Thur.	Wed.	Tues.	Mon.	Sun.

CYCLE OF THE SUN.—As the number of dominical letters, or days in the week, is seven, and as every fourth year is bissextile or leap-year, the same order of dominical letters for a specified year of the Julian Calendar only returns after 4 times 7, or 28 years, which is the period of the solar cycle. The cycle is considered as having commenced nine years before the era, so that the number or year of the cycle corresponding to any year Y of the Julian Calendar, is determined by the formula,

$$s = \left(\frac{Y+9}{28}\right)_{r'}$$

which may be stated in the following rule:—

Rule.—Add 9 to the given year; divide the sum by 28;

the quotient is the number of cycles elapsed, and the remainder is the number or year of the cycle: if there be no remainder, the number is 28, the last of the current cycle.

If preferred, the calculation may be modified thus:—

Second Rule.—Having, as before, added 9 to the year, divide by 4, and the integral quotient again by 7; then the first remainder added to 4 times the second remainder will give the number of the solar cycle. If there be no remainder to either division, the required number is 28.

Example.—Required the number of the solar cycle for the year 1942.

The year, augmented by 9, is 1951.

1951, divided by 4, gives 487, with first remainder 3;
487 „ 7 „ 69 „ second „ 4;
and adding 3 to 4 times 4, the number of the solar cycle is 19.

Otherwise the solar cycle may perhaps most conveniently be deduced by subtracting from the given year the next less number contained in the following formula:—

Cycle (s) = year − (1811, 1839, 1867, 1895, 1923, 1951, 1979, 2007, 2035, 2063, 2091, &c.).

The cycle of the sun, or more properly the *Sunday cycle*, was invented for the purpose of determining the dominical letter or letters for any given year of the Julian Calendar, by means of a short and convenient table exhibiting the same for each of the 28 years of one cycle.

But according to the Gregorian Calendar now in general use in every Christian country, with the exception of Russia, the order of the letters is necessarily interrupted by the first suppression of a centenary leap-year, and the table of dominical letters must therefore, after every such year, be reconstructed for the next following century. We have, however, found, page 140, that the complete intercalary period of 400 Gregorian years consists of 146,097 days.

As this number is divisible by 7 without a remainder, and therefore comprises exactly 20,871 weeks, it follows that the same order of dominical letters and days of the week will recur after this period of 400 years, which is therefore the complete Sunday cycle of the Gregorian Calendar. The purport of these remarks may perhaps receive further elucidation from an examination of the perpetual table of dominical letters already given, which extends through a complete cycle of 400 years, and will therefore in future calculations supersede the use of the solar cycle.

GOLDEN NUMBER. — The cycle of the moon or lunar cycle, sometimes called the Metonic cycle, after the name of its original inventor, Meton, is a period of *nineteen* years, after which the new moons fall on the same days of the Julian year, within an hour and a half. The number which any given year occupies in the current cycle was called the *golden number*, from the circumstance of its being usually marked in letters of gold in ancient calendars, and it was used for the purpose of determining the days of new moon, and of thereby fixing the date of Easter-day, on which the other moveable feasts of the ecclesiastical calendar are made to depend. The year of the birth of our Saviour is reckoned the first of the lunar cycle, and therefore the golden number for any year Y is determined by the formula,

$$g = \left(\frac{Y+1}{19}\right)_r$$

which may be expressed by the following rule:—

Rule.—Add 1 to the year; divide the sum by 19; the quotient is the number of completed cycles, and the remainder is the golden number. If 0 remains, the number is 19, the year being in that case the 19th or last of the cycle.

By this rule the following table has been calculated, and the golden number for any proposed year can be taken from it by inspection.

PERPETUAL TABLE OF

Year of the Century.	Centuries.								
	0	100	200	300	400	500	600	700	800
	1900	2000	2100	2200	2300	2400	2500	2600	2700
	3800	3900	4000	4100	4200	4300	4400	4500	4600
	5700	5800	5900	6000	6100	6200	6300	6400	6500
	7600	7700	7800	7900	8000	8100	8200	8300	8400

Golden Number (*g*).

0	19	38	57	76	95	1	6	11	16	2	7	12	17	3
1	20	39	58	77	96	2	7	12	17	3	8	13	18	4
2	21	40	59	78	97	3	8	13	18	4	9	14	19	5
3	22	41	60	79	98	4	9	14	19	5	10	15	1	6
4	23	42	61	80	99	5	10	15	1	6	11	16	2	7
5	24	43	62	81		6	11	16	2	7	12	17	3	8
6	25	44	63	82		7	12	17	3	8	13	18	4	9
7	26	45	64	83		8	13	18	4	9	14	19	5	10
8	27	46	65	84		9	14	19	5	10	15	1	6	11
9	28	47	66	85		10	15	1	6	11	16	2	7	12
10	29	48	67	86		11	16	2	7	12	17	3	8	13
11	30	49	68	87		12	17	3	8	13	18	4	9	14
12	31	50	69	88		13	18	4	9	14	19	5	10	15
13	32	51	70	89		14	19	5	10	15	1	6	11	16
14	33	52	71	90		15	1	6	11	16	2	7	12	17
15	34	53	72	91		16	2	7	12	17	3	8	13	18
16	35	54	73	92		17	3	8	13	18	4	9	14	19
17	36	55	74	93		18	4	9	14	19	5	10	15	1
18	37	56	75	94		19	5	10	15	1	6	11	16	2

As the lunar months in the construction of the calendar must necessarily be estimated in integral days, and as the mean value of the lunar synodical month is over $29\frac{1}{2}$ days, it is evident that the calendar lunations must consist mainly of 30 and 29 days alternately, but that on the whole there should be rather more of the former than the latter. Now 19 ordinary years of 365 days make 6935 days; these are distributed into 235 calendar lunations in the following manner.

The lunations are made to consist of 30 and 29 days alternately, so that each lunar year of 12 lunations thus

GOLDEN NUMBERS.

Year of the Century.	Centuries.									
	900	1000	1100	1200	1300	1400	1500	1600	1700	1800
	2800	2900	3000	3100	3200	3300	3400	3500	3600	3700
	4700	4800	4900	5000	5100	5200	5300	5400	5500	5600
	6600	6700	6800	6900	7000	7100	7200	7300	7400	7500
	8500	8600	8700	8800	8900	9000	9100	9200	&c.	&c.
	Golden Number (g).									
0 19 38 57 76 95	8	13	18	4	9	14	19	5	10	15
1 20 39 58 77 96	9	14	19	5	10	15	1	6	11	16
2 21 40 59 78 97	10	15	1	6	11	16	2	7	12	17
3 22 41 60 79 98	11	16	2	7	12	17	3	8	13	18
4 23 42 61 80 99	12	17	3	8	13	18	4	9	14	19
5 24 43 62 81	13	18	4	9	14	19	5	10	15	1
6 25 44 63 82	14	19	5	10	15	1	6	11	16	2
7 26 45 64 83	15	1	6	11	16	2	7	12	17	3
8 27 46 65 84	16	2	7	12	17	3	8	13	18	4
9 28 47 66 85	17	3	8	13	18	4	9	14	19	5
10 29 48 67 86	18	4	9	14	19	5	10	15	1	6
11 30 49 68 87	19	5	10	15	1	6	11	16	2	7
12 31 50 69 88	1	6	11	16	2	7	12	17	3	8
13 32 51 70 89	2	7	12	17	3	8	13	18	4	9
14 33 52 71 90	3	8	13	18	4	9	14	19	5	10
15 34 53 72 91	4	9	14	19	5	10	15	1	6	11
16 35 54 73 92	5	10	15	1	6	11	16	2	7	12
17 36 55 74 93	6	11	16	2	7	12	17	3	8	13
18 37 56 75 94	7	12	17	3	8	13	18	4	9	14

comprises 354 days, which is 11 days short of the ordinary calendar year of 365 days. To correct this accumulating deficiency, 6 embolismic months of 30 days each are introduced in the course of the cycle of 19 years, and one of 29 days is added at the termination of the cycle. In this way the 235 lunations become divided thus:

$$\frac{\begin{array}{l}120 \text{ calendar lunations of} \quad 30 \text{ days} = 3600 \text{ days}\\ 115 \quad\quad\quad\quad\text{,,}\quad\quad\quad\text{,,}\quad\quad\quad\quad 29 \quad\text{,,} = 3335 \text{ ,,}\end{array}}{235 \quad\quad\quad\text{,,}\quad\quad\quad\text{,,}\quad\quad\quad 30 \text{ and } 29 \text{ ,,} = 6935 \text{ days.}}$$

Furthermore, in every bissextile year the intercalary day is also added to the days of the lunation in which it happens to be included, making the same 30 instead of 29 days, so that the 235 lunations thus distributed will then accurately measure 19 Julian years. Thus, as the intercalary day of the Julian Calendar occurs uniformly once in every four years without interruption, the average number of such days, distributed in periods of 19 years, will be $4\frac{3}{4}$ days, which, added to the 6935 days, give $6939\frac{3}{4}$ days for the mean value of the entire cycle so formed, and this is exactly equal to 19 mean Julian years of $365\frac{1}{4}$ days.

There are two objections, in point of accuracy, to the permanency of this lunar-solar period. In the first place, the Julian year, employed as the basis of calculation, does not correctly represent either the mean solar or the Gregorian year; as compared with the latter, the accumulated error after any stated epoch will be represented by the augmentation of the requisite correction for reducing the old to the new style. In the second place, the period of $6939\frac{3}{4}$ days is nearly an hour and a half in excess of 235 mean astronomical lunations, and therefore the correct time of new moon will in successive cycles happen so much earlier, and will retrograde a day in every 308 years. For the purpose of correcting and adjusting the errors whenever they amount to one day, by an established system of calculation, Lilius introduced another set of numbers called Epacts.

Epact.—The epact for any year is a number designed to represent the age of the moon on the first day, that is, on the 1st day of January, of that year.

Suppose, for the first year of a lunar cycle, a new moon to happen on the 1st day of January; then the age on the day of new moon being 0, the epact for that year will be 0. Now the civil year containing 365 days, and the lunar year only 354 days, the new moon will at the end of the year have retrograded 11 days, and this will be the same if the civil year be a bissextile of 366 days, because in that case the intercalary day is also included in the lunations, causing the lunar year to consist of 355 days. It therefore follows that the moon's age on the 1st of January of the second year of the cycle, or the epact of the second year, will be 11 days. Similarly the epact for the third year will be 22 days. Another addition of 11 would give 33 for the fourth year; but in consequence of the insertion of the embolismic month of 30 days, the epact for the fourth year is reduced to 3. In like manner the epacts of the following years are deduced by successively adding 11 and rejecting 30 whenever the sum exceeds that number, excepting at the termination of the cycle, where the last embolismic month being only 29 days, the same number is deducted, and we again have 0 for the epact of the first year of the next cycle. The order of the epacts throughout each cycle is therefore as follows:—

Year of the Cycle, or Golden Number (g).																		
1	2	3	4	5	6	7	8	9	10	11	12	13	14	15	16	17	18	19
*	11	22	3	14	25	6	17	28	9	20	1	12	23	4	15	26	7	18
Epact (ϵ).																		

This table will exhibit the epacts correctly from the year 1700 to the year 1900, the mathematical relations being,

$$g = \left(\frac{Y+1}{19}\right)_r, \quad \epsilon = \left(\frac{11(g-1)}{30}\right)_r.$$

But it has been explained under the article *Golden Number*, that in the course of centuries the astronomical new moon will deviate from the preceding deductions, from two causes, viz. the small error of the cycle of 235 calendar lunations as compared with 235 mean astronomical lunations, and the gradual shifting of dates on account of the difference between the Julian and Gregorian styles. We now proceed to investigate the principles on which these irregularities are calculated and adjusted.

To determine the error on account of the inaccurate measure of the lunations, we have

$$\left. \begin{array}{l} \text{19 mean Julian years,} \\ \text{each 365·25 days} \end{array} \right\} \; . \; . \; 6939{\cdot}75 \quad \text{days}$$

$$\left. \begin{array}{l} \text{235 astronomical lunations,} \\ \text{each 29·5305887 days} \end{array} \right\} \; 6939{\cdot}68834 \quad \text{,,}$$

$$\overline{\phantom{6939{\cdot}68834}}$$

$$0{\cdot}06166 \text{ day.}$$

The excess of the established period of the lunar cycle or the Julian Calendar over the astronomical lunations is therefore 0·06166 day, or about 1 h. 28·8 min. Thus after every cycle of 19 years the times of new moon will happen 1 h. 28·8 min. earlier than in the preceding cycle, and therefore the age of the moon will become periodically *increased* by the same quantity, which will amount to a day in about 308 years. In the construction of the calendar it has been assumed to amount to 8 days in 25 centuries, and, when computed from the year 1700, to be determined by the formula,

$$a = \left(\frac{c-17}{25}\right)_w, \quad \text{correction} = \left(\frac{c-a}{3}\right)_w - 5.$$

Thus after a period of 25 centuries,

c will be augmented by 25,

a ,, ,, ,, 1,

$c - a$,, ,, ,, 24;

therefore the correction $\left(\dfrac{c-a}{3}\right)_w - 5$ will give exactly 8 days in every 25 centuries, and this reduces the lunar error to less than a day in 270,000 years.

To obtain the correction on account of difference of style, if c, as before, denote the number of completed centuries in the proposed year Y, we have ascertained that the number of days' difference between the old and new styles will then have amounted to $c - \left(\dfrac{c}{4}\right)_w - 2$. When $c = 17$ it is 11 days for the year 1700; therefore, from 1700 to the given year Y, the divergence on account of style will be $c - \left(\dfrac{c}{4}\right)_w - 13$, and the age of the moon or epact for the year will thereby be *diminished* by the number of days represented by this last formula, which expresses, in fact, the number of centenary years passed over that are not made bissextile.

For the complete correction of the epact or moon's age, it will hence be requisite to add $\left(\dfrac{c-a}{3}\right)_w - 5$, and to subtract $c - \left(\dfrac{c}{4}\right)_w - 13$, or to apply the difference of these corrections, viz. $8 + \left(\dfrac{c}{4}\right)_w + \left(\dfrac{c-a}{3}\right)_w - c$. Thus in the new style or Gregorian Calendar, the general formulæ for determining the epact for any year Y are

$$a = \left(\dfrac{c-17}{25}\right)_w$$

$$e = \varepsilon + 8 + \left(\dfrac{c}{4}\right)_w + \left(\dfrac{c-a}{3}\right)_w - c$$

$$= \left(\dfrac{11(g-1)}{30}\right)_r + 8 + \left(\dfrac{c}{4}\right)_w + \left(\dfrac{c-a}{3}\right)_w - c.$$

Should the calculation of this expression come out negative, an embolismic month of 30 days must be added.

Note.—Should the result be $\dfrac{24}{25}$, change it into $\dfrac{25}{26}$ whenever the golden number exceeds 11.

Example.—Required the epact for the year 1942.

Here $Y = 1942$ and $c = 19$.

$Y + 1 = 1942 + 1 = 1943$, which on being divided by 19 leaves as remainder the golden number $g = 5$.

$11(g-1) = 11 \times 4 = 44$, which divided by 30 leaves as remainder $\varepsilon = 14$.

$c - 17 = 2$, which divided by 25, the whole number of the quotient is $a = 0$: (this will always be 0 until the year 4200.)

Therefore the required epact $= \epsilon + 8 + \left(\dfrac{c}{4}\right)_w + \left(\dfrac{c-a}{3}\right)_w - c = 14 + 8 + 4 + 6 - 19 = 13$.

By first taking out the golden number from the table, page 158, the epact for any given year may be obtained by inspection from the following table, in the column under the completed centuries.

TABLE OF EPACTS.

Golden Number (g).	Last Comp'eted Century (c).									
	1500 1600	1700 1800	1900 2000 2100	2200 2400	2300 2500	2600 2700 2800	2900 3000	3100 3200 3300	3400 3600	3500 3700	3800 3900 4000
1	1	*	29	28	27	26	25	24	23	22	21
2	12	11	10	9	8	7	6	5	4	3	2
3	23	22	21	20	19	18	17	16	15	14	13
4	4	3	2	1	*	29	28	27	26	25	24
5	15	14	13	12	11	10	9	8	7	6	5
6	26	25	24	23	22	21	20	19	18	17	16
7	7	6	5	4	3	2	1	*	29	28	27
8	18	17	16	15	14	13	12	11	10	9	8
9	29	28	27	26	25	24	23	22	21	20	19
10	10	9	8	7	6	5	4	3	2	1	*
11	21	20	19	18	17	16	15	14	13	12	11
12	2	1	*	29	28	27	26	25	24	23	22
13	13	12	11	10	9	8	7	6	5	4	3
14	24	23	22	21	20	19	18	17	16	15	14
15	5	4	3	2	1	*	29	28	27	26	25
16	16	15	14	13	12	11	10	9	8	7	6
17	27	26	25	24	23	22	21	20	19	18	17
18	8	7	6	5	4	3	2	1	*	29	28
19	19	18	17	16	15	14	13	12	11	10	9

NUMBER OF DIRECTION.—The number of direction is the number of days that Easter-day falls later than the 21st of March. Easter, as ordained by the Council of Nice, is the first Sunday after the first full moon which happens upon or next after the 21st day of March; and if the full moon happens on a Sunday, then Easter-day is the Sunday after. This last condition was introduced to avoid the celebration of Easter at the same time as the Jewish Passover; notwithstanding which, this coincidence will sometimes happen, and will next occur in the year 1903. The moon on which Easter immediately depends is called the *paschal moon*, and the full moon is defined to be the 14th day of the moon, that is, 13 days after the preceding day of new moon.

Now the epact, e, is the age of the moon on January 1; and therefore January $(31 - e)$ is a day of new moon. And as the months January and February together comprise the same number of days as two alternate lunations of 29 and 30 days, it follows that March $(31 - e)$ must likewise be a day of new moon. Adding 13 days, the 14th day of this moon will fall on March $(44 - e)$, and this will be upon or later than the 21st day of March, and therefore be the paschal full moon, provided e be less than 24. When e is 24 or greater than 24 the next following moon will be the paschal moon, and the date so found will require to be increased by 29 or 30 days respectively as the period of the current lunation. The reason of this distinction is, that the epacts 24 and 25 are made to occupy the same day in the calendar whenever the lunation is required to pass from 29 to 30 days, which is the case in April. The number of days from March 21 to the day of the paschal full moon, which for uniformity we shall designate the *Paschal Direction* and denote by P, is therefore thus determined:—

$$\text{When } e < 24, \quad P = 23 - e;$$
$$\text{,, } \quad e = 24, \quad P = 28;$$
$$\text{,, } \quad e > 24, \quad P = 53 - e.$$

Next, to find the Sunday following the paschal full moon, if L denote the dominical number, $L + 4 + 7m$ days after March 21 will be a Sunday, and the number of days which intervene between the day of the paschal full moon and this

Sunday will be $L + 3 + 7m - P$. Therefore the number of days which intervene between the paschal full moon and the immediately following Sunday, or Easter, will be the least positive remainder of $L + 3 + 7m - P$ when divided by 7; and, denoting these intervening days by p,

when $P = 23 - e$, $p = \left(\dfrac{L + e + 7m - 20}{7}\right)_r = \left(\dfrac{L + e + 1}{7}\right)_r$

„ $P = 28$, $\quad p = \left(\dfrac{L + 3}{7}\right)_r$

„ $P = 53 - e$, $p = \left(\dfrac{L + e + 7m - 50}{7}\right)_r = \left(\dfrac{L + e - 1}{7}\right)_r$.

Hence if N be the number of direction, $N = P + 1 + p$, and we obtain the following general formulæ for its computation:—

When $e < 24$, $N = 24 - e + \left(\dfrac{L + e + 1}{7}\right)_r$

„ $e = 24$, $N = 29 \quad + \left(\dfrac{L + 3}{7}\right)_r$

„ $e > 24$, $N = 54 - e + \left(\dfrac{L + e - 1}{7}\right)_r$.

Example.—Find by calculation the number of direction for the year 1942.

The dominical number, page 153, has been found to be $L = 4$; and, page 164, the epact to be $e = 13$

Therefore, e being less than 24,

$$N = 24 - e + \left(\dfrac{L + e + 1}{7}\right)_r$$

$$= 24 - 13 + \left(\dfrac{4 + 13 + 1}{7}\right)_r$$

$$= 11 + \left(\dfrac{18}{7}\right)_r = 11 + 4 = 15.$$

This calculation however may in all cases be dispensed with by entering the following table with the epact and dominical letter.

Perpetual Table for finding the NUMBER OF DIRECTION (*N*) *from the Epact and Dominical Letter.*

(For Leap Years use the *second* letter.)

Epact (*e*).	Dominical Letter.						
	A	B	C	D	E	F	G
*	26	27	28	29	30	24	25
1	26	27	28	29	23	24	25
2	26	27	28	22	23	24	25
3	26	27	21	22	23	24	25
4	26	20	21	22	23	24	25
5	19	20	21	22	23	24	25
6	19	20	21	22	23	24	18
7	19	20	21	22	23	17	18
8	19	20	21	22	16	17	18
9	19	20	21	15	16	17	18
10	19	20	14	15	16	17	18
11	19	13	14	15	16	17	18
12	12	13	14	15	16	17	18
13	12	13	14	15	16	17	11
14	12	13	14	15	16	10	11
15	12	13	14	15	9	10	11
16	12	13	14	8	9	10	11
17	12	13	7	8	9	10	11
18	12	6	7	8	9	10	11
19	5	6	7	8	9	10	11
20	5	6	7	8	9	10	4
21	5	6	7	8	9	3	4
22	5	6	7	8	2	3	4
23	5	6	7	1	2	3	4
24	33	34	35	29	30	31	32
25	33	34	35	29	30	31	32
26	33	34	28	29	30	31	32
27	33	27	28	29	30	31	32
28	26	27	28	29	30	31	32
29	26	27	28	29	30	31	25

EASTER-DAY.—The date of Easter-day is obtained by simply adding the number of direction to March 21. It is therefore March ($N + 21$), or April ($N - 10$); and by

MEASURES OF TIME.

employing the foregoing values of N we deduce the following formulæ for its determination:—

When $e < 24$, Easter is $\begin{Bmatrix}\text{March } (45-e) \\ \text{April } (14-e)\end{Bmatrix} + \left(\dfrac{L+e+1}{7}\right)_r$

„ $e = 24$, „ „ April $19 + \left(\dfrac{L+3}{7}\right)_r$

„ $e > 24$ „ „ April $(44-e) + \left(\dfrac{L+e-1}{7}\right)_r.$

By entering the following table with the epact and the dominical letter the date of Easter-day may always be ascertained by inspection.

Perpetual Table for determining EASTER-DAY *from the Epact and Dominical Letter.*

Epact (e).	Dominical letter. (For Leap Years use the *second* letter.)						
	A	B	C	D	E	F	G
•	Apr. 16	Apr. 17	Apr. 18	Apr. 19	Apr. 20	Apr. 14	Apr. 15
1	„ 16	„ 17	„ 18	„ 19	„ 13	„ 14	„ 15
2	„ 16	„ 17	„ 18	„ 12	„ 13	„ 14	„ 15
3	„ 16	„ 17	„ 11	„ 12	„ 13	„ 14	„ 15
4	„ 16	„ 10	„ 11	„ 12	„ 13	„ 14	„ 15
5	„ 9	„ 10	„ 11	„ 12	„ 13	„ 14	„ 15
6	„ 9	„ 10	„ 11	„ 12	„ 13	„ 14	„ 8
7	„ 9	„ 10	„ 11	„ 12	„ 13	„ 7	„ 8
8	„ 9	„ 10	„ 11	„ 12	„ 6	„ 7	„ 8
9	„ 9	„ 10	„ 11	„ 5	„ 6	„ 7	„ 8
10	„ 9	„ 10	„ 4	„ 5	„ 6	„ 7	„ 8
11	„ 9	„ 3	„ 4	„ 5	„ 6	„ 7	„ 8
12	„ 2	„ 3	„ 4	„ 5	„ 6	„ 7	„ 8
13	„ 2	„ 3	„ 4	„ 5	„ 6	„ 7	„ 1
14	„ 2	„ 3	„ 4	„ 5	„ 6	Mar. 31	„ 1
15	„ 2	„ 3	„ 4	„ 5	Mar. 30	„ 31	„ 1
16	„ 2	„ 3	„ 4	Mar. 29	„ 30	„ 31	„ 1
17	„ 2	„ 3	Mar. 28	„ 29	„ 30	„ 31	„ 1
18	„ 2	Mar. 27	„ 28	„ 29	„ 30	„ 31	„ 1
19	Mar. 26	„ 27	„ 28	„ 29	„ 30	„ 31	„ 1
20	„ 26	„ 27	„ 28	„ 29	„ 30	„ 31	Mar. 25
21	„ 26	„ 27	„ 28	„ 29	„ 30	„ 24	„ 25
22	„ 26	„ 27	„ 28	„ 29	„ 23	„ 24	„ 25
23	„ 26	„ 27	„ 22	„ 23	„ 24	„ 25	
24	Apr. 23	Apr. 24	Apr. 25	Apr. 19	Apr. 20	Apr. 21	Apr. 22
25	„ 23	„ 24	„ 25	„ 19	„ 20	„ 21	„ 22
26	„ 23	„ 24	„ 18	„ 19	„ 20	„ 21	„ 22
27	„ 23	„ 17	„ 18	„ 19	„ 20	„ 21	„ 22
28	„ 16	„ 17	„ 18	„ 19	„ 20	„ 21	„ 22
29	„ 16	„ 17	„ 18	„ 19	„ 20	„ 21	„ 15

MEASURES OF TIME.

INDICTION.—The Roman Indiction is a mode of measuring time by a cycle of 15 years, formerly used by the Romans for some time after the Emperor Constantine, but the precise time of its first adoption has not been ascertained with certainty, beyond the fact that the year 313 of the Christian era was a first year of a cycle of indiction.

To find the indiction, we must therefore observe the following rule :—

Rule.—Add 3 to the year; divide the sum by 15, and the remainder will be the Indiction. If there be no remainder, the Indiction is 15. Or divide the sum by 30, and diminish the remainder by 15 if greater than 15.

DIONYSIAN PERIOD.—The Dionysian is a period of 532 years, formed from the product of the lunar cycle 19 and the solar cycle 28, and invented by Dionysius Exiguus about the time of the Council of Nice, to include all the varieties of the new moons and dominical letters; so that after every 532 years they were expected to recur in the same order. This would have been very convenient for fixing the date of Easter and of the other days of the calendar by a table calculated for the years of one cycle ; but as the measure of the lunar cycle was supposed to be exact, which is not the case, and as the Sunday cycle is now interrupted at the centenary years that are not bissextile, the Dionysian period is no longer used in such calculations.

To find the year of the Dionysian Period :—

Rule.—Add 457 to the year of Christ ; divide the sum by 532, and the remainder will be the number required. Or for any year, from the present time up to the year 2203,

$$\text{Year of the Dionysian} = 129 + (\text{Year} - 1800),$$
$$\text{or} = \text{Year} - (1671, 2128, \&\text{c.}).$$

When divided by 28 the remainder is the solar cycle ; when divided by 19 the remainder is the lunar cycle or golden number.

MEASURES OF TIME.

JULIAN PERIOD.—The Julian Period is a large cycle of 7980 years, formed by multiplying together the lunar cycle of 19 years, the solar cycle of 28, and the indiction of 15, and its commencement goes back 4714 years beyond the Christian era.

To determine the number for any year.

Rule.—Add 4713 to the given year; divide the sum by 7980; the remainder will be the year of the Julian Period. Or for any year from the present time up to the year 3267,

Year of the Julian Period = 6513 + (Year − 1800).

If the year of the Julian Period be divided by 19, the remainder will be the *golden number;* if it be divided by 28, the remainder will be the *solar cycle;* and if it be divided by 15, the remainder will be the *indiction* for the corresponding year. Also, if it be divided by 532, the remainder will be the year of the Dionysian.

MOON'S AGE.—The age of the moon on any given date may be approximately deduced by adding to the epact the day of the month and the number below for the month, rejecting 30's if necessary.

Jan.	Feb.	Mar.	Apr.	May	June	July	Aug.	Sept.	Oct.	Nov.	Dec.
0	1	0	1	2	3	4	5	7	7	9	9

MOVEABLE FEASTS.—These are in general made to depend on the date of Easter-day. The following are some of the principal Sundays:—

Septuagesima Sunday is 9 ⎫
Shrove Sunday „ 7 ⎬ weeks before Easter.
1 Sunday in Lent „ 6 ⎪
Midlent Sunday „ 3 ⎭

Rogation Sunday is 5 ⎫
Whit-Sunday „ 7 ⎬ weeks after Easter.
Trinity Sunday „ 8 ⎭

Advent Sunday is the nearest Sunday to November 30, whether before or after.

Also,

First day of Lent is 3 days after Shrove Sunday.
Good Friday „ 2 „ before Easter.
Ascension-day „ 4 „ after Rogation Sunday.

The number of days which *intervene* between Epiphany (January 6) and Septuagesima Sunday is $10 + N$ in ordinary years, and $11 + N$ in bissextile years, N denoting the number of direction. Therefore, as the Epiphany Sundays are included in this interval,

$$\text{Sundays after Epiphany} = \left(\frac{10 + N}{7}\right)_w \text{ in ordinary years}$$

$$\text{„ „ „} = \left(\frac{11 + N}{7}\right)_w \text{ „ bissextile „}$$

Also the number of days *intervening* between Trinity Sunday and November 27, the earliest possible date of Advent Sunday, $= 194 - N$. Therefore,

$$\text{Sundays after Trinity} = \left(\frac{194 - N}{7}\right)_w$$

$$= 22 + \left(\frac{40 - N}{7}\right)_w.$$

To determine the elements of the Christian calendar, for any given year, it is only requisite to take out, by inspection,

the Dominical Letter from the table, page 154
„ Golden Number „ „ „ 158
„ Epact „ „ „ 164
„ Number of Direction „ „ „ 167

When the number of direction has thus been ascertained, the moveable feasts and other articles of the calendar will be shown by the following tables.

MEASURES OF TIME.

Table of the MOVEABLE FEASTS, *&c., according to the Number of Direction.*

Number of Direction (N.)	Dominical Letter.	Sundays after Epiphany.	Septuagesima Sunday.	Shrove Sunday.	First Day of Lent.	1 Sunday in Lent.	Midlent Sunday.
1	D, ED	1	Jan. 18, 19	Feb. 1, 2	Feb. 4, 5	Feb. 8, 9	Mar. 1
2	E, FE	1	,, 19, 20	,, 2, 3	,, 5, 6	,, 9, 10	,, 2
3	F, GF	1, 2	,, 20, 21	,, 3, 4	,, 6, 7	,, 10, 11	,, 3
4	G, AG	2	,, 21, 22	,, 4, 5	,, 7, 8	,, 11, 12	,, 4
5	A, BA	2	,, 22, 23	,, 5, 6	,, 8, 9	,, 12, 13	,, 5
6	B, CB	2	,, 23, 24	,, 6, 7	,, 9, 10	,, 13, 14	,, 6
7	C, DC	2	,, 24, 25	,, 7, 8	,, 10, 11	,, 14, 15	,, 7
8	D, ED	2	,, 25, 26	,, 8, 9	,, 11, 12	,, 15, 16	,, 8
9	E, FE	2	,, 26, 27	,, 9, 10	,, 12, 13	,, 16, 17	,, 9
10	F, GF	2, 3	,, 27, 28	,, 10, 11	,, 13, 14	,, 17, 18	,, 10
11	G, AG	3	,, 28, 29	,, 11, 12	,, 14, 15	,, 18, 19	,, 11
12	A, BA	3	,, 29, 30	,, 12, 13	,, 15, 16	,, 19, 20	,, 12
13	B, CB	3	,, 30, 31	,, 13, 14	,, 16, 17	,, 20, 21	,, 13
14	C, DC	3	,, 31, F 1	,, 14, 15	,, 17, 18	,, 21, 22	,, 14
15	D, ED	3	Feb. 1, 2	,, 15, 16	,, 18, 19	,, 22, 23	,, 15
16	E, FE	3	,, 2, 3	,, 16, 17	,, 19, 20	,, 23, 24	,, 16
17	F, GF	3, 4	,, 3, 4	,, 17, 18	,, 20, 21	,, 24, 25	,, 17
18	G, AG	4	,, 4, 5	,, 18, 19	,, 21, 22	,, 25, 26	,, 18
19	A, BA	4	,, 5, 6	,, 19, 20	,, 22, 23	,, 26, 27	,, 19
20	B, CB	4	,, 6, 7	,, 20, 21	,, 23, 24	,, 27, 28	,, 20
21	C, DC	4	,, 7, 8	,, 21, 22	,, 24, 25	,, 28, 29	,, 21
22	D, ED	4	,, 8, 9	,, 22, 23	,, 25, 26	Mar. 1	,, 22
23	E, FE	4	,, 9, 10	,, 23, 24	,, 26, 27	,, 2	,, 23
24	F, GF	4, 5	,, 10, 11	,, 24, 25	,, 27, 28	,, 3	,, 24
25	G, AG	5	,, 11, 12	,, 25, 26	,, 28, 29	,, 4	,, 25
26	A, BA	5	,, 12, 13	,, 26, 27	Mar. 1	,, 5	,, 26
27	B, CB	5	,, 13, 14	,, 27, 28	,, 2	,, 6	,, 27
28	C, DC	5	,, 14, 15	,, 28, 29	,, 3	,, 7	,, 28
29	D, ED	5	,, 15, 16	Mar. 1	,, 4	,, 8	,, 29
30	E, FE	5	,, 16, 17	,, 2	,, 5	,, 9	,, 30
31	F, GF	5, 6	,, 17, 18	,, 3	,, 6	,, 10	,, 31
32	G, AG	6	,, 18, 19	,, 4	,, 7	,, 11	Apr. 1
33	A, BA	6	,, 19, 20	,, 5	,, 8	,, 12	,, 2
34	B, CB	6	,, 20, 21	,, 6	,, 9	,, 13	,, 3
35	C, DC	6	,, 21, 22	,, 7	,, 10	,, 14	,, 4

*** Second entries in the above table are for bissextile years.

LAW TERMS.—The Law Terms and Returns were formerly regulated by statute 1 William IV. cap. 70, passed the 22nd of July,

MEASURES OF TIME. 173

Table (continued) of the MOVEABLE FEASTS, &c.

Good Friday.	Easter Day.	Rogation Sunday.	Ascension.	Whit-Sunday.	Trinity Sunday.	Sundays after Trinity.	Advent Sunday.
Mar. 20	Mar. 22	Apr. 26	Apr. 30	May 10	May 17	27	Nov. 29
,, 21	,, 23	,, 27	May 1	,, 11	,, 18	27	,, 30
,, 22	,, 24	,, 28	,, 2	,, 12	,, 19	27	Dec. 1
,, 23	,, 25	,, 29	,, 3	,, 13	,, 20	27	,, 2
,, 24	,, 26	,, 30	,, 4	,, 14	,, 21	27	,, 3
,, 25	,, 27	May 1	,, 5	,, 15	,, 22	26	Nov. 27
,, 26	,, 28	,, 2	,, 6	,, 16	,, 23	26	,, 28
,, 27	,, 29	,, 3	,, 7	,, 17	,, 24	26	,, 29
,, 28	,, 30	,, 4	,, 8	,, 18	,, 25	26	,, 30
,, 29	,, 31	,, 5	,, 9	,, 19	,, 26	26	Dec. 1
,, 30	Ap 1	,, 6	,, 10	,, 20	,, 27	26	,, 2
,, 31	,, 2	,, 7	,, 11	,, 21	,, 28	26	,, 3
Apr. 1	,, 3	,, 8	,, 12	,, 22	,, 29	25	Nov. 27
,, 2	,, 4	,, 9	,, 13	,, 23	,, 30	25	,, 28
,, 3	,, 5	,, 10	,, 14	,, 24	,, 31	25	,, 29
,, 4	,, 6	,, 11	,, 15	,, 25	June 1	25	,, 30
,, 5	,, 7	,, 12	,, 16	,, 26	,, 2	25	Dec. 1
,, 6	,, 8	,, 13	,, 17	,, 27	,, 3	25	,, 2
,, 7	,, 9	,, 14	,, 18	,, 28	,, 4	25	,, 3
,, 8	,, 10	,, 15	,, 19	,, 29	,, 5	24	Nov. 27
,, 9	,, 11	,, 16	,, 20	,, 30	,, 6	24	,, 28
,, 10	,, 12	,, 17	,, 21	,, 31	,, 7	24	,, 29
,, 11	,, 13	,, 18	,, 22	June 1	,, 8	24	,, 30
,, 12	,, 14	,, 19	,, 23	,, 2	,, 9	24	Dec. 1
,, 13	,, 15	,, 20	,, 24	,, 3	,, 10	24	,, 2
,, 14	,, 16	,, 21	,, 25	,, 4	,, 11	24	,, 3
,, 15	,, 17	,, 22	,, 26	,, 5	,, 12	23	Nov. 27
,, 16	,, 18	,, 23	,, 27	,, 6	,, 13	23	,, 28
,, 17	,, 19	,, 24	,, 28	,, 7	,, 14	23	,, 29
,, 18	,, 20	,, 25	,, 29	,, 8	,, 15	23	,, 30
,, 19	,, 21	,, 26	,, 30	,, 9	,, 16	23	Dec. 1
,, 20	,, 22	,, 27	,, 31	,, 10	,, 17	23	,, 2
,, 21	,, 23	,, 28	June 1	,, 11	,, 18	23	,, 3
,, 22	,, 24	,, 29	,, 2	,, 12	,, 19	22	Nov. 27
,, 23	,, 25	,, 30	,, 3	,, 13	,, 20	22	,, 28

1830. The following is an abstract of clause VI. of the Act:—
"Hilary Term shall begin on the 11th and end on the 31st

day of January; Easter Term shall begin on the fifteenth day of April, and end on the eighth day of May; Trinity Term shall begin on the twenty-second day of May, and end on the twelfth day of June; and Michaelmas Term shall begin on the second and end on the twenty-fifth day of November; and that the Essoign and General Return Days of each Term shall, until further provision be made by Parliament, be as follows; that is to say, the first Essoign or General Return day for every Term shall be the fourth day before the day of the commencement of the Term, both days being included in the computation; the second Essoign day shall be the fifth day of the Term; the third shall be the fifteenth day of the Term; and the fourth and last shall be the nineteenth day of the Term, the first day of the Term being already included in the computation; with the same relation to the commencement of each Term as they now bear, and shall be distinguished by the day of the Term on which they respectively fall, the Monday being in all cases substituted for the Sunday when it shall happen that the day would fall on Sunday, except always that in Easter Term there shall be but four Returns instead of five, the last being omitted; provided that in case the day of the month on which any Term according to the Act aforesaid is to end shall fall to be on a Sunday, then the Monday next after such day shall be deemed and taken to be the last day of the Term; and that if the whole or any number of the days[1] intervening *between* the *Thursday before and the Wednesday next after Easter-day* shall fall within Easter Term, there shall be no Sittings in Banc on any of such intervening days, but the Term shall in such case be prolonged and continue for such number of *days of business* as shall be equal to the number

---

[1] The intervening days, exclusive of Easter-day, are
        Good Friday,
        Saturday,
        Easter Monday,
        Easter Tuesday.

MEASURES OF TIME. 175

of the intervening days before mentioned exclusive of Easter-day, and the commencement of the ensuing Trinity Term shall in such case be postponed and its continuance prolonged for an equal number of *days of business.*"

The wording of the Act is somewhat confused and obscure, and its correct interpretation and practical application require some little consideration. In order to obviate this inconvenience, as these terms are not yet entirely obsolete, we annex the following table, in which the dates of the commencement and ending of the several terms are made to depend simply on the Number of Direction.

Table of the OLD LAW TERMS, *according to the Number of Direction.*

No. of Direction (*N*).	Name.	Begins.	Ends.	No. of days.
1, 2, 3, 5, 8, 9, 10, 12, 15, 16, 17, 19, 22	Hilary Term	Jan. 11	Jan. 31	21
	Easter „	Apr. 15	May 8	24
	Trinity „	May 22	June 12	22
	Michaelmas „	Nov. 2	Nov. 25	24
4, 11, 18	Hilary Term	Jan. 11	Jan. 31	21
	Easter „	Apr. 15	May 8	24
	Trinity „	May 22	June 12	22
	Michaelmas „	Nov. 2	Nov. 26	25
6, 13, 20	Hilary Term	Jan. 11	{ Jan. 31 Feb. 1 (bis.)	21 22 }
	Easter „	Apr. 15	May 9	25
	Trinity „	May 22	June 13	23
	Michaelmas „	Nov. 2	Nov. 25	24
7, 14, 21	Hilary Term	Jan. 11	{ Feb. 1 Jan. 31 (bis.)	22 21 }
	Easter „	Apr. 15	May 8	24
	Trinity „	May 22	June 12	22
	Michaelmas „	Nov. 2	Nov. 25	24
23	Hilary Term	Jan. 11	Jan. 31	21
	Easter „	Apr. 15	May 9	25
	Trinity „	May 23	June 13	22
	Michaelmas „	Nov. 2	Nov. 25	24
24	Hilary Term	Jan. 11	Jan. 31	21
	Easter „	Apr. 15	May 10	26
	Trinity „	May 24	June 14	22
	Michaelmas „	Nov. 2	Nov. 25	24

MEASURES OF TIME.

Table of the OLD LAW TERMS, *according to the Number of Direction.*

No. of Direction (N).	Name.		Begins.	Ends.	No. of days.
25	Hilary	Term	Jan. 11	Jan. 31	21
	Easter	"	Apr. 15	May 10	26
	Trinity	"	May 24	June 14	22
	Michaelmas	"	Nov. 2	Nov. 26	25
26	Hilary	Term	Jan. 11	Jan. 31	21
	Easter	"	Apr. 15	May 11	27
	Trinity	"	May 25	June 15	22
	Michaelmas	"	Nov. 2	Nov. 25	24
27, 34	Hilary	Term	Jan. 11	{Jan. 31 / Feb. 1 (bis.)}	{21 / 22}
	Easter	"	Apr. 15	May 12	28
	Trinity	"	May 26	June 16	22
	Michaelmas	"	Nov. 2	Nov. 25	24
28, 35	Hilary	Term	Jan. 11	{Feb. 1 / Jan. 31 (bis.)}	{22 / 21}
	Easter	"	Apr. 15	May 13	29
	Trinity	"	May 27	June 17	22
	Michaelmas	"	Nov. 2	Nov. 25	24
29, 30, 31	Hilary	Term	Jan. 11	Jan. 31	21
	Easter	"	Apr. 15	May 13	29
	Trinity	"	May 27	June 17	22
	Michaelmas	"	Nov. 2	Nov. 25	24
32	Hilary	Term	Jan. 11	Jan. 31	21
	Easter	"	Apr. 15	May 12	28
	Trinity	"	May 26	June 16	22
	Michaelmas	"	Nov. 2	Nov. 26	25
33	Hilary	Term	Jan. 11	Jan. 31	21
	Easter	"	Apr. 15	May 12	28
	Trinity	"	May 26	June 16	22
	Michaelmas	"	Nov. 2	Nov. 25	24

LAW SITTINGS.—On the 1st of November, 1875, the Judicature Act, 1873, came into force, and the division of the legal year into terms was abolished so far as relates to the administration of justice. The Rules of Court provide that there shall be four sittings in every year of the Court of Appeal and High Court of Justice, namely :—

MEASURES OF TIME. 177

1. MICHAELMAS SITTINGS, beginning October 24, ending December 21.
2. HILARY SITTINGS, beginning January 11, ending Wednesday before Easter.
3. EASTER SITTINGS, beginning Tuesday after Easter week, ending Friday before Whit Sunday.
4. TRINITY SITTINGS, beginning Tuesday after Whitsun week, ending August 12.

There are also Law Terms in use observed by the Inns of Court. These terms are—

Hilary.—Same as old Law Term, January 11 to January 31 or February 1.
Easter.—The first 28 days of the Easter Law Sittings.
Trinity.—The first 21 days of the Trinity Law Sittings.
Michaelmas.—Same as old Law Term, November 2 to November 25 or November 26.

UNIVERSITY TERMS.—The University Terms may also be obtained simply from the Number of Direction by means of the following formulary table :—

	N.					Oxford Act.	Cambridge Commencement.
	2	9	16	23	30	July 1	June 17
	3	10	17	24	31	,, 2	,, 18
	4	11	18	25	32	,, 3	,, 19
	5	12	19	26	33	,, 4	,, 20
	6	13	20	27	34	,, 5	,, 21
	7	14	21	28	35	,, 6	,, 22
1	8	15	22	29		,, 7	,, 23

OXFORD UNIVERSITY TERMS.

When the beginning or ending of a Term falls on a festival, the rule is to take the day *after*, excepting the *ending* of Easter Term, which is to be the day *before*. The festivals, *besides Sundays*, which interfere are—

Annunciation . . . . March 25.
St. Mark . . . . April 25.
Accession Chas. II. . . . . . . May 29.
St. Barnabas . . . . June 11.

The Act, *first* Tuesday in July.

Hilary. Term begins Jan. 14   } ends Mar. 13 + N
         (if Sunday) Jan. 15   } or Mar. 15 + N (if N = 12)
Easter   ,,   ,,   Mar. 24 + N } ,, May 7 + N
         (if N = 1, 32) Mar. 25 + N } or May 6 + N (if N = 35)
Trinity   ,,   ,,   May 8 + N }
         (if N = 34) May 10 + N } ,, The Act + 4
Michaelmas ,,   ,,   Oct. 10   } ,,   Dec. 17 }
         (if Sunday) Oct. 11   }   (if Sunday) Dec. 18 }

CAMBRIDGE UNIVERSITY TERMS *before* 1884.
Commencement, Tuesday *before* Midsummer.

Hilary   Term begins Jan. 13   ends Mar. 12 + N
Easter   ,,   ,,   Mar. 26 + N ,, Comm. + 3
Michaelmas ,,   ,,   Oct. 1   ,, Dec. 16
Hilary   Term divides ...... Feb. 11 + ½ N }
         (bis)   ...... Feb. 11 + ½ (N + 1) }
Easter   ,,   divides ...... May ½ (N + Comm. − 1)
Michaelmas ,,   ,, ... Nov. 8, noon.
(See page 207.)

To facilitate the construction of the Gregorian Calendar for any proposed year, the following table is inserted, showing the dominical letter, golden number, epact, number of direction, and Easter Sunday for three centuries, namely, from 1800 to 2100. It is only requisite to take the number of direction from the last column but one, and, with it, to enter the table on pages 172 and 173, and everything becomes immediately known without any calculation.

Elements of the GREGORIAN CALENDAR *for Three Centuries,*
1800 *to* 2100.

Year.	Dominical Letter.	Golden Number (g).	Epact (e).	No. of Direction (N).	Easter Sunday.
1800	E	15	4	23	April 13.
1801	D	16	15	15	April 5.
1802	C	17	26	28	April 18.
1803	B	18	7	20	April 10.
1804	AG	19	18	11	April 1.

MEASURES OF TIME. 179

Year.	Dominical Letter.	Golden Number (g).	Epact (e).	No. of Direction (N).	Eas'er Sunday.
1805	F	1	*	24	April 14.
1806	E	2	11	16	April 6.
1807	D	3	22	8	March 29.
1808	CB	4	3	27	April 17.
1809	A	5	14	12	April 2.
1810	G	6	25	32	April 22.
1811	F	7	6	24	April 14.
1812	ED	8	17	8	March 29.
1813	C	9	28	28	April 18.
1814	B	10	9	20	April 10.
1815	A	11	20	5	March 26.
1816	GF	12	1	24	April 14.
1817	E	13	12	16	April 6.
1818	D	14	23	1	March 22.
1819	C	15	4	21	April 11.
1820	BA	16	15	12	April 2.
1821	G	17	26	32	April 22.
1822	F	18	7	17	April 7.
1823	E	19	18	9	March 30.
1824	DC	1	*	28	April 18.
1825	B	2	11	13	April 3.
1826	A	3	22	5	March 26.
1827	G	4	3	25	April 15.
1828	FE	5	14	16	April 6.
1829	D	6	25	29	April 19.
1830	C	7	6	21	April 11.
1831	B	8	17	13	April 3.
1832	AG	9	28	32	April 22.
1833	F	10	9	17	April 7.
1834	E	11	20	9	March 30.
1835	D	12	1	29	April 19.
1836	CB	13	12	13	April 3.
1837	A	14	23	5	March 26.
1838	G	15	4	25	April 15.
1839	F	16	15	10	March 31.

MEASURES OF TIME.

Year.	Dominical Letter.	Golden Number (g).	Epact (e).	No. of Direction (N).	Easter Sunday.
1840	ED	17	26	29	April 19.
1841	C	18	7	21	April 11.
1842	B	19	18	6	March 27.
1843	A	1	*	26	April 16.
1844	GF	2	11	17	April 7.
1845	E	3	22	2	March 23.
1846	D	4	3	22	April 12.
1847	C	5	14	14	April 4.
1848	BA	6	25	33	April 23.
1849	G	7	6	18	April 8.
1850	F	8	17	10	March 31.
1851	E	9	28	30	April 20.
1852	DC	10	9	21	April 11.
1853	B	11	20	6	March 27.
1854	A	12	1	26	April 16.
1855	G	13	12	18	April 8.
1856	FE	14	23	2	March 23.
1857	D	15	4	22	April 12.
1858	C	16	15	14	April 4.
1859	B	17	26	34	April 24.
1860	AG	18	7	18	April 8.
1861	F	19	18	10	March 31.
1862	E	1	30	30	April 20.
1863	D	2	11	15	April 5.
1864	CB	3	22	6	March 27.
1865	A	4	3	26	April 16.
1866	G	5	14	11	April 1.
1867	F	6	25	31	April 21.
1868	ED	7	6	22	April 12.
1869	C	8	17	7	March 28.
1870	B	9	28	27	April 17.
1871	A	10	9	19	April 9.
1872	GF	11	20	10	March 31.
1873	E	12	1	23	April 13.
1874	D	13	12	15	April 5.

MEASURES OF TIME.

Year.	Dominical Letter.	Golden Number (g).	Epact (e).	No. of Direction (N).	Easter Sunday.
1875	C	14	23	7	March 28.
1876	BA	15	4	26	April 16.
1877	G	16	15	11	April 1.
1878	F	17	26	31	April 21.
1879	E	18	7	23	April 13.
1880	DC	19	18	7	March 28.
1881	B	1	*	27	April 17.
1882	A	2	11	19	April 9.
1883	G	3	22	4	March 25.
1884	FE	4	3	23	April 13.
1885	D	5	14	15	April 5.
1886	C	6	25	35	April 25.
1887	B	7	6	20	April 10.
1888	AG	8	17	11	April 1.
1889	F	9	28	31	April 21.
1890	E	10	9	16	April 6.
1891	D	11	20	8	March 29.
1892	CB	12	1	27	April 17.
1893	A	13	12	12	April 2.
1894	G	14	23	4	March 25.
1895	F	15	4	24	April 14.
1896	ED	16	15	15	April 5.
1897	C	17	26	28	April 18.
1898	B	18	7	20	April 10.
1899	A	19	18	12	April 2.
1900	G	1	29	25	April 15.
1901	F	2	10	17	April 7.
1902	E	3	21	9	March 30.
1903	D	4	2	22	April 12.
1904	CB	5	13	13	April 3.
1905	A	6	24	33	April 23.
1906	G	7	5	25	April 15.
1907	F	8	16	10	March 31.
1908	ED	9	27	29	April 19.
1909	C	10	8	21	April 11.

MEASURES OF TIME.

Year.	Dominical Letter.	Golden Number (g).	Epact (e).	No. of Direction (N).	Easter Sunday.
1910	B	11	19	6	March 27.
1911	A	12	*	26	April 16.
1912	GF	13	11	17	April 7.
1913	E	14	22	2	March 23.
1914	D	15	3	22	April 12.
1915	C	16	14	14	April 4.
1916	BA	17	26	33	April 23.
1917	G	18	6	18	April 8.
1918	F	19	17	10	March 31.
1919	E	1	29	30	April 20.
1920	DC	2	10	14	April 4.
1921	B	3	21	6	March 27.
1922	A	4	2	26	April 16.
1923	G	5	13	11	April 1.
1924	FE	6	24	30	April 20.
1925	D	7	5	22	April 12.
1926	C	8	16	14	April 4.
1927	B	9	27	27	April 17.
1928	AG	10	8	18	April 8.
1929	F	11	19	10	March 31.
1930	E	12	*	30	April 20.
1931	D	13	11	15	April 5.
1932	CB	14	22	6	March 27.
1933	A	15	3	26	April 16.
1934	G	16	14	11	April 1.
1935	F	17	26	31	April 21.
1936	ED	18	6	22	April 12.
1937	C	19	17	7	March 28.
1938	B	1	29	27	April 17.
1939	A	2	10	19	April 9.
1940	GF	3	21	3	March 24.
1941	E	4	2	23	April 13.
1942	D	5	13	15	April 5.
1943	C	6	24	35	April 25.
1944	BA	7	5	19	April 9.

MEASURES OF TIME.

Year.	Dominical Letter.	Golden Number (g).	Epact (e).	No. of Direction (N).	Easter Sunday.
1945	G	8	16	11	April 1.
1946	F	9	27	31	April 21.
1947	E	10	8	16	April 6.
1948	DC	11	19	7	March 28.
1949	B	12	*	27	April 17.
1950	A	13	11	19	April 9.
1951	G	14	22	4	March 25.
1952	FE	15	3	23	April 13.
1953	D	16	14	15	April 5.
1954	C	17	26	28	April 18.
1955	B	18	6	20	April 10.
1956	AG	19	17	11	April 1.
1957	F	1	29	31	April 21.
1958	E	2	10	16	April 6.
1959	D	3	21	8	March 29.
1960	CB	4	2	27	April 17.
1961	A	5	13	12	April 2.
1962	G	6	24	32	April 22.
1963	F	7	5	24	April 14.
1964	ED	8	16	8	March 29.
1965	C	9	27	28	April 18.
1966	B	10	8	20	April 10.
1967	A	11	19	5	March 26.
1968	GF	12	*	24	April 14.
1969	E	13	11	16	April 6.
1970	D	14	22	8	March 29.
1971	C	15	3	21	April 11.
1972	BA	16	14	12	April 2.
1973	G	17	26	32	April 22.
1974	F	18	6	24	April 14.
1975	E	19	17	9	March 30.
1976	DC	1	29	28	April 18.
1977	B	2	10	20	April 10.
1978	A	3	21	5	March 26.
1979	G	4	2	25	April 15.

Year.	Dominical Letter.	Golden Number (g).	Epact (e).	No. of Direction (N).	Easter Sunday.
1980	FE	5	13	16	April 6.
1981	D	6	24	29	April 19.
1982	C	7	5	21	April 11
1983	B	8	16	13	April 3.
1984	AG	9	27	32	April 22
1985	F	10	8	17	April 7.
1986	E	11	19	9	March 30.
1987	D	12	*	29	April 19.
1988	CB	13	11	13	April 3.
1989	A	14	22	5	March 26.
1990	G	15	3	25	April 15.
1991	F	16	14	10	March 31.
1992	ED	17	26	29	April 19.
1993	C	18	6	21	April 11.
1994	B	19	17	13	April 3.
1995	A	1	29	26	April 16.
1996	GF	2	10	17	April 7.
1997	E	3	21	9	March 30.
1998	D	4	2	22	April 12.
1999	C	5	13	14	April 4.
2000	BA	6	24	33	April 23.
2001	G	7	5	25	April 15.
2002	F	8	16	10	March 31.
2003	E	9	27	30	April 20.
2004	DC	10	8	21	April 11.
2005	B	11	19	6	March 27
2006	A	12	*	26	April 16.
2007	G	13	11	18	April 8.
2008	FE	14	22	2	March 23.
2009	D	15	3	22	April 12.
2010	C	16	14	14	April 4.
2011	B	17	26	34	April 24.
2012	AG	18	6	18	April 8.
2013	F	19	17	10	March 31.
2014	E	1	29	30	April 20.

MEASURES OF TIME.

Year.	Dominical Letter.	Golden Number (g).	Epact (e).	No. of Direction (N).	Easter Sunday.
2015	D	2	10	15	April 5.
2016	CB	3	21	6	March 27.
2017	A	4	2	26	April 16.
2018	G	5	13	11	April 1.
2019	F	6	24	31	April 21.
2020	ED	7	5	22	April 12.
2021	C	8	16	14	April 4.
2022	B	9	27	27	April 17.
2023	A	10	8	19	April 9.
2024	GF	11	19	10	March 31.
2025	E	12	*	30	April 20.
2026	D	13	11	15	April 5.
2027	C	14	22	7	March 28.
2028	BA	15	3	26	April 16.
2029	G	16	14	11	April 1.
2030	F	17	26	31	April 21.
2031	E	18	6	23	April 13.
2032	DC	19	17	7	March 28.
2033	B	1	29	27	April 17.
2034	A	2	10	19	April 9.
2035	G	3	21	4	March 25.
2036	FE	4	2	23	April 13.
2037	D	5	13	15	April 5.
2038	C	6	24	35	April 25.
2039	B	7	5	20	April 10.
2040	AG	8	16	11	April 1.
2041	F	9	27	31	April 21.
2042	E	10	8	16	April 6.
2043	D	11	19	8	March 29.
2044	CB	12	*	27	April 17.
2045	A	13	11	19	April 9.
2046	G	14	22	4	March 25.
2047	F	15	3	24	April 14.
2048	ED	16	14	15	April 5.
2049	C	17	26	28	April 18.

MEASURES OF TIME.

Year.	Dominical Letter.	Golden Number (g).	Epact (e).	No of Direction (N).	Easter Sunday.
2050	B	18	6	20	April 10.
2051	A	19	17	12	April 2.
2052	GF	1	29	31	April 21.
2053	E	2	10	16	April 6.
2054	D	3	21	8	March 29.
2055	C	4	2	28	April 18.
2056	BA	5	13	12	April 2.
2057	G	6	24	32	April 22.
2058	F	7	5	24	April 14.
2059	E	8	16	9	March 30.
2060	DC	9	27	28	April 18.
2061	B	10	8	20	April 10.
2062	A	11	19	5	March 26.
2063	G	12	*	25	April 15.
2064	FE	13	11	16	April 6.
2065	D	14	22	8	March 29.
2066	C	15	3	21	April 11.
2067	B	16	14	13	April 3.
2068	AG	17	26	32	April 22.
2069	F	18	6	24	April 14.
2070	E	19	17	9	March 30.
2071	D	1	29	29	April 19.
2072	CB	2	10	20	April 10.
2073	A	3	21	5	March 26.
2074	G	4	2	25	April 15.
2075	F	5	13	17	April 7.
2076	ED	6	24	29	April 19.
2077	C	7	5	21	April 11.
2078	B	8	16	13	April 3.
2079	A	9	27	33	April 23.
2080	GF	10	8	17	April 7.
2081	E	11	19	9	March 30.
2082	D	12	*	29	April 19.
2083	C	13	11	14	April 4.
2084	BA	14	22	5	March 26.

MEASURES OF TIME.

Year.	Dominical Letter	Golden Number (g).	Epact (e).	No. of Direction (N).	Easter Sunday.
2085	G	15	3	25	April 15.
2086	F	16	14	10	March 31.
2087	E	17	26	30	April 20.
2088	DC	18	6	21	April 11.
2089	B	19	17	13	April 3.
2090	A	1	29	26	April 16.
2091	G	2	10	18	April 8.
2092	FE	3	21	9	March 30.
2093	D	4	2	22	April 12.
2094	C	5	13	14	April 4.
2095	B	6	24	34	April 24.
2096	AG	7	5	25	April 15.
2097	F	8	16	10	March 31.
2098	E	9	27	30	April 20.
2099	D	10	8	22	April 12.
2100	C	11	19	7	March 28.

HEBREW CALENDAR.

A paper on the computation of the Hebrew Calendar was given by Mr. Herschell Fillipowski, in the "Lady and Gentleman's Diary," for the year 1850. As the Diary is out of print the following introductory extract may be acceptable :—

"Of all calendars known the Hebrew Calendar is the most complicated. It is the more remarkable on account of its astronomical system as laid down by the Hebrews, A.D. 358, and which, to the present time, agrees in its results with the revolutions of both luminaries.

"The Hebrew Calendar is composed either of twelve or thirteen months, the former being called a *common* the latter an *embolismic* year. The months receive alternately thirty and twenty-nine days ; thus, in a common year the number of days amount to 354, whilst in an embolismic one they amount to 384 ; in both these cases they are called *regular* years. The month of intercalation is always the twelfth in the year, and invariably receives thirty days; it is then called ADAR THE FIRST, whilst in a common year there is only one month of *Adar*.

"The year begins with the month of TISHRI, it being the seventh in the year, and is fixed on or immediately after the day of the mean new moon of that month (which always takes place during the month of September), provided the new moon does not occur within the last six hours of the day (the day commences and terminates at 6 p.m.).

"The new year is not to be fixed on either of the following days—Sunday, Wednesday, or Friday; accordingly, if the new moon happens to take place on either of these three days, though within the first eighteen hours of the day, the new year is postponed respectively to the next following day. These regulations cause, occasionally, either the eighth month, *Heshvon*, to receive thirty days instead of twenty-nine, thus increasing the year with one day; or the ninth month, *Kislev*, to receive twenty-nine days instead of thirty, by which the year is made one day shorter.

"The time of one mean lunar revolution is assumed by the Hebrews to be 29 days, 12 hours, and $44\frac{1}{18}$ minutes, differing from the modern astronomers by only half a second. With this they have measured, without exception, all lunar months ever since the introduction of their calendar, the results of which seem to correspond closely with those of modern computation."

The Jews date their Calendar from the Creation, which is considered by them to have occurred 3760 years and 3 months before the commencement of the Christian era. The year is luni-solar, and, according as it is ordinary or embolismic, consists of twelve or thirteen lunar months, each of which has 29 or 30 days. It is occasionally made a day more or less than the mean value in order that certain festivals may fall on proper days of the week for their due observance. The days of the respective months, according to the number comprised in the different years, are distributed as follows:—

MEASURES OF TIME.

Month.	Ordinary year.			Embolismic year.		
	Imperfect.	Common.	Perfect.	Imperfect.	Common.	Perfect.
Tisri	30	30	30	30	30	30
Hesvan	29	29	30	29	29	30
Kislev	29	30	30	29	30	30
Tebet	29	29	29	29	29	29
Sebat	30	30	30	30	30	30
Adar	29	29	29	30	30	30
(Veadar)	..	..	..	(29)	(29)	(29)
Nisan	30	30	30	30	30	30
Yiar	29	29	29	29	29	29
Sivan	30	30	30	30	30	30
Tamuz	29	29	29	29	29	29
Ab	30	30	30	30	30	30
Elul	29	29	29	29	29	29
No. of days in the year	353	354	355	383	384	385

The intercalary month, Veadar, is introduced that Passover, the 15th day of Nisan, may be kept at its proper season, which is the full moon of the vernal equinox, or that which takes place after the sun has entered the sign Aries. The distribution of the embolismic years is determined by a cycle of 19 years, according to the following rule.

Divide the Hebrew year by 19, the quotient is the number of the last completed cycle, and the remainder is the year of the current cycle; should it be 3, 6, 8, 11, 14, 17, or 19 (0), the year is embolismic; if any other it is ordinary.

Or if Y denote the year, and

$$R = \left(\frac{7Y + 13}{19}\right)_r;$$

the year is embolismic when $R > 11$.

The calendar is constructed by assuming the mean lunation to be 29 days 12 hours 44 min. $3\frac{1}{3}$ sec., and that the year commences on, or immediately after, the new moon following the autumnal equinox. The mean solar year is also assumed to be 365 days 5 hours 55 min. $25\frac{25}{57}$ sec.[1], so that a cycle of nineteen of such years contains 6939 days 16 hours 33 min. $3\frac{1}{3}$ sec., the exact measure of 235 of the assumed lunations. The year 5606 was the first of a cycle, and the computed new moon answering to the 1st of Tisri for that year was 1845, Oct. 1, 15h. 42m. $43\frac{1}{3}$s., according to Lindo, adopting the civil mode of reckoning from the previous midnight. The future times of new moon are consequently deduced by successively adding 29 days 12h. 44m. $3\frac{1}{3}$s. to this date.

Or to compute the times of new moon which belong to the commencement of successive years, we must in passing from an ordinary year, deduce the new moon of the following year by subtracting the interval that twelve lunations fall short of the corresponding Gregorian year of 365 or 366 days; and for an embolismic year we must add the excess of thirteen lunations over the Gregorian year; that is, to get the new moon of Tisri for the year immediately following any given year Y, we must,

for an ordinary year, subtract $\left\{\begin{array}{c}10\\11\end{array}\right\}$ days 15h. 11m. 20s.,

„ embolismic „ add $\left\{\begin{array}{c}18\\17\end{array}\right\}$ days 21h. 32m. $43\frac{1}{3}$s.,

the second mentioned number of days being used whenever the year Y is divisible by 4 without a remainder, or, more correctly, when the following or new Gregorian year is bissextile.

Thus, by knowing which of the years are embolismic, from their ordinal position in the cycle, according to the rule before stated, the times of the commencement of successive years may be carried on indefinitely without much trouble.

[1] This being 6 min. 37 sec. in excess of the mean solar year, the dates of commencement of future Jewish years so calculated will advance forward from the equinox a day in error in every 218 years. The lunations it will be observed, are estimated with greater precision.

We annex a few by way of example, and have distinguished by a * the years which are embolismic and bissextile. The hours are counted from the previous midnight, according to the civil reckoning.

Year				date		d.	h.	m.	s.
Year 5606,	year of cycle	1,	date	1845	Oct.	1	15	42	43½
5607	,,	,,	2,	,,	1846	Sept. 21	0	31	23½
5608	,,	,,	*3,	,,	1847	,,	10	9	20	3½
5609	,,	,,	4,	,,	*1848	,,	28	6	52	46¾
5610	,,	,,	5,	,,	1849	,,	17	15	41	·26¾
5611	,,	,,	*6,	,,	1850	,,	7	0	30	6¾
5612	,,	,,	7,	,,	1851	,,	25	22	2	50
5613	,,	,,	*8,	,,	*1852	,,	14	6	51	30
5614	,,	,,	9,	,,	1853	Oct.	3	4	24	13½
&c.				&c.				&c.	

But it must be observed, for the reasons before assigned, to avoid certain festivals falling on incompatible days of the week, that the year must not begin on a Sunday, Wednesday, or Friday, so that if the computed conjunction falls on one of these days the new year is to be fixed on the day after. The following conditions also require to be attended to :—

If the computed new moon be after 12h. the following day is to be taken, and if that happen to be Sunday, Wednesday, or Friday, it must be postponed one day more.

If in an ordinary year the new moon is on Tuesday, as late as 9h. 11m. 20s., it is not to be observed thereon; and, as it may not be held on Wednesday, it is to be postponed to Thursday.

If in a year immediately following an embolismic year the new moon is on Monday, as late as 15h. 30m. 52s., the new year is to be fixed on Tuesday.

The number of days contained in any given Jewish year, and the day of the week on which it commences, may be readily calculated by means of the following tables, the use of which will be best explained by an example.

Example.—Required the character of the Jewish yeai 5640.

Having found the next less year in the table of completed cycles, the calculation is as follows:—

Proposed year 5640		
296 cycles 5624	Number	2525,480
Year of cycle.. 16	($R = 11$)	2278,271
	Sum	4803,751

Referring to the third table in the column under $R = 11$, the number next less to this sum is 4743,996, and it stands opposite to "354 Thur.," showing that the year is ordinary, contains 354 days, and begins on a Thursday.

By the second table the approximate date of commencement is 17 Sept., and the corresponding year of our Lord is $5640 - 3761 = 1879$. Now, referring to the table, page 146, the Dominical letter for this year is E, and 17 Sept. is Wednesday. Therefore, as the Jewish year has been found to begin on a Thursday, the true date of commencement is 18 Sept. 1879.

The Gregorian epact being the age of the moon of Tebet at the beginning of the year, it represents the day of Tebet which corresponds to Jan. 1; and the approximate date of Tisri 1 may be otherwise deduced by subtracting the epact

from $\left\{ \begin{array}{l} \text{Sept. 24} \\ \text{Oct. 24} \end{array} \right\}$ after an $\left\{ \begin{array}{l} \text{ordinary} \\ \text{embolismic} \end{array} \right\}$ year.

The result so obtained would in general be more accurate than the Jewish calculation, and it may differ a day from the latter, as fractions of a day are omitted in these computations. The difference may, however, be adjusted as before by means of the day of the week.

MEASURES OF TIME.

Completed Cycles.	Year.	Number.
295	5605	1200,895
296	5624	2525,480
297	5643	402,705
298	5662	1727,290
299	5681	3051,875
300	5700	929,100
301	5719	2253,685
302	5738	130,910
303	5757	1455,495
304	5776	2780,080
305	5795	657,305
306	5814	1981,890
307	5833	3306,475
308	5852	1183,700
309	5871	2508,285
310	5890	385,510
311	5909	1710,095
312	5928	3034,680
313	5947	911,905
314	5966	2236,490
315	5985	113,715
316	6004	1438,300
317	6023	2762,885
318	6042	640,110
319	6061	1964,695
320	6080	3289,280

Year of Cycle.	R.	Number.	Approximate Commencement (Tisri 1).
1	1	722,076	2 Oct. *
2	8	2872,800	21 Sept.
3	15	1576,164	10 „
4	3	1033,315	29 „
5	10	3184,039	18 Sept.
6	17	1887,403	8 „
7	5	1344,554	27 „
8	12	47,918	15 „
9	0	2952,429	4 Oct.
10	7	1655,793	23 Sept.
11	14	359,157	12 „
12	2	3263,668	1 Oct.
13	9	1967,032	20 Sept.
14	16	670,396	9 „
15	4	127,547	28 Sept.
16	11	2278,271	17 „
17	18	981,635	6 „
18	6	438,786	25 „
19	13	2589,510	14 „

* The corresponding year of our Lord is obtained by subtracting 3761 from the Jewish year, and *vice versâ;* and the approximate date of commencement may be adjusted to the accurate date by means of the true day of the week taken from the next table.

	Ordinary Years.				Embolismic Years.	
Character.	$R = 0 \ldots 4$	$R = 5, 6$	$R = 7 \ldots 11$	$R = 12 \ldots 18$	Character.	
353 Mon.	0	0	0	0	383 Mon.	
355 Mon.	311,676	311,676	311,676	542,849	385 Mon.	
354 Tues.	934,591	934,591	984,960	984,960	384 Tues.	
354 Thur.	1296,636	1296,636	1296,636	1477,440	383 Thur.	
355 Thur.	2281,596	2281,596	2281,596	1839,485	385 Thur.	
353 Sat.	2462,400	2462,400	2462,400	2462,400	383 Sat.	
355 Sat.	2593,272	2774,076	2774,076	3005,249	385 Sat.	
353 Mon.	3447,360	3447,360	3447,360	3447,360	383 Mon.	
355 Mon.	3759,036	3759,036	3759,036	3990,209	385 Mon.	
354 Tues.	4381,951	4381,951	4432,320	4432,320	384 Tues.	
354 Thur.	4743,996	4743,996	4743,996	4924,800	383 Thur.	
355 Thur.	5728,956	5728,956	5728,956	5286,845	385 Thur.	
353 Sat.	5909,760	5909,760	5909,760	5909,760	383 Sat.	
355 Sat.	6040,632	6221,436	6221,436	6452,609	385 Sat.	

K

MEASURES OF TIME.

The annexed table exhibits the principal fasts and festivals, and it is followed by a table of the dates, &c. of Jewish years up to the year 2072, and the completion of the 307th cycle. When the date of commencement and number of days contained in a year are known, the number of days contained in the several months are shown in the table, page 189, and the construction of a detailed calendar for that year is obvious.

Principal Days of the JEWISH CALENDAR.

Tebet	1	1st of Tebet.
,,	10 Fast	Fast; Siege of Jerusalem.
Sebat	1	1st of Sebat.
Adar	1	1st of Adar.
,,	13, or 11 if Tisri 1 be Thursday	Fast of Esther — In Embol Years substitute *Veadar*, and same days.
,,	14 } Purim	Purim
,,	15	Second Day
Veadar	1	1st of Veadar (if Embolismic Year).
Nisan	1	1st of Nisan.
,,	15 } Passover	Passover.
,,	16	Second Day.
Yiar	1	1st of Yiar.
Sivan	1	1st of Sivan.
,,	6 } Sebuot	Pentecost.
,,	7	Second Day.
Tamuz	1	1st of Tamuz.
,,	17, or 18 if Tisri 1 be Monday	Fast; Taking of Jerusalem.
Ab	1	1st of Ab.
,,	9, or 10 if Tisri 1 be Monday	Fast; Destruction of the Temple.
Elul	1	1st of Elul.
TISRI	1 New Year	1st of Tisri (Yr .... begins).
,,	2	Second Day.
,,	3, or 4 if Tisri 1 be Thursday	Fast of Guedaliah.
,,	10 Kipur	Fast of Expiation.
,,	15 } Tabernacle.	Feast of Tabernacles.
,,	16	Second Day.
,,	21 Hosana Raba	Last day of the Festival.
,,	22 } 8th day	Feast of the 8th day.
,,	23	Rejoicing of the Law.
Hesvan	1	1st of Hesvan.
Kislev	1	1st of Kislev.
	25 Hanuca	Dedication of the Temple.
Tebet	1	1st of Tebet.
	&c.	&c.

Table of HEBREW YEARS.

Jewish Year.	Number of Days.	Commencement (1st of Tisri).			Jewish Year.	Number of Days.	Commencement (1st of Tisri).		
5606	354	Thur.	2 Oct.	1845	5644	354	Tues.	2 Oct.	1883
07	355	Mon.	21 Sept.	1846	45	355	Sat.	20 Sept.	1884
08	383	Sat.	11 Sept.	1847	46	385	Thur.	10 Sept.	1885
09	354	Thur.	28 Sept.	1848	47	354	Thur.	30 Sept.	1886
10	355	Mon.	17 Sept.	1849	48	353	Mon.	19 Sept.	1887
11	385	Sat.	7 Sept.	1850	49	385	Thur.	6 Sept.	1888
12	353	Sat.	27 Sept.	1851	50	354	Thur.	26 Sept.	1889
13	384	Tues.	14 Sept.	1852	51	383	Mon.	15 Sept.	1890
296 Cycle. 14	355	Mon.	3 Oct.	1853	298 Cycle. 52	355	Sat.	3 Oct.	1891
15	355	Sat.	23 Sept.	1854	53	354	Thur.	22 Sept.	1892
16	383	Thur.	13 Sept.	1855	54	385	Mon.	11 Sept.	1893
17	354	Tues.	30 Sept.	1856	55	353	Mon.	1 Oct.	1894
18	355	Sat.	19 Sept.	1857	56	355	Thur.	19 Sept.	1895
19	385	Thur.	9 Sept.	1858	57	384	Tues.	8 Sept.	1896
20	354	Thur.	29 Sept.	1859	58	355	Mon.	27 Sept.	1897
21	353	Mon.	17 Sept.	1860	59	353	Sat.	17 Sept.	1898
22	385	Thur.	5 Sept.	1861	60	384	Tues.	5 Sept.	1899
23	354	Thur.	25 Sept.	1862	61	355	Mon.	24 Sept.	1900
24	383	Mon.	14 Sept.	1863	62	383	Sat.	14 Sept.	1901
5625	355	Sat.	1 Oct.	1864	5663	355	Thur.	2 Oct.	1902
26	354	Thur.	21 Sept.	1865	64	354	Tues.	22 Sept.	1903
27	385	Mon.	10 Sept.	1866	65	385	Sat.	10 Sept.	1904
28	353	Mon.	30 Sept.	1867	66	355	Sat.	30 Sept.	1905
29	354	Thur.	17 Sept.	1868	67	354	Thur.	20 Sept.	1906
30	385	Mon.	6 Sept.	1869	68	383	Mon.	9 Sept.	1907
31	355	Mon.	26 Sept.	1870	69	355	Sat.	26 Sept.	1908
32	383	Sat.	16 Sept.	1871	70	383	Thur.	16 Sept.	1909
297 Cycle. 33	354	Thur.	3 Oct.	1872	299 Cycle. 71	354	Tues.	4 Oct.	1910
34	355	Mon.	22 Sept.	1873	72	355	Sat.	23 Sept.	1911
35	383	Sat.	12 Sept.	1874	73	385	Thur.	12 Sept.	1912
36	355	Thur.	30 Sept.	1875	74	354	Thur.	2 Oct.	1913
37	354	Tues.	19 Sept.	1876	75	353	Mon.	21 Sept.	1914
38	385	Sat.	8 Sept.	1877	76	385	Thur.	9 Sept.	1915
39	355	Sat.	28 Sept.	1878	77	354	Thur.	28 Sept.	1916
40	354	Thur.	18 Sept.	1879	78	355	Mon.	17 Sept.	1917
41	383	Mon.	6 Sept.	1880	79	383	Sat.	7 Sept.	1918
42	355	Sat.	24 Sept.	1881	80	354	Thur.	25 Sept.	1919
43	383	Thur.	14 Sept.	1882	81	385	Mon.	13 Sept.	1920

Table of HEBREW YEARS.

Jewish Year.	Number of Days.	Commencement (1st of Tisri).			Jewish Year.	Number of Days.	Commencement (1st of Tisri).		
5682	355	Mon.	3 Oct.	1921	5720	355	Sat.	3 Oct.	1959
83	353	Sat.	23 Sept.	1922	21	354	Thur.	22 Sept.	1960
84	384	Tues.	11 Sept.	1923	22	383	Mon.	11 Sept.	1961
85	355	Mon.	29 Sept.	1924	23	355	Sat.	29 Sept.	1962
86	355	Sat.	19 Sept.	1925	24	354	Thur.	19 Sept.	1963
87	383	Thur.	9 Sept.	1926	25	385	Mon.	7 Sept.	1964
88	354	Tues.	27 Sept.	1927	26	353	Mon.	27 Sept.	1965
89	385	Sat.	15 Sept.	1928	27	385	Thur.	15 Sept.	1966
300 Cycle. 90	353	Sat.	5 Oct.	1929	302 Cycle. 28	354	Thur.	5 Oct.	1967
91	354	Tues.	23 Sept.	1930	29	355	Mon.	23 Sept.	1968
92	385	Sat.	12 Sept.	1931	30	383	Sat.	13 Sept.	1969
93	355	Sat.	1 Oct.	1932	31	354	Thur.	1 Oct.	1970
94	354	Thur.	21 Sept.	1933	32	355	Mon.	20 Sept.	1971
95	383	Mon.	10 Sept.	1934	33	383	Sat.	9 Sept.	1972
96	355	Sat.	28 Sept.	1935	34	355	Thur.	27 Sept.	1973
97	354	Thur.	17 Sept.	1936	35	354	Tues.	17 Sept.	1974
98	385	Mon.	6 Sept.	1937	36	385	Sat.	6 Sept.	1975
99	353	Mon.	26 Sept.	1938	37	353	Sat.	25 Sept.	1976
5700	385	Thur.	14 Sept.	1939	38	384	Tues.	13 Sept.	1977
5701	354	Thur.	3 Oct.	1940	5739	355	Mon.	2 Oct.	1978
02	355	Mon.	22 Sept.	1941	40	355	Sat.	22 Sept.	1979
03	383	Sat.	12 Sept.	1942	41	383	Thur.	11 Sept.	1980
04	354	Thur.	30 Sept.	1943	42	354	Tues.	29 Sept.	1981
05	355	Mon.	18 Sept.	1944	43	355	Sat.	18 Sept.	1982
06	383	Sat.	8 Sept.	1945	44	385	Thur.	8 Sept.	1983
07	354	Thur.	26 Sept.	1946	45	354	Thur.	27 Sept.	1984
08	385	Mon.	15 Sept.	1947	46	383	Mon.	16 Sept.	1985
301 Cycle. 09	355	Mon.	4 Oct.	1948	303 Cycle. 47	355	Sat.	4 Oct.	1986
10	353	Sat.	24 Sept.	1949	48	354	Thur.	24 Sept.	1987
11	384	Tues.	12 Sept.	1950	49	383	Mon.	12 Sept.	1988
12	355	Mon.	1 Oct.	1951	50	355	Sat.	30 Sept.	1989
13	355	Sat.	20 Sept.	1952	51	354	Thur.	20 Sept.	1990
14	383	Thur.	10 Sept.	1953	52	385	Mon.	9 Sept.	1991
15	354	Tues.	28 Sept.	1954	53	353	Mon.	28 Sept.	1992
16	355	Sat.	17 Sept.	1955	54	355	Thur.	16 Sept.	1993
17	385	Thur.	6 Sept.	1956	55	384	Tues.	6 Sept.	1994
18	354	Thur.	26 Sept.	1957	56	355	Mon.	25 Sept.	1995
19	383	Mon.	15 Sept.	1958	57	383	Sat.	14 Sept.	1996

MEASURES OF TIME. 197

Table of HEBREW YEARS.

Jewish Year.	Number of Days.	Commencement (1st of Tisri).			Jewish Year.	Number of Days.	Commencement (1st of Tisri).		
5758	354	Thur.	2 Oct.	1997	5796	354	Thur.	4 Oct.	2035
59	355	Mon.	21 Sept.	1998	97	353	Mon.	22 Sept.	2036
60	385	Sat.	11 Sept.	1999	98	385	Thur.	10 Sept.	2037
61	353	Sat.	30 Sept.	2000	99	354	Thur.	30 Sept.	2038
62	354	Tues.	18 Sept.	2001	5800	355	Mon.	19 Sept.	2039
63	385	Sat.	7 Sept.	2002	01	383	Sat.	8 Sept.	2040
64	355	Sat.	27 Sept.	2003	02	354	Thur.	26 Sept.	2041
65	383	Thur.	16 Sept.	2004	03	385	Mon.	15 Sept.	2042
304 Cycle. 66	354	Tues.	4 Oct.	2005	306 Cycle. 04	353	Mon.	5 Oct.	2043
67	355	Sat.	23 Sept.	2006	05	355	Thur.	22 Sept.	2044
68	383	Thur.	13 Sept.	2007	06	384	Tues.	12 Sept.	2045
69	354	Tues.	30 Sept.	2008	07	355	Mon.	1 Oct.	2046
70	355	Sat.	19 Sept.	2009	08	353	Sat.	21 Sept.	2047
71	385	Thur.	9 Sept.	2010	09	384	Tues.	8 Sept.	2048
72	354	Thur.	29 Sept.	2011	10	355	Mon.	27 Sept.	2049
73	353	Mon.	17 Sept.	2012	11	355	Sat.	17 Sept.	2050
74	385	Thur.	5 Sept.	2013	12	383	Thur.	7 Sept.	2051
75	354	Thur.	25 Sept.	2014	13	354	Tues.	24 Sept.	2052
76	385	Mon.	14 Sept.	2015	14	385	Sat.	13 Sept.	2053
5777	353	Mon.	3 Oct.	2016	5815	355	Sat.	3 Oct.	2054
78	354	Thur.	21 Sept.	2017	16	354	Thur.	23 Sept.	2055
79	385	Mon.	10 Sept.	2018	17	383	Mon.	11 Sept.	2056
80	355	Mon.	30 Sept.	2019	18	355	Sat.	29 Sept.	2057
81	353	Sat.	19 Sept.	2020	19	354	Thur.	19 Sept.	2058
82	384	Tues.	7 Sept.	2021	20	383	Mon.	8 Sept.	2059
83	355	Mon.	26 Sept.	2022	21	355	Sat.	25 Sept.	2060
84	383	Sat.	16 Sept.	2023	22	385	Thur.	15 Sept.	2061
305 Cycle. 85	355	Thur.	3 Oct.	2024	307 Cycle. 23	354	Thur.	5 Oct.	2062
86	354	Tues.	23 Sept.	2025	24	353	Mon.	24 Sept.	2063
87	385	Sat.	12 Sept.	2026	25	385	Thur.	11 Sept.	2064
88	355	Sat.	2 Oct.	2027	26	354	Thur.	1 Oct.	2065
89	354	Thur.	21 Sept.	2028	27	355	Mon.	20 Sept.	2066
90	383	Mon.	10 Sept.	2029	28	383	Sat.	10 Sept.	2067
91	355	Sat.	28 Sept.	2030	29	354	Thur.	27 Sept.	2068
92	354	Thur.	18 Sept.	2031	30	355	Mon.	16 Sept.	2069
93	383	Mon.	6 Sept.	2032	31	383	Sat.	6 Sept.	2070
94	355	Sat.	24 Sept.	2033	32	355	Thur.	24 Sept.	2071
95	385	Thur.	14 Sept.	2034	33	384	Tues.	13 Sept.	2072

MAHOMETAN CALENDAR.

The Mahometan era, or era of the Hegira, employed in Turkey, Persia, Arabia, &c., is dated from the flight of Mahomet from Mecca to Medina, which was in the night of Thursday the 15th of July, A.D. 622, and it commenced on the day following. The years of the Hegira are purely lunar, and always consist of twelve lunar months commencing with the approximate new moon, without any intercalation to keep them to the same season with respect to the Sun, so that they retrograde through all the seasons in about $32\frac{1}{2}$ years. They are also partitioned into cycles of 30 years, 19 of which are common years of 354 days each, and the other 11 are intercalary years having an additional day appended to the last month. The mean length of the year is therefore $354\frac{11}{30}$ days or 354 days 8 hours 48 min., which divided by 12 gives $29\frac{191}{380}$ days or 29 days 12 hours 44 min. as the time of a mean lunation, and this differs from the astronomical mean lunation by only 3 seconds. This small error will not amount to a day in less than 2300 years.

To find if a year is intercalary or common, divide it by 30; the quotient will be the number of completed cycles and the remainder will be the year of the current cycle; if this last be one of the numbers 2, 5, 7, 10, 13, 16, 18, 21, 24, 26, 29, the year is intercalary and consists of 355 days; if it be any other number, the year is ordinary.

Or if Y denote the number of the year, and

$$R = \left(\frac{11\,Y + 14}{30}\right)_r;$$

the year is intercalary when $R < 11$.

Also the number of intercalary years from the year 1 up to the year Y inclusive $= \left(\dfrac{11\,Y + 14}{30}\right)_w$; and the same, exclusive of the year Y, $= \left(\dfrac{11\,Y + 3}{30}\right)_w$.

MEASURES OF TIME.

To find the day of the week on which any year of the Hegira begins, we observe that the year 1 began on a Friday, and that after every common year of 354 days, or 50 weeks and 4 days, the day of the week must necessarily become postponed 4 days, besides the additional day of each intercalary year.

Hence if $w = 1$ indicate Sun. | 2 Mon. | 3 Tues. | 4 Wed. | 5 Thur. | 6 Frid. | 7 Sat.

the day of the week on which the year Y commences will be

$$w = 2 + 4\left(\frac{Y}{7}\right)_r + \left(\frac{11\,Y + 3}{30}\right)_w \text{ (rejecting sevens)};$$

which admits of being reduced to the more convenient formula

$$w = 7 - \left(\frac{Y}{7}\right)_r + 3\left(\frac{11\,Y + 3}{30}\right)_r \text{ (rejecting sevens),}$$

the values of which obviously circulate in a period of 7 times 30 or 210 years.

Let C denote the number of completed cycles, and y the year of the cycle; then $Y = 30\,C + y$, and

$$w = 5\left(\frac{C}{7}\right)_r + 6\left(\frac{y}{7}\right)_r + 3\left(\frac{11\,y + 3}{30}\right)_r \text{ (rejecting sevens).}$$

From this formula the following table has been constructed:—

Year of the Current Cycle (y).				Number of the Period of Seven Cycles $= \left(\frac{C}{7}\right)_r$						
				0	1	2	3	4	5	6
0	8			Mon.	Sat.	Thur.	Tues.	Sun.	Frid.	Wed.
1	9	17	25	Frid.	Wed.	Mon.	Sat.	Thur.	Tues.	Sun.
*2	*10	*18	*26	Tues.	Sun.	Frid.	Wed.	Mon.	Sat.	Thur.
3	11	19	27	Sun.	Frid.	Wed.	Mon.	Sat.	Thur.	Tues.
4	12	20	28	Thur.	Tues.	Sun.	Frid.	Wed.	Mon.	Sat.
*5	*13	*21	*29	Mon.	Sat.	Thur.	Tues.	Sun.	Frid.	Wed.
6	14	22	30	Sat.	Thur.	Tues.	Sun.	Frid.	Wed.	Mon.
*7	15	23		Wed.	Mon.	Sat.	Thur.	Tues.	Sun.	Frid.
	*16	*24		Sun.	Frid.	Wed.	Mon.	Sat.	Thur.	Tues.

To find from this table the day of the week on which any year of the Hegira commences, the rule to be observed will be as follows:—

Rule.—Divide the year of the Hegira by 30; the quotient is the number of cycles, and the remainder is the year of the current cycle. Next divide the number of cycles by 7, and the second remainder will be the Number of the Period, which being found at the top of the table, and the year of the cycle on the left hand, the required day of the week is immediately shown.

The intercalary years of the cycle are distinguished by an asterisk.

For the computation of the Christian date, the ratio of a mean year of the Hegira to a solar year is

$$\frac{\text{Year of Hegira}}{\text{Mean solar year}} = \frac{354\frac{11}{30}}{365\cdot 24222} = 0\cdot 970224.$$

The year 1 began 16 July, 622, Old Style, or 19 July, 622, according to the New or Gregorian Style. Now the day of the year answering to the 19th of July is 200 which, in parts of the solar year, is 0·5476, and the number of years elapsed $= Y - 1$. Therefore, as the intercalary days are distributed with considerable regularity in both calendars, the date of commencement of the year Y expressed in Gregorian years is

$$0\cdot 970224\,(Y-1) + 622\cdot 5476,$$
$$\text{or } 0\cdot 970224\,Y + 621\cdot 5774.$$

This formula gives the following rule for calculating the date of the commencement of any year of the Hegira, according to the Gregorian or New Style.

Rule.—Multiply 970224 by the year of the Hegira, cut off six decimals from the product, and add 621·5774. The

MEASURES OF TIME. 201

sum will be the year of the Christian era, and the day of the year will be found by multiplying the decimal figures by 365.

The result may sometimes differ a day from the truth as the intercalary days do not occur simultaneously; but as the day of the week can always be accurately obtained from the foregoing table, the error, if any, can be readily adjusted.

Example.—Required the date on which the year 1362 of the Hegira begins.

```
    970224
      1362
    ───────
   1940448
   5821344
   2910672
    970224
   ───────
  1321·445088
    621·5774
   ───────
   1943·0225
        365
   ───────
      1125
      1350
       675
   ───────
     8·2125
```

Thus the date is the 8th day, or 8 January, of the year 1943.

To find, as a test, the accurate day of the week, the proposed year of the Hegira divided by 30 gives 45 cycles and remainder 12, the year of the current cycle.

Also 45 divided by 7 leaves a remainder 3 for the number of the period.

Therefore, referring to 3 at the top of the table and 12 on the left, the required day is Friday.

The tables, pages 154-5, show that 8 Jan. 1943 is a Friday; therefore the date is exact.

For any other date of the Mahometan year it is only requisite to know the names of the consecutive months, and the length of each; these are,

Muharram	. . . 30	Shaaban	. . . 29
Saphar	 29	Ramadân	. . . 30
Rabia I.	 30	Shawall	 29
Rabia II.	. . . 29	Dulkaada	. . . 30
Jomada I.	. . . 30	Dulheggia	. . . 29 ⎫
Jomada II.	. . . 29	and, in intercalary	⎬
Rajab	 30	years, 30 days.	⎭

The ninth month, Ramadân, is the month of Abstinence observed by the Turks.

The Turkish calendar may evidently be carried on indefinitely by successive addition, observing only to allow for the additional day that occurs in the bissextile and intercalary years; but for any remote date the computation according to the preceding rules will be most efficient, and such computation may be usefully employed as a check on the accuracy of any considerable extension of the calendar by induction alone.

The following table shows the dates of commencement of Mahometan years for 1845 up to 2047, or from the 43rd to the 49th cycle inclusive, which form the whole of the seventh period of seven cycles. Throughout the next period of seven cycles, and all other like periods, the days of the week will recur in exactly the same order.

All the tables hitherto published, of this kind, which extend beyond the year 1900 of the Christian era, are erroneous, not excepting the celebrated French work, *L'Art de vérifier les Dates*, so justly regarded as the greatest authority in chronological matters. The errors have probably arisen from a continued excess of 10 in the discrimination of the intercalary years, and they have been faithfully transcribed by other authors.

MEASURES OF TIME. 203

Table of MAHOMETAN YEARS.

43rd Cycle.			44th Cycle.		
Year of Hegira.	Commencement (1st of Muharram).		Year of Hegira.	Commencement (1st of Muharram).	
1261	Frid.	10 Jan. 1845	1291	Wed.	18 Feb. 1874
1262*	Tues.	30 Dec. 1845	1292*	Sun.	7 Feb. 1875
1263	Sun.	20 Dec. 1846	1293	Frid.	28 Jan. 1876
1264	Thur.	9 Dec. 1847	1294	Tues.	16 Jan. 1877
1265*	Mon.	27 Nov. 1848	1295*	Sat.	5 Jan. 1878
1266	Sat.	17 Nov. 1849	1296	Thur.	26 Dec. 1878
1267*	Wed.	6 Nov. 1850	1297*	Mon.	15 Dec. 1879
1268	Mon.	27 Oct. 1851	1298	Sat.	4 Dec. 1880
1269	Frid.	15 Oct. 1852	1299	Wed.	23 Nov. 1881
1270*	Tues.	4 Oct. 1853	1300*	Sun.	12 Nov. 1882
1271	Sun.	24 Sept. 1854	1301	Frid.	2 Nov. 1883
1272	Thur.	13 Sept. 1855	1302	Tues.	21 Oct. 1884
1273*	Mon.	1 Sept. 1856	1303*	Sat.	10 Oct. 1885
1274	Sat.	22 Aug. 1857	1304	Thur.	30 Sept. 1886
1275	Wed.	11 Aug. 1858	1305	Mon.	19 Sept. 1887
1276*	Sun.	31 July 1859	1306*	Frid.	7 Sept. 1888
1277	Frid.	20 July 1860	1307	Wed.	28 Aug. 1889
1278*	Tues.	9 July 1861	1308*	Sun.	17 Aug. 1890
1279	Sun.	29 June 1862	1309	Frid.	7 Aug. 1891
1280	Thur.	18 June 1863	1310	Tues.	26 July 1892
1281*	Mon.	6 June 1864	1311*	Sat.	15 July 1893
1282	Sat.	27 May 1865	1312	Thur.	5 July 1894
1283	Wed.	16 May 1866	1313	Mon.	24 June 1895
1284*	Sun.	5 May 1867	1314*	Frid.	12 June 1896
1285	Frid.	24 April 1868	1315	Wed.	2 June 1897
1286*	Tues.	13 April 1869	1316*	Sun.	22 May 1898
1287	Sun.	3 April 1870	1317	Frid.	12 May 1899
1288	Thur.	23 Mar. 1871	1318	Tues.	1 May 1900
1289*	Mon.	11 Mar. 1872	1319*	Sat.	20 April 1901
1290	Sat.	1 Mar. 1873	1320	Thur.	10 April 1902

MEASURES OF TIME.

Table of MAHOMETAN YEARS.

45th Cycle.			46th Cycle.		
Year of Hegira.	Commencement (1st of Muharram).		Year of Hegira.	Commencement (1st of Muharram).	
1321	Mon.	30 Mar. 1903	1351	Sat.	7 May 1932
1322*	Frid.	18 Mar. 1904	1352*	Wed.	26 April 1933
1323	Wed.	8 Mar. 1905	1353	Mon.	16 April 1934
1324	Sun.	25 Feb. 1906	1354	Frid.	5 April 1935
1325*	Thur.	14 Feb. 1907	1355*	Tues.	24 Mar. 1936
1326	Tues.	4 Feb. 1908	1356	Sun.	14 Mar. 1937
1327*	Sat.	23 Jan. 1909	1357*	Thur.	3 Mar. 1938
1328	Thur.	13 Jan. 1910	1358	Tues.	21 Feb. 1939
1329	Mon.	2 Jan. 1911	1359	Sat.	10 Feb. 1940
1330*	Frid.	22 Dec. 1911	1360*	Wed.	29 Jan. 1941
1331	Wed.	11 Dec. 1912	1361	Mon.	19 Jan. 1942
1332	Sun.	30 Nov. 1913	1362	Frid.	8 Jan. 1943
1333*	Thur.	19 Nov. 1914	1363*	Tues.	28 Dec. 1943
1334	Tues.	9 Nov. 1915	1364	Sun.	17 Dec. 1944
1335	Sat.	28 Oct. 1916	1365	Thur.	6 Dec. 1945
1336*	Wed.	17 Oct. 1917	1366*	Mon.	25 Nov. 1946
1337	Mon.	7 Oct. 1918	1367	Sat.	15 Nov. 1947
1338*	Frid.	26 Sept. 1919	1368*	Wed.	3 Nov. 1948
1339	Wed.	15 Sept. 1920	1369	Mon.	24 Oct. 1949
1340	Sun.	4 Sept. 1921	1370	Frid.	13 Oct. 1950
1341*	Thur.	24 Aug. 1922	1371*	Tues.	2 Oct. 1951
1342	Tues.	14 Aug. 1923	1372	Sun.	21 Sept. 1952
1343	Sat.	2 Aug. 1924	1373	Thur.	10 Sept. 1953
1344*	Wed.	22 July 1925	1374*	Mon.	30 Aug. 1954
1345	Mon.	12 July 1926	1375	Sat.	20 Aug. 1955
1346*	Frid.	1 July 1927	1376*	Wed.	8 Aug. 1956
1347	Wed.	20 June 1928	1377	Mon.	29 July 1957
1348	Sun.	9 June 1929	1378	Frid.	18 July 1958
1349*	Thur.	29 May 1930	1379*	Tues.	7 July 1959
1350	Tues.	19 May 1931	1380	Sun.	26 June 1960

MEASURES OF TIME. 205

Table of MAHOMETAN YEARS.

47th Cycle.			48th Cycle.		
Year of Hegira.	Commencement	(1st of Muharram).	Year of Hegira.	Commencement	(1st of Muharram).
1381	Thur.	15 June 1961	1411	Tues.	24 July 1990
1382*	Mon.	4 June 1962	1412*	Sat.	13 July 1991
1383	Sat.	25 May 1963	1413	Thur.	2 July 1992
1384	Wed.	13 May 1964	1414	Mon.	21 June 1993
1385*	Sun.	2 May 1965	1415*	Frid.	10 June 1994
1386	Frid.	22 April 1966	1416	Wed.	31 May 1995
1387*	Tues.	11 April 1967	1417*	Sun.	19 May 1996
1388	Sun.	31 Mar. 1968	1418	Frid.	9 May 1997
1389	Thur.	20 Mar. 1969	1419	Tues.	28 April 1998
1390*	Mon.	9 Mar. 1970	1420*	Sat.	17 April 1999
1391	Sat.	27 Feb. 1971	1421	Thur.	6 April 2000
1392	Wed.	16 Feb. 1972	1422	Mon.	26 Mar. 2001
1393*	Sun.	4 Feb. 1973	1423*	Frid.	15 Mar. 2002
1394	Frid.	25 Jan. 1974	1424	Wed.	5 Mar. 2003
1395	Tues.	14 Jan. 1975	1425	Sun.	22 Feb. 2004
1396*	Sat.	3 Jan. 1976	1426*	Thur.	10 Feb. 2005
1397	Thur.	23 Dec. 1976	1427	Tues.	31 Jan. 2006
1398*	Mon.	12 Dec. 1977	1428*	Sat.	20 Jan. 2007
1399	Sat.	2 Dec. 1978	1429	Thur.	10 Jan. 2008
1400	Wed.	21 Nov. 1979	1430	Mon.	29 Dec. 2008
1401*	Sun.	9 Nov. 1980	1431*	Frid.	18 Dec. 2009
1402	Frid.	30 Oct. 1981	1432	Wed.	8 Dec. 2010
1403	Tues.	19 Oct. 1982	1433	Sun.	27 Nov. 2011
1404*	Sat.	8 Oct. 1983	1434*	Thur.	15 Nov. 2012
1405	Thur.	27 Sept. 1984	1435	Tues.	5 Nov. 2013
1406*	Mon.	16 Sept. 1985	1436*	Sat.	25 Oct. 2014
1407	Sat.	6 Sept. 1986	1437	Thur.	15 Oct. 2015
1408	Wed.	26 Aug. 1987	1438	Mon.	3 Oct. 2016
1409*	Sun.	14 Aug. 1988	1439*	Frid.	22 Sept. 2017
1410	Frid.	4 Aug. 1989	1440	Wed.	12 Sept. 2018

Table of MAHOMETAN YEARS.

49th Cycle.

Year of Hegira.	Commencement (1st of Muharram).			Year of Hegira.	Commencement (1st of Muharram).		
1441	Sun.	1 Sept.	2019	1456*	Tues.	21 Mar.	2034
1442*	Thur.	20 Aug.	2020	1457	Sun.	11 Mar.	2035
1443	Tues.	10 Aug.	2021	1458*	Thur.	28 Feb.	2036
1444	Sat.	30 July	2022	1459	Tues.	17 Feb.	2037
1445*	Wed.	19 July	2023	1460	Sat.	6 Feb.	2038
1446	Mon.	8 July	2024	1461*	Wed.	26 Jan.	2039
1447*	Frid.	27 June	2025	1462	Mon.	16 Jan.	2040
1448	Wed.	17 June	2026	1463	Frid.	4 Jan.	2041
1449	Sun.	6 June	2027	1464*	Tues.	24 Dec.	2041
1450*	Thur.	25 May	2028	1465	Sun.	14 Dec.	2042
1451	Tues.	15 May	2029	1466*	Thur.	3 Dec.	2043
1452	Sat.	4 May	2030	1467	Tues.	22 Nov.	2044
1453*	Wed.	23 April	2031	1468	Sat.	11 Nov.	2045
1454	Mon.	12 April	2032	1469*	Wed.	31 Oct.	2046
1455	Frid.	1 April	2033	1470	Mon.	21 Oct.	2047

Table of EPOCHS of the PRINCIPAL ERAS AND PERIODS.

Name.	Christian Date of Commencement.	Name.	Christian Date of Commencement.
Grecian Mundane era	1 Sept 5598 B.C.	Sidonian era . .	Oct. 110 B.C.
Civil era of Constantinople .	1 Sept. 5508 „	Cæsarean era of Antioch . .	1 Sept. 48 „
Alexandrian era .	29 Aug. 5502 „	Julian year . .	1 Jan. 45 „
Ecclesiastical era of Antioch .	1 Sept. 5492 „	Spanish era . .	1 Jan. 38 „
Julian Period .	1 Jan. 4713 „	Actian era . .	1 Jan. 30 „
Mundane era .	Oct. 4008 „	Augustan era .	14 Feb. 27 „
Jewish Mundane era	Oct. 3761 „	Vulgar Christian era	1 Jan. 1 A.D.
Era of Abraham	1 Oct. 2015 „	Destruction of Jerusalem . .	1 Sept. 69 „
Era of the Olympiads . . .	1 July 776 „	Era of Maccabees	24 Nov. 166 „
Roman era . .	24 April 753 „	„ „ Dioclesian	17 Sept. 284 „
Era of Nabonassar	26 Feb. 747 „	„ „ Ascension	12 Nov. 295 „
Metonic Cycle .	15 July 432 „	„ „ the Armenians . . .	7 July 552 „
Grecian or Syro-Macedonian era	1 Sept. 312 „	Mahometan era of the Hegira .	16 July 622 „
Tyrian era . .	19 Oct. 125 „	Persian era of Yezdegird . . .	16 June 632 „

CAMBRIDGE UNIVERSITY TERMS, AFTER 1883:

Up to the end of the year 1883 the Cambridge University Terms were, in substance, determined by the process we briefly formulated on page 178. About that time a new scheme was recommended by the Council of the Senate and ordered to be adopted for all subsequent years. In order to bring this new scheme into the most convenient practical use it is here formulated in like manner, and made to depend on the number of direction:—

Commencement, Tuesday *before* Midsummer.

Lent Term begins Jan. 8, ends $\begin{cases} N = 1 \ldots 9 & \text{March } (N+18) \\ 10 \ldots 26 & \text{,, } 27 \\ 27 \ldots 35 & \text{,, } (N+1) \end{cases}$

Easter Term begins $\begin{cases} N = 1 \ldots 9 & \text{April } (N+9) \\ 10 \ldots 26 & \text{,, } 18 \\ 27 \ldots 35 & \text{,, } (N-8) \end{cases}$ ends June 24.

Michaelmas Term begins October 1, ends December 19.

$\left.\begin{array}{l}\text{From Lent Term ends}\\\text{to Easter Term begins}\end{array}\right\} = 22 \text{ days.}$

An improved Table for finding the date of Easter Day is given overleaf. By its use the Table of Epacts on page 164 is not wanted, and the operations require less consideration.

EASTER-DAY *found from the* GOLDEN

Epact.	1500 1600	1700 1800	1900 2000 2100	2200 2400	2300 2500	2600 2700 2800	2900 3000	3100 3200 3300	3400 3600	3500 3700	3800 3900 4000
* 1 2	1 12	1 12	12 4	4 15	4 15	15 7	7 18	7 18	18 10	10	10 2
3 4 5	4 15	4 15	15 7	7 18	7 18	18 10	10	10 2	2 13	2 13	13 5
6 7 8	7 18	7 18	18 10	10	10 2	2 13	2 13	13 5	5 16	5 16	16 8
9 10 11	10	10 2	2 13	2 13	13 5	5 16	5 16	16 8	8 19	8 19	19 11
12 13 14	2 13	13 5	5 16	5 16	16 8	8 19	8 19	19 11	11	11 3	3 14
15 16 17	5 16	16 8	8 19	8 19	19 11	11	11 3	3 14	3 14	14 6	6 17
18 19 20	8 19	19 11	11	11 3	3 14	3 14	14 6	6 17	6 17	17 9	9
21 22 23	11 3	3 14	3 14	14 6	6 17	6 17	17 9	9	9 1	1 12	1 12
24 25 26	14 6	6 17	6 17	17 9	9	9 1	1 12	1 12	12 4	4 15	4 15
27 28 29	17 9	9	9 1	1 12	1 12	12 4	4 15	4 15	15 7	7 18	7 18

Directions.—Under the century find the golden number; Dominical Letter the date of Easter Day is shown.

MEASURES OF TIME.

NUMBER and DOMINICAL LETTER only.

	Dominical Letter. (For Leap Years use the *second* letter).					
	A	B	C	D	E	F
	Apr. 16	Apr. 17	Apr. 18	Apr. 19	Apr. 20	Apr. 14
	,, 16	,, 17	,, 18	,, 19	,, 13	,, 14
	,, 16	,, 17	,, 18	,, 12	,, 13	,, 14
	,, 16	,, 17	,, 11	,, 12	,, 13	,, 14
	,, 16	,, 10	,, 11	,, 12	,, 13	,, 14
	,, 9	,, 10	,, 11	,, 12	,, 13	,, 14
	,, 9	,, 10	,, 11	,, 12	,, 13	,, 14
	,, 9	,, 10	,, 11	,, 12	,, 13	,, 7
	,, 9	,, 10	,, 11	,, 12	,, 6	,, 7
	,, 9	,, 10	,, 11	,, 5	,, 6	,, 7
	,, 9	,, 10	,, 4	,, 5	,, 6	,, 7
	,, 9	,, 3	,, 4	,, 5	,, 6	,, 7
	,, 2	,, 3	,, 4	,, 5	,, 6	,, 7
	,, 2	,, 3	,, 4	,, 5	,, 6	,, 7
	,, 2	,, 3	,, 4	,, 5	,, 6	Mar. 31
	,, 2	,, 3	,, 4	,, 5	Mar. 30	,, 31
	,, 2	,, 3	,, 4	Mar. 29	,, 30	,, 31
	,, 2	,, 3	Mar. 28	,, 29	,, 30	,, 31
	,, 2	Mar. 27	,, 28	,, 29	,, 30	,, 31
	Mar. 26	,, 27	,, 28	,, 29	,, 30	,, 31
	,, 26	,, 27	,, 28	,, 29	,, 30	,, 31
	,, 26	,, 27	,, 28	,, 29	,, 30	,, 24
	,, 26	,, 27	,, 28	,, 29	,, 23	,, 24
	,, 26	,, 27	,, 28	,, 22	,, 23	,, 24
	Apr. 23	Apr. 24	Apr. 25	Apr. 19	Apr. 20	Apr. 21
	,, 23	,, 24	,, 25	,, 19	,, 20	,, 21
	,, 23	,, 24	,, 18	,, 19	,, 20	,, 21
	,, 23	,, 17	,, 18	,, 19	,, 20	,, 21
	,, 16	,, 17	,, 18	,, 19	,, 20	,, 21
	,, 16	,, 17	,, 18	,, 19	,, 20	,, 21

then on the same horizontal line and under the
The Epact is also shown on the left in the first column.

APPENDIX.

THE valuable miscellaneous information conveyed in what follows has for the most part been obtained from the Reports by the Board of Trade on their proceedings and business under the Weights and Measures Act, 1878. This is regarded as the principal Act in relation to weights and measures, and its chief enactments have already been briefly stated on pages 26-8, 50 and 53.

Imperial weights and measures are now legally in force in the following colonies, &c. :—

Antigua.	Jamaica.	Singapore.
Barbadoes.	Malta.	South Australia.
Bermuda.	Natal.	St. Christopher.
British Guiana.	Nevis.	St. Helena.
British Honduras.	New Brunswick.	St. Vincent.
Canada.	New South Wales.	Tobago.
Cape of Good Hope.	New Zealand.	Trinidad.
Cyprus.	Nova Scotia.	Vancouver's Island.
Dominica.	Queensland.	Victoria.
Grenada.	Sierra Leone.	Western Australia.
Hong Kong.		

In Germany, by a law promulgated on 11th July, 1884, it was declared that the "meter" was then to be adopted as the basis of all weight and measure, and that the weight of distilled water contained in a cubic decimeter, in a vacuum at the temperature of 4° centigrade, was to be the unit of weight, a "kilogramm."

On this metric basis the other values of the French metric system (pp. 75-6) were likewise adopted, and under all but identical denominations. The several parts and multiples, including the half and quarter, of these metric denominations were also to be used.

CHINA.

Weights and Measures legally in use.

The subject of the legal use of weights and measures in China is attended with some difficulty, as each shopkeeper recognises no standard but his own, and the Penal Code is only strictly enforced in matters affecting the revenue. To provide against abuses certain to arise, the framers of the various foreign Treaties usually inserted clauses defining the values of the units of length and weight. In the list these Treaty equivalents are given wherever they exist.

Linear Measure.

		ENGLISH VALUE.
10 fên (lines)	= 1 ts'un ('punto,' or inch)	1·41 inches
10 ts'un	= 1 ch'ih ('covid,' or foot)	14·1 ,,
10 ch'ih	= 1 chang (rod)	141 ,,
10 chang	= 1 yin	1410 ,,
Ch'ih of the Board of Works		12·5 ,,

Itinerary Measure.

5 ch'ih ('covids')	= 1 pu* (pace)	about 66 inches
360 pu	= 1 li	,, 660 yards
10 li	= 1 t'ang-hsün	,, 6600 ,,
250 li	= 1 tu (or degree)	,, 94 miles

Land Measure.

5 ch'ih ('covids')	= 1 kung (bow)	about 66 inches
2 kung	= 1 chang	,, 11 feet
1 square chang	= 1 ching	,, 121 sq. feet
15 ching	= 1 chüo	,, 1815 ,,
4 chüo	= mou	,, 7260 ,,

Cubic Measure.

| 100 cubic ch'ih ('covids') = 1 fang or ma | about 113 cubic feet |

---

* The *pu* or *pace* is here estimated to be the distance from right foot-print to right foot-print of a man walking steadily; sometimes it has half this value, and then it represents the more familiar distance from right foot-print to left foot-print.

Capacity.

		ENGLISH VALUE.
2 yo = 1 ho (gill)	about	0·2 pint
10 ho (gills) = 1 shêng (pint)	,,	2 pints
10 shêng = 1 tou (peck)	,,	20 ,,
5 tou = 1 hu (bushel)	,,	100 ,,

The number of *tou* in a *hu* varies.

Weight Avoirdupois.

24 chu	= 1 liang	1¼ oz. av.
16 liang ('taels,' or ounces)	= 1 chin (' catty ')	1¼ lbs.
100 chin (' catties ')	= 1 tan (' picul ')	133⅓ ,,

Silver Weight.

10 ssŭ	= 1 hao (thousandths)	·000013 oz. av.
10 hao	= 1 li (hundredths—' cash')	·00013 ,,
10 li	= 1 fên (tenth—' candreen')	·0013 ,,
10 fên	= 1 ch'ien ('mace ')	·013 ,,
10 ch'ien	= 1 *liang* (' tael ')	1·3 ,,

JAPAN.

WEIGHTS AND MEASURES LEGALLY IN USE.

Artisan's Measure.

10 Shi	= 1 Mo	·001193 inch
10 Mo	= 1 Rin	·01193 ,,
10 Rin	= 1 Bu	·1193 ,,
10 Bu	= 1 Sun	1·1931 ,,
10 Sun	= 1 Shaku*	11·9305 inches
10 Shaku	= 1 Jô	9·9421 feet

Dry Goods Measure.

(One-fourth more than Artisan's Measure.)

10 Rin	= 1 Bu	·14913 inch
10 Bu	= 1 Sun	1·4913 ,,
10 Sun	= 1 Shaku	14·9132 inches
10 Shaku	= 1 Jô	12·4276 feet

Itinerary Measure.

6 Shaku = 1 Ken	5·9653 feet
60 Ken = 1 Chô	357·9163 ,,
36 Chô = 1 Ri	4295 yards, or 2·4403 miles
Kai-Ri, or Sea Mile = 16·975 Chô	2025 ,, 1·1506 ,,

* The shaku = $\frac{10}{33}$ mètre.

APPENDIX.

Superficial or Land Measure.

		ENGLISH VALUE.
36 square Shaku	= 1 Tsubo	3·9538 square yards
30 Tsubo	= 1 Se	118·6149 ,,
10 Se	= 1 Tan	39·2115 square poles
10 Tan	= 1 Chō	2·4507 acres

Capacity.

10 Kei	= 1 Sai	·000318 pint.
10 Sai	= 1 Shō	·003176 ,,
10 Shō	= 1 Shaku	·03176 ,,
10 Shaku	= 1 Gō	·3176 ,,
10 Gō	= 1 Shō	1·5881 quarts.
10 Shō	= 1 To	3·9703 gallons.
10 To	= 1 Koku	4·9629 bushels.

Weight.

10 Shi	= 1 Mo	·05797 grain.
10 Mo	= 1 Rin	·5797 ,,
10 Rin	= 1 Fun	5·7972 grains.
10 Fun	= 1 Momme	57·972 ,,
100 Momme	= 1 Hyaku-me	0·8282 lbs. av.
1000 Momme	= 1 Kwam-me	8·2817 ,,
160 Momme	= 1 Kin	1·3251 ,,
100 Kin	= 1 Hiyak-kin	132·5073 ,,

The Imperial Standard Yard.—An accurate copy ("P. C. VI."), of the same form and material as the Imperial Standard Yard, has been made, as required by the Act, 1878. No two primary standards are really alike, but these two primary standards are as much alike as it is possible to make them. (Standards Office, 10th July, 1886.)

The Indian Seer.—A question having arisen as to the true weight of the Indian seer to be marked on weighing-machines, the information given by the Board of Trade was that the seer of the British Government is equal to 80 tolos, or to 32·914 ounces avoirdupois, the tola, the legal weight of the rupee, being equal to 180 grains.

Illegal Denominations of Weight.—There are no Board of Trade standards of 3, 5, 6, 8, 9, 10, or 11 lbs., and the use in trade of weights of these denominations would appear, therefore, to be illegal, as the inspectors could not stamp them.

Wire and Sheet Gauges.

The wire manufacturer has his special instrument, the wire gauge, for accurately measuring the diameter of the wire he draws. This gauge is generally a rectangular steel plate, having around its edges a series of notches graduated to all the sizes required. Up to a recent date the greatest confusion existed in the free use of wire and metal gauges, the various sizes of which were of a very promiscuous character; indeed, the gauges most generally used were severally known by the common appellation of the Birmingham Wire Gauge (B. W. G.). Yet there were many varieties all bearing this name and differing materially from each other, almost every firm, in the absence of a definite standard, being accustomed to employ its own gauge as the correct one. Also wire-drawers as a body were, in consequence thereof, less careful, knowing that disputes could not be settled by gauges.

But since the establishment of telegraphy and self-acting machines for working wires into nails, pins, rivets, screws, &c., exactitude both in diameter and "cast" have become indispensable, and the wire manufacturer's ingenuity is being strained to meet the increasing demands for wires having accurate diameters, and, as a check, weighing so much per given length.

The necessities of some trades require an extended series of sizes with small differences between them; others a smaller series with greater differences; while some, again, deal only with the thicker sizes, and others only with the finer. An effective standard gauge should be sufficiently comprehensive to include all these purposes, so that the sizes of any special trade may be obtained from it, although only a portion of such standard be used. The standard gauge should, therefore, have a large range of sizes from which each trade may select such as may suit it, or it may consist of a more limited series of primary sizes readily divisible into half and quarter sizes, according to the very convenient arrangement adopted in the construction of the gauge patented by Mr. Hughes.

Imperial Standard Wire Gauge.

The serious complications often arising from the use of so many different wire gauges have been effectually remedied by the recent introduction of a standard wire gauge by the Board of Trade. It is designated the Imperial Standard Wire Gauge, and was legalised by an Order in Council of 23rd August, 1883, and came into operation as the only legal wire gauge on 1st March, 1884. From and after that date it therefore became necessary to use in trade the particular equivalent sizes in parts of an inch, from 0·500 to 0·001 inch, as stated in the Order referred to.

The Weights and Measures Act of 1878 specially provides that a contract or dealing shall not be invalid or open to objection on the ground that the weights or measures expressed or referred to therein are weights and measures of the metric system.

The Order in Council, by which the new gauge is legalised, does not alter or amend this law, and contracts for wire, &c., may continue, therefore, to be made in millimètres, or other metric measures as above provided.

The following table contains all the measurements represented, or intended to be represented, by the said Imperial Standard Wire Gauge, together with the equivalents expressed in millimètres:—

TABLE I.

IMPERIAL STANDARD WIRE GAUGE.

Number of Standard Gauge.	Diameter.		Number of Standard Gauge.	Diameter.	
	In parts of an inch.	In millimètres.		In parts of an inch.	In parts of millimètres.
No.	Inch.	Millimètres.	No.	Inch.	Millimètre.
7·0	0·500	12·700	23	0·024	0·610
6·0	·464	11·785	24	·022	·559
5·0	·432	10·973	25	·020	·508
4·0	·400	10·160	26	·018	·457
3·0	·372	9·449	27	·0164	·4166
2·0	·348	8·839	28	·0148	·3759
1·0	·324	8·229	29	·0136	·3454
1	·300	7·620	30	·0124	·3150
2	·276	7·010	31	·0116	·2946
3	·252	6·401	32	·0108	·2743
4	·232	5·893	33	·0100	·2540
5	·212	5·385	34	·0092	·2337
6	·192	4·877	35	·0084	·2134
7	·176	4·470	36	·0076	·1930
8	·160	4·064	37	·0068	·1727
9	·144	3·658	38	·0060	·1524
10	·128	3·251	39	·0052	·1321
11	·116	2·946	40	·0048	·1219
12	·104	2·642	41	·0044	·1118
13	·092	2·337	42	·0040	·1016
14	·080	2·032	43	·0036	·0914
15	·072	1·829	44	·0032	·0813
16	·064	1·626	45	·0028	·0711
17	·056	1·422	46	·0024	·0610
18	·048	1·219	47	·0020	·0508
19	·040	1·016	48	·0016	·0406
20	·036	0·914	49	·0012	·0305
21	·032	·813	50	0·0010	0·0254
22	·028	·711			

To make the standard wire-gauge additionally useful as a plate-gauge, Table II. has been calculated, showing the corresponding weight in pounds avoirdupois of each superficial square foot of metal plate.

TABLE II.

Showing the weight of a *superficial square foot* of wrought iron for all thicknesses of the IMPERIAL STANDARD WIRE GAUGE.

At 62° Fahr.

Density—7·8; Weight of cubic inch of water = 252·458 grains.

Descriptive Number.	Weight of Square Foot.	Descriptive Number.	Weight of Square Foot.
No.	lbs.	No.	lbs.
7·0	20·254	23	0·972
6·0	18·796	24	·891
5·0	17·500	25	·810
4·0	16·203	26	·729
3·0	15·069	27	·664
2·0	14·097	28	·600
1·0	13·125	29	·551
1	12·153	30	·502
2	11·180	31	·470
3	10·208	32	·437
4	9·398	33	·405
5	8·588	34	·373
6	7·778	35	·340
7	7·130	36	·308
8	6·481	37	·275
9	5·833	38	·243
10	5·185	39	·211
11	4·699	40	·194
12	4·213	41	·178
13	3·727	42	·162
14	3·241	43	·146
15	2·917	44	·130
16	2·593	45	·113
17	2·268	46	·097
18	1·944	47	·081
19	1·620	48	·065
20	1·458	49	·049
21	1·296	50	·041
22	1·134		

Tables III. and IV., which also relate to the Imperial Standard Wire Gauge, supply other important values of great practical use to engineers and others.

TABLE III.

IMPERIAL STANDARD WIRE GAUGE.

Table of Sectional Areas, Weights, Lengths, and Breaking Strains of Iron Wire; issued by the Iron and Steel Wire Manufacturers' Association.

Size on Wire Gauge.	Sectional Area in square inches.	Weight of		Length of Cwt.	Breaking Strains.	
		100 yards.	Mile.		Annealed.	Bright.
No.	inch.	lbs.	lbs.	yards.	lbs.	lbs.
7·0	·1963	193·4	3404	58	10470	15700
6·0	·1691	166·5	2930	67	9017	13525
5·0	·1466	144·4	2541	78	7814	11725
4·0	·1257	123·8	2179	91	6702	10052
3·0	·1087	107·1	1885	105	5796	8694
2·0	·0951	93·7	1649	120	5072	7608
1·0	·0824	81·2	1429	138	4397	6595
1	·0707	69·6	1225	161	3770	5655
2	·0598	58·9	1037	190	3190	4785
3	·0499	49·1	864	228	2660	3990
4	·0423	41·6	732	269	2254	3381
5	·0353	34·8	612	322	1883	2824
6	·0290	28·5	502	393	1544	2316
7	·0243	24·0	422	467	1298	1946
8	·0201	19·8	348	566	1072	1608
9	·0163	16·0	282	700	869	1303
10	·0129	12·7	223	882	687	1030
11	·0106	10·4	183	1077	564	845
12	·0085	8·4	148	1333	454	680
13	·0066	6·5	114	1723	355	532
14	·0050	5·0	88	2240	268	402
15	·0041	4·0	70	2800	218	326
16	·0032	3·2	56	3500	172	257
17	·0025	2·4	42	4667	131	197
18	·0018	1·8	32	6222	97	145
19	·0013	1·2	21	9333	67	100
20	·0010	1·0	18	11200	55	82

TABLE IV.

Table of WEIGHTS and RESISTANCES of COPPER and IRON WIRE, compiled by Mr. C. H. Barney, Sec. of the National Telephone Exchange Association, New York.

Number of standard Gauge.	Pure Copper Wire.				Galvanis'd iron wire.	
	Weight, sp. gr., 8·90		Resistance at 59·9° Fahr.		Weight, sp. gr., 7·74.	Resistance at 59·9° Fahr.
	lbs. per 1000 feet.	lbs. per mile.	Ohms per 1000 feet.	Ohms per mile.	lbs. per mile.	Ohms per mile.
7·0	756·28					
6·0	651·31					
5·0	564·44					
4·0	484·03					
3·0	418·63					
2·0	366·36					
1·0	317·54	1676·61	·095774	·50558		
1	272·27	1437·58	·11171	·58982		
2	230·44	1216·72	·13199	·69690		
3	192·11	1014·34	·15832	·83592		
4	162·83	859·74	·18680	·98630	747·68	6·6884
5	135·96	717·87	·22370	1·18113	624·30	8·0100
6	111·52	588·82	·27273	1·4400	572·07	9·7656
7	93·71	494·78	·32458	1·7137	430·29	11·6222
8	77·445	408·89	·39273	2·0736	355·60	14·0625
9	62·730	331·21	·48486	2·5600	279·34	17·3611
10	49·565	261·70	·61365	3·2400	227·59	21·9726
11	40·707	214·93	·74718	3·9451	186·92	26·7539
12	32·720	172·76	·92955	4·9080	150·24	33·2840
13	25·605	135·19	1·1878	6·2715	117·57	42·5331
14	19·361	102·22	1·5709	8·2943	88·90	56·2500
15	15·683	82·80	1·9394	10·2400	72·01	69·4444
16	12·391	65·42	2·4546	12·9602	56·89	87·8906
17	9·4869		3·2060			
18	6·9700		4·3637			
19	4·8403		6·2837			
20	3·9206		7·7577			
21	3·0978		9·8184			
22	2·3708		12·824			
23	1·7425		17·455			
24	1·4642		20·773			
25	1·2100		25·135			
26	0·98015		31·031			
27	·81365		37·381			
28	·66263		45·900			
29	·55953		54·358			
30	·46515		65·388			
31	·40707		74·718			
32	·35286		86·197			
33	·30252		100·541			
34	·25605		118·79			
35	·21346		142·49			
36	·17473		174·07			
37	·13988		216·94			
38	·10891		279·28			
39	·08180		371·82			
40	·06970		436·37			

Resistance $\left(\begin{array}{c}\text{copper}\\ \text{iron}\end{array}\right)$ wire increases about $\left(\begin{array}{c}0·21\\ 0·35\end{array}\right)$ per cent. for each additional degree Fahr. The diameters of wire are already given in Table I.

Tables V., VI. and VII., which appertain to three other prominent gauges, may be of some use to those interested in the subject.

TABLE V.

STUBS' IRON WIRE GAUGE, 1879.

No.	Diameter.	No.	Diameter.	No.	Diameter.	No.	Diameter.
	Inch.		Inch.		Inch.		Inch.
4·0	·454	7	·180	17	·058	27	·016
3·0	·425	8	·165	18	·049	28	·014
2·0	·380	9	·148	19	·042	29	·013
1·0	·340	10	·134	20	·035	30	·012
1	·300	11	·120	21	·032	31	·010
2	·284	12	·109	22	·028	32	·009
3	·259	13	·095	23	·025	33	·008
4	·238	14	·083	24	·022	34	·007
5	·220	15	·072	25	·020	35	·005
6	·203	16	·065	26	·018	36	·004

Stubs' iron wire gauge has been somewhat extensively used in America and elsewhere, under the mistaken name of the Birmingham wire gauge (B.W.G.).

Table VI. is that of the Lancashire gauge for steel wire and pinion wire. It will be observed that the "letter gauge" is a continuation of the steel wire gauge.

TABLE VI.

STEEL WIRE GAUGES.

	Diameter.	No.	Diameter.	No.	Diameter.	No.	Diameter.
	Inch.		Inch.		Inch.		Inch.
Z	·413	1	·227	28	·139	55	·050
Y	·404	2	·219	29	·134	56	·045
X	·397	3	·212	30	·127	57	·042
W	·386	4	·207	31	·120	58	·041
V	·377	5	·204	32	·115	59	·040
U	·368	6	·201	33	·112	60	·039
T	·358	7	·199	34	·110	61	·038
S	·348	8	·197	35	·108	62	·037
R	·339	9	·194	36	·106	63	·036
Q	·332	10	·191	37	·103	64	·035
P	·323	11	·188	38	·101	65	·033
O	·316	12	·185	39	·099	66	·032
N	·302	13	·182	40	·097	67	·031
M	·295	14	·180	41	·095	68	·030
L	·290	15	·178	42	·092	69	·029
K	·281	16	·175	43	·088	70	·027
J	·277	17	·172	44	·085	71	·026
I	·272	18	·168	45	·081	72	·024
H	·266	19	·164	46	·079	73	·023
G	·261	20	·161	47	·077	74	·022
F	·257	21	·157	48	·075	75	·020
E	·250	22	·155	49	·072	76	·018
D	·246	23	·153	50	·069	77	·016
C	·242	24	·151	51	·066	78	·015
B	·238	25	·148	52	·063	79	·014
A	·234	26	·146	53	·058	80	·013
		27	·143	54	·055		

TABLE VII.

A SHEET AND HOOP IRON GAUGE issued by the South Staffordshire Ironmasters' Association for the use of sheet and hoop iron makers, 1st March, 1884.

Parts of inch.	No. on gauge.	Thickness.	Parts of inch.	No. on gauge.	Thickness.	Parts of inch.	No. on gauge.	Thickness.
1	15°	1·000		7	·1764	$\frac{1}{64}$	28	·015625
	14°	0·9583		8	·1570		29	·0139
	13°	·9167		9	·1398		30	·0123
							31	·0110
$\frac{7}{8}$	12°	·8750	$\frac{1}{8}$	10	·1250		32	·0098
	11°	·8333		11	·1113		33	·0087
	10°	·7917		12	·0991		34	·0077
				13	·0882		35	·0069
$\frac{3}{4}$	9°	·7500		14	·0785		36	·0061
	8°	·7083		15	·0699		37	·0054
	7°	·6666					38	·0048
			$\frac{1}{16}$	16	·0625		39	·0043
$\frac{5}{8}$	6°	·6250		17	·0556		40	·00386
	5°	·5883		18	·0495		41	·00343
	4°	·5416		19	·0440		42	·00306
				20	·0392		43	·00272
$\frac{1}{2}$	3°	·5000		21	·0349		44	·00242
	2°	·4452					45	·00215
	1°	·3964	$\frac{1}{32}$	22	·0312		46	·00192
	1	·3532		23	·0278		47	·00170
	2	·3147		24	·0247		48	·00152
	3	·2804		25	·0220		49	·00135
				26	·0196		50	·00120
$\frac{1}{4}$	4	·2500		27	·0174		51	·00107
	5	·2225					52	·00095
	6	·1981						

By an Order in Council, dated 26th June, 1884, it was provided that in the several places therein named the maximum fee hereafter to be taken by the local inspectors of weights and measures on the verification and stamping of wire-gauge measures should be as follows:—

Measures from 0·500 to 0·001 inch in the form of wire-gauge plates; for each notch, or for each internal gauge or separate size, from half-an-inch to one-thousandth of an inch, 0^s 0$\frac{1}{4}$^{d}.

In reply to inquiries as to the use of gauges other than the new standard wire gauge, it has been stated by the Board of Trade that, in their view of the matter, a contract can only be maintained when made according to denominations of standards duly legalised under the Act. The Board have, however, no power to put an authoritative construction on the Act.

With reference to gauges used for flat metals, for sheet and hoop iron, it has been represented that a standard flat metal gauge would be desirable. The Board have, however, no power under the Act to make a standard applicable to sheet and hoop iron only. All that the Board could do, if after consideration they should think it advisable, would be to cause to be made a standard flat metal gauge applicable to any metal, whether iron, brass, tin, &c.

A new WEIGHTS AND MEASURES ACT, 1889 (52 & 53 Vict., cap. 21), came into operation on the 1st January, 1890. In the preamble the Weights and Measures Act, 1878, is referred to as the principal Act. Some of the chief enactments are in substance the following:—

1. That every weighing instrument used for trade shall be verified and stamped by an inspector of weights and measures with a stamp of verification under this Act.

2. Every person who, after twelve months from the commencement of this Act, uses, or has in his possession for use for trade, any weighing instrument not stamped as

required by this Act, shall be liable to a fine not exceeding £2, or in the case of a second offence £5;

4. And imprisonment in case of fraud.

6. The Board of Trade shall from time to time cause such new denominations of standards for the measurement of electricity, temperature, pressure, or gravities as appear to them to be required.

9. Every local authority having power to appoint inspectors of weights and measures shall, with the approval of the Board of Trade, make for the guidance of inspectors appointed or employed by that authority or person, and with the like approval, amend or rescind general regulations to be observed in the verification, stamping, and the inspection of weights, measures, and weighing and measuring instruments.

11. The Board of Trade shall provide for the holding of examinations for the purpose of ascertaining whether persons acting or appointed to act as inspectors of weights and measures possess sufficient practical knowledge for the proper performance of their duties as such, and for the grant of certificates to persons who satisfactorily pass such examinations.

20. (1) All coal shall be sold by weight only, except where by the written consent of the purchaser it is sold by boat-load, or by waggons or tubs delivered from the colliery into the works of purchaser.

(2) If any person sells coals otherwise than is required by this section he shall be liable to a fine not exceeding £5 for every such sale.

21. Weight ticket or note to be given on delivery of coal over 2 cwt.

24. Penalty, on deficiency in weight of coal on small sales, not exceeding £5.

25. Weighing instrument to be kept in place where coal is sold by retail.

27. Any seller or purchaser of coal may require that any

coal, or vehicle used for the carriage of coal in bulk, be weighed or re-weighed by a properly stamped weighing instrument.

39. This Act may be cited as the Weights and Measures Act, 1889; and the principal Act and this Act may be cited together as the Weights and Measures Acts, 1878 and 1889.

The Board of Trade, in pursuance of Section 9, have, at the commencement of the new Act, also caused to be issued a book entitled, "Model Regulations with respect to Inspectors and the Inspection of Weights, Measures, and Weighing and Measuring Instruments," which may be purchased, either directly or through any bookseller, from Messrs. Eyre & Spottiswoode, East Harding Street, E.C. The Board of Trade are prepared to approve of these regulations if made by local authorities, or to take into consideration any other or additional regulations which a local authority may desire to make for a particular district.

Every shop or place having weights, measures, or weighing instruments liable to inspection, should be visited by the inspector once at least in every year.

No weights should be stamped which are found to differ from the inspector's standards by more than the following:—

Avoirdupois weights.	Allowance in excess only.	
	Iron weights.	Brass weights.
56 lbs.	50 grains	20 grains
28 ,,	40 ,,	15 ,,
14 ,,	30 ,,	10 ,,
7 ,,	20 ,,	5 ,,
4 ,,	20 ,,	5 ,,
2 ,,	10 ,,	5 ,,
1 lb.	10 ,,	2 ,,
8 oz.	5 ,,	2 ,,
4 ,,	5 ,,	2 ,,
2 ,,	2 ,,	1 ,,
1 ,, and under	2 ,,	1 ,,

Apothecaries' Weights.	Allowance in excess only.
10, 8, or 6 oz. troy	0·5 grain
4, 2, or 1 oz. ,,	0·2 ,,
4, 2, or 1 drachms	0·1 ,,
2, 1½, 1 or ½ scruples	0·05 ,,
6, 5, 4, or 3 grains	0·02 ,,
2, 1, or ½ grains	0·01 ,,

No measures of capacity to be stamped which differ from the standards by more than the following:—

Capacity.	Liquid measure.	Dry measure.
Bushel of 8 gallons	5 fluid ounces	} 20 cubic inches
Half bushel of 4 gallons	4 ,,	
Peck or 2 gallons	3 ,,	} 15 ,,
Gallon	2 ,,	
Half gallon	1 ,,	} 10 ,,
Quart	1 ,,	
Pint	4 fluid drachms	} 5 ,,
Half pint	3 ,,	
Gill	2 ,,	} 2 ,,
Half gill	1 ,,	
Quarter gill	½ ,,	

1 fluid ounce = 437½ grains.

On an ordinary measure of length no larger amount of error should be allowed than the following:—

	End measures.		Line measures.	
	Error in excess.	Error in deficiency.	Error in excess.	Error in deficiency.
Wood:				
Yard or foot	$\frac{1}{16}$ in.	$\frac{1}{32}$ in.	$\frac{1}{32}$ in.	$\frac{1}{64}$ in.
Inch	$\frac{1}{32}$,,	$\frac{1}{32}$,,	$\frac{1}{64}$,,	$\frac{1}{64}$,,
Metal:				
Yard or foot	0·05 in.	0·02 in.	0·02 in.	0·01 in.
Inch	0·02 ,,	0·01 ,,	0·005 ,,	0·002 ,,

On other measures of length such amounts of error may be allowed as the local authority may approve.

Mr. Herbage, in 1886, afforded the Standards Department an opportunity of weighing some sovereigns of the years 1864, 1865, and 1866, which he had taken from circulation:—

100 sovereigns of the year 1864 weighed only 12,226 grains.
,,   ,,   1865   ,,   12,232 ,,
,,   ,,   1866   ,,   12,232 ,,

Mr. Herbage is of opinion that the "life" of a gold coin should be legally limited to a duration of twenty years.

Ascertained weights of standards used by some scale-makers for adjusting bankers' weights.

The 500 sovereigns weight = 61,435 grains.
,, 300   ,,   ,,   36,861 ,,
,, 200   ,,   ,,   24,574 ,,
,, 100   ,,   ,,   12,287 ,,
,,   50   ,,   ,,   6,143·5 ,,
,,   30   ,,   ,,   3,686·1 ,,
,,   20   ,,   ,,   2,457·4 ,,
,,   10   ,,   ,,   1,228·7 ,,
,,   5   ,,   ,,   614·35 ,,
,,   3   ,,   ,,   368·61 ,,
,,   2   ,,   ,,   245·74 ,,
,,   1   ,,   ,,   122·87 ,,
,,   ½   ,,   ,,   61·44 ,,

On pages 27-8 is given a list of standards in the custody of the Board of Trade at the time of the passing of the Act, 1878. They have now the following additional standards:—

Length: Quarter of a yard, eighth of a yard, sixteenth of a yard, and nail.
Liquid Measures: 2, 3, 4, &c., to 25 gallons.
Dry Measure: 4 bushels.
Weight: Cental or 100 lbs.; 480 grains or 20 pennyweights.

INDEX.

ABASSI (money), 94
   Abyssinia (Africa), 57
Accini (weight), 92
Achtel (capacity), 59, 77, 80, 82, 118
Achterli (capacity), 109
Acker (surface), 83
Acre (surface), 51
Adarme (weight), 106
Adoulies (money), 73
Aguirage (capacity), 80
Aix-la-Chapelle; *see* Prussia, 97
Aleppo, Syria, 93
Alexandria; *see* Egypt, 74
Algiers; *see* France, 75
Alicante; *see* Spain, 106
Almud (capacity), 113
Almude (capacity), 95, 106
Aln (length), 71, 107
Alqueire (capacity), 64, 95, 96
Altona; *see* Denmark, 71
America, United States, 115
American dollar (money), 65
Amole (capacity), 102
Amsterdam; *see* Holland, 84
Ancient attic or olympic foot, 79
Ancient Greek foot, 79
Ancient Pythic foot, 79
Ancient foot, 99
Ancient keramion or metretes (capacity), 79
Ancona; *see* Roman States, 100
Anker (capacity), 64, 71, 80, 82, 87, 97, 101, 104
Anna (money), 72, 74
Antwerp; *see* Belgium, 62
Appendix, 211

Arabia, 58
Archangel; *see* Russia, 101
Archimede's foot, 105
Archine (length), 101
Ardeb (capacity), 57, 74
Are (surface), 75
Arish (length), 94
Arpent (surface), 66
Arragon; *see* Spain, 106
Arratel or pound, 96
Arrōba (weight), 96, 106
Artuba (capacity), 94
As (weight), 108
Ass (weight), 61
Assay reports, 39
Assays of gold and silver, 30
Athens; *see* Greece, 79
Atomi (length), 93, 102, 116
Augsburg; *see* Bavaria, 61
Augsburg ell, 61
Augsburg foot, 61
Augsburg mark (weight), 62
Augsburg mass (capacity), 61
Augsburg muid (capacity), 61
Augsburg pounds (weights), 62
Augsburg scheffels (capacities), 62
Aum of hock, &c. (capacity), 55
Aune, large ell, 76, 112
Austria, 59
Azumbre (capacity), 106

BADEN, Germany, 60
   Bahar (weight), 58, 68
Bajocchi (money), 100

230 INDEX.

Bamboo, or dha (length), 63
Banco (money), 81
Bankers' weights for sovereigns, 228
Barbadoes; *see* West Indies, 117
Barcelona; *see* Spain, 106
Barge (weight), 54
Barile (capacity), 86, 88, 90, 91, 99, 102, 113, 114
Barleycorn (length), 2
Barrel (capacity), 52
Baryd (length), 58
Basle or Bâle, 112
Bat (weight), 105
Batavia, 86
Batman (weight), 93, 94
Bavaria, 61
Bavarian eimer (capacity), 61
Bavarian ell, 61
Bavarian foot, 61
Bavarian pound, 62
Bavarian scheffel (capacity), 62
Becher (capacity), 59, 61, 67, 109, 110, 112
Belgium, 62
Benda (weight), 80
Benda-offa (weight). 80
Bengal, 72
Bengal coss or mile, 72
Bergen; *see* Sweden and Norway, 107
Berkowitz (weight), 101
Berlin; *see* Prussia, 97
Bermudas; *see* West Indies, 117
Berne, Switzerland, 109
Biolca (surface), 90, 93
Birmingham wire gauge, 215, 221
Bissextile year, 144
Bissextile year, rule for, 145, 146
Bit, or threepenny-piece, 56
Boccale (capacity), 90, 99, 100, 103, 114
Bohemia, 63
Bologna; *see* Roman States, 100
Bombay, East Indies, 73
Bonn; *see* Prussia, 97
Bordeaux; *see* France, 75
Boston; *see* United States, 115
Botte (capacity), 91, 99
Bozze (capacity), 113
Braça, or fathom, 95
Braccio (length), 90, 99, 100, 102, 114, 116

Braccio di legno (length), 93
Brasse, small ell, 112
Brazil, South America, 64
Brazil, rate of exchange, 64
Brazilian dollar, 96
Bremen, Germany, 64
Brenta (capacity), 103
British bronze coins, 35
British coins in use, 56
British coins not in use, 56
British gold coins, 35
British Islands; *see* Great Britain, 50
British North America, 65
British moneys in North America, 65
British money table, 56
British money values (are named as "sterling"), 56
British silver coins, 35
Bronze coins, British, 35
Brunswick, Germany, 66
Brussels; *see* Belgium, 62
Bu (length), 213
Bushel (capacity), 51
Bussoli (capacity), 115
Butt (capacity), 52, 55
Butt of sherry, 55

CABLE'S length, 51
Cadiz; *see* Spain, 106
Caffiso (capacity), 88
Cahiz (capacity), 106
Cairo; *see* Egypt, 74
Calcutta, 72
Calendar, how determined, 171
Calendar month, 144
Calendar, Gregorian, 145
Cambridge University terms, 178, 207
Canada; *see* British America, 65
Canada (liquid measure), 95
Canary Islands (Atlantic), 67
Candereen " fun " (money), 69
Candia (Mediterranean), 68
Candy (weight), 63, 68, 73
Canna (length), 88, 90, 91, 99, 102, 105, 114
Canne (length), 99
Cantar (weight), 113
Cantaro (capacity), 95, 106

INDEX. 231

Cantaro (weight), 68, 74, 79, 88, 93, 114
Canton; *see* China, 69
Cape of Good Hope, 68
Cape Verd Islands; *see* Portugal, 95
Capichas (capacity), 94
Caraffi (capacity), 91
Carat (weight), 32—33, 53
Carats-fine, 33
Carats (weight), 100, 103
Carga (capacity), 68
Carlino (money), 92
Carro (capacity), 91, 103
Cash, "le" (money), 69
Cassel; *see* Hesse Cassel, 83
Castile, Spain, 106
Castilian foot, 67
Catena (length), 99
Catty (weight), 63, 69, 105
Caveer current (money), 58
Cavezzo (length), 90
Cent (money), 56, 85
Centass (weight), 61
Centenas (money), 107
Centesimi (money), 115, 117
Centiare (surface), 75
Centigramme (weight), 76
Centilitre (capacity), 75
Centime (money), 62, 77
Centimes, or lepta (money), 79
Centimètre (length), 75
Centinajo (weight), 86
Centner (weight), 12, 67, 98
Ceylon; *see* East Indies, 72
Ceylon, coin in circulation, 68
Chain (length), 50, 51
Chaldron (weight), 55
Chang, "rod" (length), 212
Chang (weight), 105
Char (capacity), 111
Chequee (weight), 114
Chenicas (capacity), 94
Cheüng (length), 69
Ch'ien, "mace" (weight), 213
Ch'ih ("covid" or foot), 212
Chik (length), 69
Chilo (capacity), 86
China, 69, 212
Chinese mile, "li," 69
Ching (surface), 212
Chittack (weight), 72
Chô (length), 213

Chô (surface), 214
Christiana; *see* Sweden and Norway, 107
Chu (weight), 213
Chunam (capacity), 73
Chüo (surface), 212
Cobido, or covid (length), 58
Coblentz; *see* Prussia, 97
Coins, particulars relating thereto, 28
Coins, computation of values of, 39
Cologne, Prussia, 70, 97
Cologne mark (weight), 81, 98
Columbo, coin in circulation, 68
Commassees (money), 58
Company's rupee (money), 72
Constantinople; *see* Turkey, 113
Conto of rees (money), 96
Coomb (capacity), 51
Copas (capacity), 106
Copec (money), 101
Copello (capacity), 103
Copenhagen; *see* Denmark, 71
Coppi (capacity), 116
Corba (capacity), 100
Corn gallon, 5, 50
Corsica; *see* France, 75
Coss, or mile, 72
Coupe, or sack, 111, 112
Covada, or cubit (length), 95
Covid, or cobido (length), 58, 69, 73, 212
Cracow, Poland, 70
Cracow foot, 70
Cremona; *see* Venetian Lombardy, 116
Crore (money), 74
Crown (money), 56
Crusado (money), 96
Cuba, West Indies, 117
Cubic foot (capacity), 51
Cubic foot of water, 19
Cubic inch of water, 5, 19
Cubic yard (capacity), 51
Cubit (length), 2, 51, 72, 74, 79, 90
Cucchiari (capacity), 103
Cycle of the sun, 155
Cycle of the sun, rules for calculating, 155—156

DAIN (length), 63
　Damascus ; *see* Ottoman
　　Asia, 93
Danish specie dollar, 72
Dantzic ; *see* Prussia, 97
Days of the week, table of, 155
Décagramme (weight), 76
Décalitre (capacity), 75
Decare (surface), 75
Décamètre (length), 75
Décastère (capacity), 75
Decigramme (weight), 76
Decilitre (capacity), 75
Decimal coinage, 36, 56
Decimas (money), 78, 107
Decime (money), 77
Decimètre (length), 75
Decimi (length), 99
Décistère (capacity), 75
Dedo, or finger (length), 95, 106
Degree of equator (length), 51
Degree of meridian (length), 51
Demerara ; *see* West Indies, 117
Denaro (weight), 93, 102, 103, 114, 115, 116, 117
Denmark, 71
Derah (length), 74
Dernier (weight), 111
Dha (length), 63
Diamond carats (weight), 53
Diamond grains (weight), 53
Dicotoli (capacity), 86
Difference of style, 148—149
Digit (length), 2
Dinars (money), 94
Dinars-bisti (money), 94
Dionysian period, 169
Dionysian period, rule for, 169
Dirhem (weight), 57, 94
Doli (weight), 101
Dollar (money), 66, 72, 78, 89, 104, 115, 117
Dollar of Mexico, 89
Dollar of South America, 89
Dollar of Spain, 89
Dominical letter, 149
Dominical letter, rule for, 153
Dominical letters, table of, 154
Doubloon (money), 78, 117
Drachm (weight and capacity), 53
Drachma (money), 79
Drachma (weight), 74, 108, 110

Drachme (weight), 60, 92, 98
Dracma (weight), 106
Dram (weight), 53, 113, 114
Dramma (weight), 92, 115
Dreah (length), 113
Dresden, 103
Drittel (capacity), 82
Drömt (capacity), 87
Ducat (money), 60, 92
Ducatoon (money), 85
Duim (length), 84
Duro (money), 107

EAST Indies, 72
　Easter day, rule for, 148, 164—165
Easter day, table for finding, 168, 208
Easter day, &c., for three centuries, 178—187
Easter Sunday ; *see* Easter day
Ecklein (capacity), 118
Eggēba (weight), 80
Egypt, Africa, 74
Eich-mass (capacity), 77
Eimer (capacity), 59, 80, 82, 87, 97, 104, 109, 111, 112, 118
Ell (length), 68, 102, 109, 110, 111, 112, 113
Ell, Scotch, 51
Elle (length), 59, 64, 66, 80, 82, 84, 87, 97, 103, 110
Elle, Stuttgard, 118
Elsinore ; *see* Denmark, 71
England ; *see* Great Britain, 50
English coins, table of, 32
English ell (length), 51
Epact, 160
Epact, formula for, 163
Epacts, table of, 164
Epochs, Eras and Periods, 206
Errors allowed in measures of capacity, 227
Errors allowed in measures of length, 227
Errors allowed in measures of weight, 226
Estadal (length), 106
Exchange, principles of, 45
Exchange in Brazil, 64

INDEX. 233

FALKLAND Islands; *see* Great Britain, 50
Famn (length), 107
Fanega (capacity), 106
Fang or ma (capacity), 212
Fanga (capacity), 96
Farsakh (length), 58
Farthing (money), 56
Fass (capacity), 66, 81, 87, 104
Fathom (length), 2, 51, 72, 99
Feddan al risach (surface), 74
Feld acker (surface), 109
Fên (length), 69, 212
Fên (weight), 213
Ferlini (weight), 90
Ferlino (weight), 100
Fiasco (capacity), 90, 114
Finland; *see* Russia, 101
Firkin (capacity), 52
Five-frank piece, 77
Five-lire piece, 93
Fjerding (capacity), 71, 107
Flanders; *see* Belgium, 62
Florence, 114
Florin (money), 56, 60, 62, 85, 113
Florin of Batavia, 86
Florin of Nuremberg, 62
Fluctuations of exchange, 48
Fod (length), 71
Foglietta (capacity), 99, 100
Foot (length), 2, 50, 100, 111, 112, 113
Fortin (capacity), 114
Fot (length), 107
Fourpenny-piece, or groat, 56
Franc (money), 62, 77
France, 75
Frankfort-on-Maine, 77
Frank, Swiss, 112
Frazil (weight), 58
Frederick (money), 98
Fuder (capacity), 59, 60, 64, 66, 77, 80, 82, 83, 84, 87, 97, 104, 118
Fun (weight), 214
Funt, or pound, 101
Furlong (length), 50
Fuss (length), 59, 60, 64, 80, 82, 83, 87, 97, 103, 109, 110, 118
Futtermassel (capacity), 59

GALLON (capacity), 11, 51—52, 53
Garce (capacity), 73
Garnetz (capacity), 101
Gasab (length), 74
Gauges, wire and sheet, 215
Geira (surface), 95
Geneva, 111
Genoa, 102
Germany, 78, 211
Gescheid (capacity), 77, 84
Gibraltar, 78
Gill (capacity), 51, 52
Giornata (surface), 103
Girth (length), 2
Glass (capacity), 60
Gō (capacity), 214
Gold, assays of, 39
Gold coins, wear of, 228
Gold crown, 96
Gold doubloon, 79
Gold ducat, 78, 81, 85, 98, 101, 104, 108, 110, 111
Gold eagle, 115
Golden number, 157
Golden number, rule for, 157
Golden numbers, table of, 158—159
Gold imperial, 101
Gold pistole, 107, 110, 117
Gold rosini, 115
Gold rupee, 94
Gold rusponi, 115
Gold ryder, 85
Gold sequin, 93, 100, 115, 117
Gold tical, 105
Gombette (capacity), 102
Grain (weight), 52, 53
Grains (weight), 76, 77, 108, 111, 112
Gramme (weight), 76
Gran (weight), 60, 92, 98, 104, 110
Grani (money), 88, 92, 105
Grani (weight), 93, 99, 100, 102, 115, 116, 117
Grano (weight), 103, 106
Granotini (weight), 103
Grao (length), 95
Graos (weight), 96
Great Britain, 50
Greece, 79
Gregorian calendar, 145

Gregorian calendar for three centuries, 178—187
Groat, fourpenny piece, 56
Gros (weight), 76, 77, 112
Groschen (money), 98
Grosso (weight), 116
Grot (money), 65
Grote (money), 63
Gudda (capacity), 58
Guerze (length), 94
Guilder (money), 85, 113
Guinea, Africa, 79
Gulden (money), 86
Gulden, or florin, 60
Guz (length), 58, 73
Guz, or yard, 72

HALF-CROWN, 56
Half-pence, an inch in diameter, 36
Half-penny, 56
Half-sovereign, 56
Hall mark, &c., 33
Hamburg, Germany, 80
Hand (length), 50
Hand, or moot (length), 72
Hanover, Germany, 82
Hao (weight), 213
Hap, or pecul (weight), 105
Hath (length), 73
Haut, or cubit (length), 72
Havannah, or Havana; see West Indies, 117
Hayti, or Haiti; see West Indies, 117
Hebrew calendar, 187
Hebrew calendar, principal days of, 194
Hebrew calendar, table of months, 189
Hebrew years, table of, 195—197
Hectare (surface), 75
Hectogramme (weight), 76
Hectolitre (capacity), 75
Hectomètre (length), 75
Heller (weight), 67, 78
Hesse-Cassel, Germany, 83
Hesse-Darmstadt, Germany, 83
Himt (capacity), 67, 81, 82, 83
Hindoostan; see East Indies, 72
Ho (gill), 213
Hogshead (capacity), 52

Hogshead of claret, &c., 55
Holland, 84
Holstein; see Denmark, 71
Hu (bushel), 213
Hundredweight, 53
Hungary; see Austria, 59
Hwŭh (capacity), 69
Hyaku-me (weight), 214

IKJE (length), 86
Illegal denominations of weight, 214
Immi (capacity), 109, 110, 111, 118
Imperial gallon, 19, 50
Imperial gold coin, 101
Imperial standard wire gauge, 216, 217
Inc (length), 86
Inch (length), 50
Indian seer (weight), 214
Indiction, 169
Indiction, rule for, 169
Inspectors, examination of, 225
Ionian Islands, 86
Ireland; see Great Britain, 50
Island of Java, 86
Italic stadium (length), 2
Italy, 86
Itinerary or road measures, 140—142
Itinerary pace (length), 51

JACKTAN (length), 79
Jamaica; see West Indies, 117
Japan, Asia, 86, 213
Jewish calendar, 187
Jewish calendar, principal days of, 194
Jewish calendar, table of months, 189
Jewish years, table of, 195—197
Jô (length), 213
Joch, or day's work, 59
Jow (length), 72
Juchart, or feld acker (surface), 109
Julian period, 170
Julian period, rule for, 170
Jungfrus (capacity), 107

K AI-RI (length), 213
  Kan (capacity), 85
Kande (capacity), 71
Kanna (capacity), 107
Kanne (capacity), 59, 80, 82, 86, 87, 104
Kappe (capacity), 107
Kasbequis (money), 94
Keel (weight), 54
Kei (capacity), 214
Kellas, or mecmedas (capacity), 58
Ken (length), 104, 213
Kharoubas (weight), 113
Kiel; *see* Germany, 78
Kila (capacity), 79
Kilderkin (capacity), 52
Killow (capacity), 114
Kilogramme (weight), 76
Kilolitre (capacity), 75
Kilomètre (length), 75
Kin (weight), 214
Kintal (weight), 114
Klafter (length), 59, 64, 83, 118
Koku (capacity), 214
Königsberg; *see* Prussia, 97
Kop (capacity), 69, 85
Köpflein (capacity) 112
Kopt (capacity), 111
Korrel (weight), 85
Kreuzer (money), 113
Kreuzer piece, 60, 62
Kronor (money), 72
Kub (length), 104
Kümpf (capacity), 84
Kung (length), 212
Kwam-me (weight), 214

L AC (money), 74
  Lana (weight), 101
Landfass (capacity), 109
Last (capacity), 51, 65, 71, 81, 82, 85, 87, 97
Law relating to the sale of coals, 225
Law sittings, 176—177
Law terms (old), 172
Law terms (Inns of Court), 177
League (length), 51
Leap year, 144
Leap year, rule for, 145—146
" Legal tender," 34

Legal weights of coins, 34
Leghorn, 114
Legua (length), 106
Leipzic, 103
Lepta, or centimes (money), 79
Li, Chinese mile, 69, 212
Li (weight), 213
Liang (weight), 213
Libbra (weight), 90, 92, 93, 99, 100, 102, 103, 105, 115
Libbra grossa (weight), 86
Libbra metrica (weight), 116
Libbra sottile (weight), 86
Libra (weight), 67, 96, 106
Liespfund (weight), 67, 81, 98
"Lieu de poste" (length), 76
Ligne (length), 76
Lille, or Lisle; *see* France, 75
Line (length), 50, 59, 64
Linea (length), 106
Linear measures, table of, 120—129
Linha (length), 95
Linie (length), 60, 87, 97, 109, 118
Linien (length), 71, 82, 83, 103, 110
Linies (length), 107
Lira (money), 86, 90
Lira Austriaca (money), 117
Lira nuova (money), 112
Lisbon; *see* Portugal, 95
Lispund (weight), 71, 108
Liter (weight), 57
Lithuania meile (length), 101
Litre (capacity), 75
Livre (weight), 76
Load (capacity), 51
Load (weight), 54
Löcher (capacity), 67
Lod (weight), 71, 108
London; *see* Great Britain, 50
Lood (weight), 85
Loth (weight), 59, 65, 67, 70, 78, 81, 82, 83, 84, 87, 97, 101, 104, 109, 110, 111
Lübeck, Germany, 87
Lucca, Italy; *see* Tuscany, 114
Lucerne, 110
Lyons; *see* France, 75

M AATJE (capacity), 85
  Mace, "tsëen" (money), 69

INDEX.

Madega (capacity), 57
Madeira, Atlantic; see Portugal, 95
Madras, East Indies, 73
Madrid, 106
Mahhub (money), 113
Mahometan calendar, 198
Mahometan months, table of, 202
Mahometan year, rule for, 200, 201
Mahometan years, table of, 203—206
Malaga; see Spain, 106
Malta, Mediterranean, 88
Malter (capacity), 61, 77, 82, 84, 97, 104, 110, 111
Mamoodi (money), 94
Maravedise (money), 107
Marc (weight), 109
Marcal (capacity), 73
Marco (weight), 96, 103, 106, 117
Mark (money), 72, 88
Mark (weight), 59, 60, 64, 65, 67, 70, 71, 76, 77, 78, 81, 82, 87, 97, 104, 108
"Mark banco" (money), 81
Mark of Cologne (weight), 60, 61
Marseilles; see France, 75
Mascais (weight), 94
Mass (capacity), 59, 60, 83, 84, 109, 110, 111, 118
Mässchen (capacity), 77, 83, 84, 97, 104
Mässlein (capacity), 61, 118
Mässli (capacity), 109, 111
Maund (weight), 58, 72, 73
Mauritius, 89
Measure (capacity), 73
Measures numerical relations of, 19
Mecklenburg-Schwerin, 89
Mecklenburg-Strelitz, 89
Mecmedas or kellas (capacity), 58
Media-tabla (weight), 80
Medida (capacity), 64
Medio (capacity), 106
Meile (length), 59, 64, 66, 80, 82, 109
Memel, see Prussia, 97
Metadella (capacity), 115
Metical (weight), 74, 113
Metretes (capacity), 79
Metric system, 26

Metrical system, 75
Mètre (length), 75, 76, 77
Mètze (capacity), 59, 83, 97, 104
Mexico, North America, 90
Mezzaruōla (capacity), 102
Mezzetta (capacity), 114, 115
Miglio (length), 91, 102, 114, 116
Mijle (length), 84
Mil (length), 107
Mil (money), 56
Milan, 116
Milan foot, 116
Milo (length), 50
Military pace (length), 51
Millet (weight), 12
Milligramme (weight), 76
Millimètre (length), 75
Millitre (capacity), 75
Milrea (money), 96
Milrea of Azores (money), 96
Milrea of Madeira (money), 96
Milreis (money), 64
Milyard (length), 11
Mina (capacity), 93, 102, 103, 115, 116
Mingel (capacity), 64
Minim (capacity), 53
Minot (capacity), 66
Mint, remedy of the, 34
Minuti (length), 91
Mistate (capacity), 68
Misure (capacity), 91
Misurelle (capacity), 91
Mo (length), 213
Mo (weight), 214
Mocha; see Arabia, 58
Modena, Italy, 90
Moeda (money), 96
Moggio (surface), 91, 115
Mohur, gold coin, 72, 74, 89
Moio (capacity), 96
Momme (weight), 214
Monkelser (length), 94
Montpelier; see France, 75
Moon's Age, 170
Moot, or hand (length), 72
Morgen (surface), 60, 64, 66, 80, 84, 97, 118
Morgen of Calenberg (surface), 82
Morocco, Africa, 90
Moscow; see Russia, 101
Mou (surface), 212
Movable Feasts, 170

Movable Feasts, table of, 172-3
Mudde or Zak (capacity), 85
Muhlmassel (capacity), 59
Mül (length), 71
Munich ; *see* Bavaria, 61
Munich eimer (capacity), 61
Munich pound (weight), 62
Munich scheffel (capacity), 62
Munster ; *see* Prussia, 97
Mural standards of length, 22
Muth (capacity), 59
Mütt (capacity), 109, 110, 111, 112
Myriagramme (weight), 76
Myriamètre (length), 75

NAIL (length), 50
Nantes, the Two Sicilies, 91
Naples, 91
Napoleon (money), 77, 89
Netherlands ; *see* Holland, 84
Neuchâtel } *see* Switzerland, 109
Neufchâtel }
Neufchâtel, 113
Neu groschen (money), 104
Neu-mass (capacity), 77
Newfoundland ; *see* Great Britain, 50
New South Wales; *see* Great Britain, 50
New York ; *see* United States, 115
Nice ; *see* Sardinia, 102
Nin (capacity), 105
Nin (length), 104
Noggin (capacity), 52
Nouslia (capacity), 58
North America, United States, 115
Norway, 107
Nössel (capacity), 66, 82, 104
Nova Scotia ; *see* British North America, 65
Nufs-orbah (capacity), 113
Number of direction, 164
Number of direction, formulas for, 166
Number of direction, table of, 167
Numerical relations of measures, 19
Nürnberg } Bavaria, 92
Nuremberg }
Nuremberg eimers (capacity), 61
Nuremberg ell (length), 61

Nuremberg foot (length), 61
Nuremberg malter (capacity), 62
Nuremberg mark (weight), 62
Nuremberg pound (weight), 62

OBOLI (money), 86
Occa (weight), 68
Ochava (weight), 106
Ochavillos (capacity), 106
Octave, &c. (capacity), 55
Odessa ; *see* Russia, 101
Ohm (capacity), 60, 64, 66, 77, 80, 82, 83, 84, 87, 97, 104, 110, 112
Oke (weight), 74, 79, 93, 114
Okie, oz. (weight), 113
Old Ale gallon, 5, 50
Old and New Measures, 20
Old Irish mile, 51
Old Law Terms, table of, 175-6
Old Scotch mile, 51
Old Wine gallon, 5, 50
Olluck (capacity), 73
Olympic foot (length), 79
Onça (weight), 96
Once (weight), 76, 77
Oncetta (money), 92
Oncia (length), 91, 93, 99, 102, 116
Oncia (money), 105
Oncia (weight), 90, 92, 93, 99, 100, 102, 103, 115, 116, 117
Ons (weight), 85
Onza (weight), 106
Oporto ; *see* Portugal, 95
Orbah, (capacity), 113
Öre (money), 72, 108
Ort (weight), 65, 71, 87
Örtchen (weight), 82
Orts (capacity), 107
Osmin (capacity), 101
Össel (capacity), 80, 97
Ostend ; *see* Belgium, 62
Ottavo (weight), 100, 103
Ottoman Asia, 93
Ottoman Empire; *see* Turkey, 113
Ounce (capacity), 53
Ounce (weight), 12, 52, 53
Outava (weight), 96
Oxford University Terms, 177-8
Oxhoft (capacity), 64, 66, 80, 82, 87, 104

PACE (length), 2
Pace itinerary, 51
Pace military, 51
Packen (weight), 101
Padua; *see* Venetian Lombardy, 116
Pagoda (weight), 73
Pagoda star, gold coin, 74
Pajak (capacity), 101
Palermo; *see* Sicily, 105
Palm (length), 2, 50, 84
Palmo (length), 88, 91, 95, 99, 102, 105, 106, 114, 116
Panilla (capacity), 106
Paolo (money), 100
Paper rouble (money), 101
Parah (capacity), 73
Paras (money), 68, 74, 113, 114
Paris; *see* France, 75
Parma, Italy, 93
Parrah (capacity), 68
Patàca (money), 96
Pataka (money), 57
Paulgaut (length), 63
Peck (capacity), 51
Pecul (weight), 69, 105
Penny (money), 56
Pennyweight (weight), 52
Perch (length), 2, 50, 51
Persia, Asia, 94
Pertica or passo (length), 91, 93, 102
Peseta (money), 107
Pesos (money), 90
Petersburg; *see* Russia, 101
Pezza (money), 88, 89
Pfennig (weight), 59, 61, 67, 70, 78, 81, 84, 87, 104, 109
Pfennig (money), 78, 81, 88, 98, 104
Pfiff (capacity), 59
Pfund, (weight), 59, 60, 61, 65, 67, 70, 78, 81, 82, 83, 84, 87, 92, 97, 98, 104, 109, 110, 111, 112
Philadelphia; *see* United States, 115
Piastre (money), 68, 74, 114
Piastre (money), 58, 107
Piastre of Selim (money), 114
Pic Turkish (length), 57
Pic or ell (length), 68, 74, 90
Pic or pike (length), 93, 113
Pice (money), 72

Pice (weight), 73
Picha (length), 79
Pié (length), 67, 93, 99, 106
Pie (money), 72
"Pied usuel" (length), 76
Piede (length), 90
Piede liprando (length), 102
Piede manuale (length), 102
Piedmont, 102
Pignate (capacity), 91
Pik (length), 113
Pike or pic (length), 93. 113
Pint (capacity), 51-2, 53
Pinta (capacity), 103, 116
Pinte (capacity), 102
Pipe (capacity), 52
Pipe of Port, &c., 55
Piso (weight), 80
Pistole, gold coin, 100, 107, 110, 117
Planke (capacity), 87
Pole (length), 50
Pole (surface), 51
Pollam (weight), 73
Pollegada, thumb or inch (length), 95
Pond (weight), 85
Pondicherry; *see* France, 75
Pontos (length), 95
Portugal, 95
Portuguese mile, 95
Post-meile (length), 97
Pote (capacity), 95
Pots (capacity), 111
Pott (capacity), 71, 112
Pottle (capacity), 51
Pouce (length), 76
Pound (weight), 12
Pound (weight), 52, 53, 79, 112
Pound or Rotl, (weight), 74
Pound or sovereign (money), 56
Prague, 63
Prague foot, 63
Presburg; *see* Austria, 59
Primas (capacity), 110
Prince Edward's Island; *see* British North America, 65
Principles of Exchange, 45
Probmetzen (capacity), 59
Prussia, 97
Pu (pace), 212
Pud (weight), 101
Puddy (capacity), 73
Pulgada (length), 106

Puncheon (capacity), 52
Puncheon of Whisky, &c., 55
Pund (weight), 71
Punkte (length), 59, 60, 87, 118
Punto (length), 93, 102, 106, 116, 212

QUART (capacity), 51—2
Quarta (capacity), 99
Quartaroli (capacity), 93
Quarter (capacity), 51
Quarter (money), 74
Quarter (weight), 53
Quartern (capacity), 52
Quarteron (capacity), 111
Quarterone (capacity), 100
Quarticini (capacity), 100
Quartier (capacity), 64, 66, 80, 82, 87, 97, 104
Quartière (capacity), 103
Quartilhos (capacity), 95
Quartillo (capacity), 106
Quartini (capacity), 103
Quärtli (capacity), 111
Quarto (capacity), 91, 102, 115
Quartucci (capacity), 99, 114
Quartuccio (capacity), 115
Quebec; *see* Great Britain, 50
Quent (weight), 109
Quentchen (weight), 59, 65, 67, 70, 78, 81, 82, 83, 84, 87, 97, 110
Quenten (weight), 111
Quentlein (weight), 104
Quintal (weight), 96
Quintin (weight), 71
Quinto (weight), 80
Qwarter (capacity), 107
Qwintin (weight), 108

RACION (capacity), 106
Rangoon, Asia, 63
Rap (money), 112
Raso or ell (length), 102
Rattel (weight), 94
Real Vellon (money), 78, 107
Reals of plate (money), 107
Rea (money), 74, 96
Reichs-mark (money), 78
Remedy of the mint, 34
Revel; *see* Russia, 101
Rhenish foot (length), 80

Ri (length), 213
Riga; *see* Russia, 101
Rigsbank (money), 72
Riksdaler (money), 108
Riksdaler banco (money), 108
Rin (length), 213
Rin (money), 86
Rin (weight), 214
Rio de Janeiro; *see* Brazil, 64
Rix-banco (money), 72
Rixdollar (money), 60, 62, 65, 68, 81, 85, 86, 98
Road or itinerary measures, 140-2
Rob (capacity), 74
Rochelle; *see* France, 75
Rod (length), 50
Rod (surface), 51
Roede (length), 84
Roëneng (length), 104
Roman scudo (money), 100
Roman States, Italy, 99
Rome, Italy, 99
Rood (surface), 51
Room (weight), 54
Rotl or pound (weight), 74
Rotolo (weight), 92, 114
Rottol (weight), 113
Rottoli (weight), 68
Rottolo (weight), 57, 88, 102, 105
Rotterdam; *see* Holland, 84
Rouen; *see* France, 75
Rubbio (capacity), 99, 100, 103
Rouble (money), 101
Runstycken (money), 108
Rupee (money), 68, 72, 74, 89
Rusponi (money), 115
Russia, 101
Ruthe (length), 60, 64, 82, 97, 103, 109, 118
Ruthen, square, 64
Ryder, gold coin, 85

SACCO (capacity), 90, 103, 115
Sachine (length), 101
Sack (weight), 54
Sai (capacity), 214
St. Domingo; *see* West Indies, 117
St. Gallen, 112
St. Petersburg; *see* Russia, 101
Salma (capacity), 88, 91, 10,
Sardinia, Italy, 102
Sarokowaja (capacity), 101

Saum (capacity), 109, 110, 112
Saundaung (length), 63
Savoy; *see* Sardinia, 102
Saxony, 103
Scheffel (capacity), 65, 67, 81, 83, 87, 97, 104, 118
Scheffel Munich (capacity), 62
Schepel (capacity), 85
Schiffpfund (weight), 67, 81, 98
Schilling (money), 63, 81, 85, 88
Schoppen (capacity), 77, 83, 84, 110, 118
Schrot (capacity), 77
Schuh, shoe or foot (length), 66, 110
Schwaren (money), 65
Scorzo (capacity), 99
Scotch ell (length), 51
Scropulo (weight), 96
Scrupel (weight), 60, 92, 97, 98, 108, 110
Scruple (weight), 53
Scrupolo (weight), 92
Scudo (money), 88, 100, 105
Se (surface), 214
Sechszehnerli (capacity), 109
Sechter (capacity), 77
Secundes (length), 109
Seer (capacity), 68
Seer (weight), 72, 73, 214
Seidel (capacity), 59
Sello (money), 96
Sen (money), 86
Sequin (money), 57
Seron (weight), 80
Sesma (length), 106
Sester (capacity), 61
Setter (capacity), 111
Seville; *see* Spain, 106
Sextarios (capacity), 94
Shaku (capacity), 214
Shaku (length), 213
Shatree (money), 94
Sheet and Hoop Iron Gauge, 223
Sheng (pint), 213
Shi (length), 213
Shi (weight), 214
Shilling (money), 56
Shing tsong (capacity), 69
Shipload (weight), 54
Shō (capacity), 214
Shoe (money), 66, 70
Siberia; *see* Russia, 101
Sicca or tola (weight), 72

Sicca rupee (money), 68
Sicilian dollar (money), 88, 89
Sicily, 105
Silver, assays of, 39
Silver crown, 111, 112
Silver ducat (money), 117
Silver rupee (money), 94
Silver tical (money), 105
Simmer (capacity), 77, 84
Simri (capacity), 118
Sixpence or tester (money), 56
Skälpund (weight), 108
Skeppund (weight), 108
Skieppe (capacity), 71
Skilling (money), 68, 72, 108
Skippund (weight), 71
Smyrna, Turkey, 93, 113
Sok (length), 104
Solar cycle, 155
Solar cycle, rules for, 155-6
Solar day, 143
Solar year, 143
Soldo (length), 114
Soldo Austriaca (money), 117
Soma (capacity), 100, 114, 116
Sovereigns, Bankers' weights for, 228
Sovereigns, wear of, 228
Spain, 106
Span (length), 2, 51, 72
Spanish dollar, 68, 86, 89
Spann (capacity), 107
Specific gravity of water, 19
Spinte (capacity), 65, 81
Square foot (surface), 51
Square measures, table of, 130-9
Square mile (surface), 51
Square yard (surface), 51
Ssŭ (weight), 213
Stadio (length), 86
Stadium Italic (length), 2
Stairo (capacity), 99
Staja (capacity), 90
Stâjo (capacity), 91, 93, 100, 103, 115
Stajolo (length), 99
Standard measures of capacity, 28
Standard measures of length, 27
Standard weights, 27-8
Standard yard, 3, 6, 214
Standards of measures and weights, 1
Standards of Gold and Silver, 39

Standards of length, mural, 22
Stǎng (length), 107
Starello (capacity), 99
Staro (capacity), 79
Steel Wire Gauges, 222
Stein (weight), 98
Step (length), 2
Stère (capacity), 75
Stiver (money), 63, 68, 85
Stockholm; *see* Sweden & Norway, 107
Stone (weight), 53
Stop (capacity), 107
Stotzen (capacity), 111
Strasbourg ; *see* Germany, 78
Streep (length), 84
Strike (capacity), 51
Stübchen (capacity), 64, 66, 71, 80, 82, 87
Stubs' wire gauge, 221
Stützte (capacity), 60
Style, difference of, 148-9
Sun (length), 213
Sundays after Epiphany, 171
Sundays after Trinity, 171
Sweden and Norway, 107
Swiss frank (money), 112
Switzerland, 109

TABLE of Assays, &c., of coins, 44
Table of Coins, 44
Table of Days of the week, 155
Table of Dominical Letters, 154
Table of Easter, &c., for three centuries, 178—187
Table of Easter Day, 168, 208
Table of English Coins, 32
Table of Epacts (measures of time), 164
Table of Epochs, Eras & Periods, 206
Table of Golden Numbers, 158—159
Table of Hebrew Months, 189
Table of Hebrew Years, 195—197
Table of Itinerary or Road Measures, 140—142
Table of Linear Measures, 120—129
Table of Mahometan Months, 202
Table of Mahometan Years, 203—206

Table of Movable Feasts, &c., 172—173
Table of Numbers of Direction, 167
Table of Old Law Terms, 175—176
Table of Square Measures, 130—139
Tables of Wire Gauges, &c., 217—223
Tael (weight), 69, 105, 213
Tael, "lëang" (money), 69
Tain (length), 63
Tan (weight), 69, 213
Tan (surface), 214
T'ang-hsün (length), 212
Tank (weight), 73
Taro (money), 88, 92, 105
Tau (capacity), 69
Tavole (surface), 90
Teman or tomand (capacity), 58
Temen (capacity), 113
Ten-franc piece, 77
Tester, or sixpence, 56
Testoon (money), 96
Thaler (money), 98
Thaler, or rix dollar (money), 60, 65
Thanan (capacity), 105
Thangsat or bucket (capacity), 105
Threepenny piece or bit, 56
Tical (weight), 63, 105
Tierce (capacity), 52
To (capacity), 214
Toise (length), 76
Tola or sicca (weight), 72
Toman (money), 94
Tomand or teman (capacity), 58
Tomin (weight), 106
Tomme (length), 71
Tomolo (capacity), 91
Ton (weight), 53, 54
Tonne (capacity), 71, 80, 87
Tonnen (capacity), 66
Tornatura (surface), 116
Tou (peck), 213
Toulon ; *see* France, 75
Trabucco (length), 102
Trapeso (weight), 92
Trieste ; *see* Austria, 59
Tripoli, Barbary, 113
Troy Weight, 52

Truss (weight), 54
Tscharkey (capacity), 101
Tschetverik (capacity), 101
Tschetverka (capacity), 101
Tschetvert (capacity), 101
Tsubo (surface), 214
Tsun (length), 69, 212
Tu (or degree), 212
Tum (length), 107
Tun (capacity), 52
Tunna (capacity), 107
Turin and Piedmont, 102
Turkey, 113
Turkish oke (weight), 93
Turkish pic (length), 93
Tuscany, Italy, 114
Tussoo (length), 73
Twenty-franc piece, 77
Twenty-lire piece, 93
Two Sicilies, The, 91

UEBA (capacity), 113
    Unglee or finger (length), 72
United States, 115
University Terms, 177
Untz (weight), 108
Unze (weight), 59, 60, 65, 70, 71,
    78, 81, 82, 83, 87, 9', 97,
    98, 104, 109, 110, 111, 112
Uzan (weight), 80

VAKIA (capacity), 58
    Valencia; *see* Spain, 106
Valuation of Coins, 39
Vara, or yard (length), 95, 106
Vat (capacity), 85
Vedro (capacity), 101
Venetian Lombardy, 116
Venice and Milan, 116
Venice foot (length), 116
Verona; *see* Venetian Lombardy, 116
Verst (length), 101
Verschoks (length), 101
Vienna, Austria, 59
Vierding (weight), 59
Vierfass (capacity), 67, 82

Vierling (capacity), 111
Viertel (capacity), 59, 60, 64, 65,
    71, 77, 80, 82, 83, 84, 87,
    97, 104, 110, 111, 118
Viertelein (capacity), 118
Vierteli (capacity), 109
Vingerhoed (capacity), 85
Vintem (money), 96
Vis (weight), 63, 73
Vouh (length), 104

WA (length), 104
    Wakca (weight), 57
Warsaw; *see* Russia, 101.
Water, specific gravity of, 19
Wear of Sovereigns, 228
Week (time), 143
Weights of bronze coins, 36
Weights and Measures Act, 1878, 26
Weights and Measures Act, 1889, 224
West Indies, 117
Wey (capacity), 51
Wigtje (weight), 85
Wire and Sheet Gauges, 215
Wire Gauges, tables of, 217
Wispel (capacity), 81, 82, 97, 104
Wurtemburg, Germany, 118

YARD (length), 50
    Year, commencement of, 146
    —147
Yen (money), 86
Yin (length), 69, 212
Yo (capacity), 213

ZAK (capacity), 85
    Zecchino (money), 93, 117
Zehnling (weight), 61
Zoll (length), 59, 60, 64, 66, 80,
    82, 83, 87, 97, 103, 109, 110, 118
Zoll-centner (weight), 78
Zolotnic (weight), 101
Zuber (capacity), 61
Zurich, Switzerland, 110

7, STATIONERS' HALL COURT, LONDON, E.C.
October, 1889.

A
CATALOGUE OF BOOKS
INCLUDING MANY NEW AND STANDARD WORKS IN
ENGINEERING, MECHANICS, ARCHITECTURE,
NATURAL AND APPLIED SCIENCE,
INDUSTRIAL ARTS, TRADE AND COMMERCE, AGRICULTURE,
GARDENING, LAND MANAGEMENT, LAW, &c.
PUBLISHED BY
CROSBY LOCKWOOD & SON.

MECHANICS, MECHANICAL ENGINEERING, etc.

New Manual for Practical Engineers.
THE PRACTICAL ENGINEER'S HAND-BOOK. Comprising a Treatise on Modern Engines and Boilers: Marine, Locomotive and Stationary. And containing a large collection of Rules and Practical Data relating to recent Practice in Designing and Constructing all kinds of Engines, Boilers, and other Engineering work. The whole constituting a comprehensive Key to the Board of Trade and other Examinations for Certificates of Competency in Modern Mechanical Engineering. By WALTER S. HUTTON, Civil and Mechanical Engineer, Author of "The Works' Manager's Handbook for Engineers," &c. With upwards of 370 Illustrations. Third Edition, Revised, with Additions. Medium 8vo, nearly 500 pp., price 18s. Strongly bound. [*Just published.*

☞ *This work is designed as a companion to the Author's* "WORKS' MANAGER'S HAND-BOOK." *It possesses many new and original features, and contains, like its predecessor, a quantity of matter not originally intended for publication, but collected by the author for his own use in the construction of a great variety of modern engineering work.*

The information is given in a condensed and concise form, and is illustrated by upwards of 370 Woodcuts; and comprises a quantity of tabulated matter of great value to all engaged in designing, constructing, or estimating for ENGINES, BOILERS *and* OTHER ENGINEERING WORK.

*** OPINIONS OF THE PRESS.

"We have kept it at hand for several weeks, referring to it as occasion arose, and we have not on a single occasion consulted its pages without finding the information of which we were in quest." —*Athenæum.*

"A thoroughly good practical handbook, which no engineer can go through without learning something that will be of service to him."—*Marine Engineer.*

"An excellent book of reference for engineers, and a valuable text-book for students of engineering."—*Scotsman.*

"This valuable manual embodies the results and experience of the leading authorities on mechanical engineering."—*Building News.*

"The author has collected together a surprising quantity of rules and practical data, and has shown much judgment in the selections he has made. . . . There is no doubt that this book is one of the most useful of its kind published, and will be a very popular compendium."—*Engineer.*

"A mass of information, set down in simple language, and in such a form that it can be easily referred to at any time. The matter is uniformly good and well chosen, and is greatly elucidated by the illustrations. The book will find its way on to most engineers' shelves, where it will rank as one of the most useful books of reference."—*Practical Engineer.*

"Full of useful information, and should be found on the office shelf of all practical engineers." —*English Mechanic.*

B

Handbook for Works' Managers.

THE WORKS' MANAGER'S HANDBOOK OF MODERN RULES, TABLES, AND DATA. For Engineers, Millwrights, and Boiler Makers; Tool Makers, Machinists, and Metal Workers; Iron and Brass Founders, &c. By W. S. HUTTON, Civil and Mechanical Engineer, Author of "The Practical Engineer's Handbook." Third Edition, carefully Revised, with Additions. In One handsome Vol., medium 8vo price 15s. strongly bound.

☞ *The Author having compiled Rules and Data for his own use in a great variety of modern engineering work, and having found his notes extremely useful, decided to publish them—revised to date—believing that a practical work, suited to the* DAILY REQUIREMENTS OF MODERN ENGINEERS, *would be favourably received.*

In the Third Edition, the following among other additions have been made, viz.: Rules for the Proportions of Riveted Joints in Soft Steel Plates, the Results of Experiments by PROFESSOR KENNEDY *for the Institution of Mechanical Engineers—Rules for the Proportions of Turbines—Rules for the Strength of Hollow Shafts of Whitworth's Compressed Steel, &c.*

*** OPINIONS OF THE PRESS.

"The author treats every subject from the point of view of one who has collected workshop notes for application in workshop practice, rather than from the theoretical or literary aspect. The volume contains a great deal of that kind of information which is gained only by practical experience, and is seldom written in books."—*Engineer.*

"The volume is an exceedingly useful one, brimful with engineers' notes, memoranda, and rules, and well worthy of being on every mechanical engineer's bookshelf."—*Mechanical World.*

"A formidable mass of facts and figures, readily accessible through an elaborate index . . . . . Such a volume will be found absolutely necessary as a book of reference in all sorts of 'works' connected with the metal trades."—*Ryland's Iron Trades Circular.*

"Brimful of useful information, stated in a concise form, Mr. Hutton's books have met a pressing want among engineers. The book must prove extremely useful to every practical man possessing a copy."—*Practical Engineer.*

"The Modernised Templeton."

THE PRACTICAL MECHANIC'S WORKSHOP COMPANION. Comprising a great variety of the most useful Rules and Formulæ in Mechanical Science, with numerous Tables of Practical Data and Calculated Results for Facilitating Mechanical Operations. By WILLIAM TEMPLETON, Author of "The Engineer's Practical Assistant," &c. &c. Fifteenth Edition, Revised, Modernised, and considerably Enlarged by WALTER S. HUTTON, C.E., Author of "The Works' Manager's Handbook," "The Practical Engineer's Handbook," &c. Fcap. 8vo, nearly 500 pp., with Eight Plates and upwards of 250 Illustrative Diagrams, 6s., strongly bound for workshop or pocket wear and tear.

☞ TEMPLETON'S "MECHANIC'S WORKSHOP COMPANION" *has been for more than a quarter of a century deservedly popular, and, as the well-worn and thumb-marked* vade mecum *of several generations of intelligent and aspiring workmen, it has had the reputation of having been the means of raising many of them in their position in life.*

In consequence of the lapse of time since the Author's death, and the great advances in Mechanical Science, the Publishers have thought it advisable to have it entirely Reconstructed and Modernised; and in its present greatly Enlarged and Improved form, they are sure that it will commend itself to the English workmen of the present day all the world over, and become, like its predecessors, their indispensable friend and referee.

A smaller type having been adopted, and the page increased in size, while the number of pages has advanced from about 330 to nearly 500, the book practically contains double the amount of matter that was comprised in the original work.

*** OPINIONS OF THE PRESS.

"In its modernised form Hutton's 'Templeton' should have a wide sale, for it contains much valuable information which the mechanic will often find of use, and not a few tables and notes which he might look for in vain in other works. This modernised edition will be appreciated by all who have learned to value the original editions of 'Templeton.'"—*English Mechanic.*

"It has met with great success in the engineering workshop, as we can testify; and there are a great many men who, in a great measure, owe their rise in life to this little book."—*Building News.*

"This familiar text-book—well known to all mechanics and engineers—is of essential service to the every-day requirements of engineers, millwrights, and the various trades connected with engineering and building. The new modernised edition is worth its weight in gold."—*Building News.* (Second Notice.)

"The publishers wisely entrusted the task of revision of this popular, valuable and useful book of Mr. Hutton, than whom a more competent man they could not have found."—*Iron.*

Stone-working Machinery.

STONE-WORKING MACHINERY, and the Rapid and Economical Conversion of Stone. With Hints on the Arrangement and Management of Stone Works. By M. POWIS BALE, M.I.M.E. Crown 8vo, 9s.

"Should be in the hands of every mason or student of stone-work."—*Colliery Guardian.*
"It is in every sense of the word a standard work upon a subject which the author is fully competent to deal exhaustively with."—*Builder's Weekly Reporter.*
"A capital handbook for all who manipulate stone for building or ornamental purposes."—*Machinery Market.*

Pump Construction and Management.

PUMPS AND PUMPING : A Handbook for Pump Users. Being Notes on Selection, Construction and Management. By M. POWIS BALE, M.I.M.E., Author of "Woodworking Machinery," "Saw Mills," &c. Crown 8vo, 2s. 6d. cloth. [*Just published.*

"The matter is set forth as concisely as possible. In fact, condensation rather than diffuseness has been the author's aim throughout; yet he does not seem to have omitted anything likely to be of use."—*Journal of Gas Lighting.*
"Thoroughly practical and simply and clearly written."—*Glasgow Herald.*

Turning.

LATHE-WORK : A Practical Treatise on the Tools, Appliances, and Processes employed in the Art of Turning. By PAUL N. HASLUCK. Third Edition, Revised and Enlarged. Crown 8vo, 5s. cloth.

"Written by a man who knows, not only how work ought to be done, but who also knows how to do it, and how to convey his knowledge to others. To all turners this book would be valuable."—*Engineering.*
"We can safely recommend the work to young engineers. To the amateur it will simply be invaluable. To the student it will convey a great deal of useful information."—*Engineer.*
"A compact, succinct, and handy guide to lathe-work did not exist in our language until Mr. Hasluck, by the publication of this treatise, gave the turner a true *vade-mecum.*"—*House Decorator.*

Screw-Cutting.

SCREW THREADS : And Methods of Producing Them. With Numerous Tables, and complete directions for using Screw-Cutting Lathes. By PAUL N. HASLUCK, Author of "Lathe-Work," &c. With Fifty Illustrations. Second Edition. Waistcoat-pocket size, price 1s. cloth.

"Full of useful information, hints and practical criticism. Taps, dies and screwing-tools generally are illustrated and their action described."—*Mechanical World.*

Smith's Tables for Mechanics, etc.

TABLES, MEMORANDA, AND CALCULATED RESULTS, FOR MECHANICS, ENGINEERS, ARCHITECTS, BUILDERS, etc. Selected and Arranged by FRANCIS SMITH. Fourth Edition, Revised and Enlarged, 250 pp., waistcoat-pocket size, 1s. 6d. limp leather.

"It would, perhaps, be as difficult to make a small pocket-book selection of notes and formulæ to suit ALL engineers as it would be to make a universal medicine; but Mr. Smith's waistcoat-pocket collection may be looked upon as a successful attempt."—*Engineer.*
"The best example we have ever seen of 250 pages of useful matter packed into the dimensions of a card-case."—*Building News.* "A veritable pocket treasury of knowledge."—*Iron.*

Engineer's and Machinist's Assistant.

THE ENGINEER'S, MILLWRIGHT'S, and MACHINIST'S PRACTICAL ASSISTANT. A collection of Useful Tables, Rules and Data. By WILLIAM TEMPLETON. 7th Edition, with Additions. 18mo, 2s. 6d. cloth.

"Occupies a foremost place among books of this kind. A more suitable present to an apprentice to any of the mechanical trades could not possibly be made."—*Building News.*
"A deservedly popular work, it should be in the 'drawer' of every mechanic."—*English Mechanic.*

Iron and Steel.

"IRON AND STEEL": A Work for the Forge, Foundry, Factory, and Office. Containing ready, useful, and trustworthy Information for Iron-masters and their Stock-takers; Managers of Bar, Rail, Plate, and Sheet Rolling Mills; Iron and Metal Founders; Iron Ship and Bridge Builders; Mechanical, Mining, and Consulting Engineers; Architects, Builders, and Draughtsmen. By CHARLES HOARE, Author of "The Slide Rule," &c. Eighth Edition, Revised and considerably Enlarged. 32mo, 6s. leather.

"One of the best of the pocket books."—*English Mechanic.*
"We cordially recommend this book to those engaged in considering the details of all kinds of iron and steel works."—*Naval Science.*

4　CROSBY LOCKWOOD & SON'S CATALOGUE.

Engineering Construction.

PATTERN-MAKING : A Practical Treatise, embracing the Main Types of Engineering Construction, and including Gearing, both Hand and Machine made, Engine Work, Sheaves and Pulleys, Pipes and Columns, Screws, Machine Parts, Pumps and Cocks, the Moulding of Patterns in Loam and Greensand, &c., together with the methods of Estimating the weight of Castings; to which is added an Appendix of Tables for Workshop Reference. By a FOREMAN PATTERN MAKER. With upwards of Three Hundred and Seventy Illustrations. Crown 8vo, 7s. 6d. cloth.

"A well-written technical guide, evidently written by a man who understands and has practised what he has written about. We cordially recommend it to engineering students, young journeymen, and others desirous of being initiated into the mysteries of pattern-making."—*Builder*.
"Likely to prove a welcome guide to many workmen, especially to draughtsmen who have lacked a training in the shops, pupils pursuing their practical studies in our factories, and to employers and managers in engineering works."—*Hardware Trade Journal*.
"More than 370 illustrations help to explain the text, which is, however, always clear and explicit, thus rendering the work an excellent *vade mecum* for the apprentice who desires to become master of his trade."—*English Mechanic*.

Dictionary of Mechanical Engineering Terms.

LOCKWOOD'S DICTIONARY OF TERMS USED IN THE PRACTICE OF MECHANICAL ENGINEERING, embracing those current in the Drawing Office, Pattern Shop, Foundry, Fitting, Turning, Smith's and Boiler Shops, &c. &c. Comprising upwards of 6,000 Definitions. Edited by A FOREMAN PATTERN-MAKER, Author of "Pattern Making." Crown 8vo, 7s. 6d. cloth.

"Just the sort of handy dictionary required by the various trades engaged in mechanical engineering. The practical engineering pupil will find the book of great value in his studies, and every foreman engineer and mechanic should have a copy. —*Building News*.
"After a careful examination of the book, and trying all manner of words, we think that the engineer will here find all he is likely to require. It will be argely used."—*Practical Engineer*.
"This admirable dictionary, although primarily intended for the use of draughtsmen and other technical craftsmen, is of much larger value as a book of reference, and will find a ready welcome in many libraries."—*Glasgow Herald*.
"One of the most useful books which can be presented to a mechanic or student."—*English Mechanic*.
"Not merely a dictionary, but, to a certain extent, also a most valuable guide. It strikes us as a happy idea to combine with a definition of the phrase useful information on the subject of which it treats."—*Machinery Market*.
"This carefully-compiled volume forms a kind of pocket cyclopædia of the extensive subject to which it is devoted. No word having connection with any branch of constructive engineering seems to be omitted. No more comprehensive work has been, so far, issued."—*Knowledge*.
"We strongly commend this useful and reliable adviser to our friends in the workshop, and to students everywhere."—*Colliery Guardian*.

Steam Boilers.

A TREATISE ON STEAM BOILERS: Their Strength, Construction, and Economical Working. By ROBERT WILSON, C.E. Fifth Edition. 12mo, 6s. cloth.

"The best treatise that has ever been published on steam boilers."—*Engineer*.
"The author shows himself perfect master of his subject, and we heartily recommend all employing steam power to possess themselves of the work."—*Ryland's Iron Trade Circular*.

Boiler Chimneys.

BOILER AND FACTORY CHIMNEYS; Their Draught-Power and Stability. With a Chapter on Lightning Conductors. By ROBERT WILSON, C.E., Author of "A Treatise on Steam Boilers," &c. Second Edition. Crown 8vo, 3s. 6d. cloth.

"Full of useful information, definite in statement, and thoroughly practical in treatment."—*The Local Government Chronicle*.
"A valuable contribution to the literature of scientific building. . . . The whole subject is very interesting and important one, and it is gratifying to know that it has fallen into such competent hands."—*The Builder*.

Boiler Making.

THE BOILER-MAKER'S READY RECKONER. With Examples of Practical Geometry and Templating, for the Use of Platers, Smiths and Riveters. By JOHN COURTNEY, Edited by D. K. CLARK, M.I.C.E. Second Edition, Revised, with Additions, 12mo, 5s. half-bound.

"No workman or apprentice should be without this book."—*Iron Trade Circular*.
"A reliable guide to the working boiler-maker."—*Iron*.
"Boiler-makers will readily recognise the value of this volume. . . . The tables are clearly printed, and so arranged that they can be referred to with the greatest facility, so that it cannot be doubted that they will be generally appreciated and much used."—*Mining Journal*.

Steam Engine.

TEXT-BOOK ON THE STEAM ENGINE. With a Supplement on Gas Engines. By T. M. GOODEVE, M.A., Barrister-at-Law, Author of "The Elements of Mechanism," &c. Tenth Edition, Enlarged. With numerous Illustrations. Crown 8vo, 6s. cloth. [*Just published.*

"Professor Goodeve has given us a treatise on the steam engine which will bear comparison with anything written by Huxley or Maxwell, and we can award it no higher praise."—*Engineer.*
"Professor Goodeve's book is ably and clearly written. It is a sound work."—*Athenæum.*
"Mr. Goodeve's text-book is a work of which every young engineer should possess himself."—*Mining Journal.*
"Essentially practical in its aim. The manner of exposition leaves nothing to be desired."—*Scotsman.*

Gas Engines.

ON GAS-ENGINES. Being a Reprint, with some Additions, of the Supplement to the *Text-book on the Steam Engine*, by T. M. GOODEVE, M.A. Crown 8vo, 2s. 6d. cloth. [*Just published.*

"Like all Mr. Goodeve's writings, the present is no exception in point of general excellence. It is a valuable little volume."—*Mechanical World.*
"This little book will be useful to those who desire to understand how the gas-engine works.'—*English Mechanic.*

Steam.

THE SAFE USE OF STEAM. Containing Rules for Unprofessional Steam-users. By an ENGINEER. Sixth Edition. Sewed, 6d.

"If steam-users would but learn this little book by heart boiler explosions would become sensations by their rarity."—*English Mechanic.*

Coal and Speed Tables.

A POCKET BOOK OF COAL AND SPEED TABLES, for Engineers and Steam-users. By NELSON FOLEY, Author of "Boiler Construction." Pocket-size, 3s. 6d. cloth; 4s. leather.

"This is a very useful book, containing very useful tables. The results given are well chosen, and the volume contains evidence that the author really understands his subject. We can recommend the work with pleasure."—*Mechanical World.*
"These tables are designed to meet the requirements of every-day use; they are of sufficient scope for most practical purposes, and may be commended to engineers and users of steam."—*Iron.*
"This pocket-book well merits the attention of the practical engineer. Mr. Foley has compiled a very useful set of tables, the information contained in which is frequently required by engineers, coal consumers and users of steam."—*Iron and Coal Trades Review.*

Fire Engineering.

FIRES, FIRE-ENGINES, AND FIRE-BRIGADES. With a History of Fire-Engines, their Construction, Use, and Management; Remarks on Fire-Proof Buildings, and the Preservation of Life from Fire; Statistics of the Fire Appliances in English Towns; Foreign Fire Systems; Hints on Fire Brigades, &c. &c. By CHARLES F. T. YOUNG, C.E. With numerous Illustrations, 544 pp., demy 8vo, £1 4s. cloth.

"To such of our readers as are interested in the subject of fires and fire apparatus, we can most heartily commend this book. It is really the only English work we now have upon the subject."—*Engineering.*
"It displays much evidence of careful research; and Mr. Young has put his facts neatly together. It is evident enough that his acquaintance with the practical details of the construction of steam fire engines, old and new, and the conditions with which it is necessary they should comply, is accurate and full."—*Engineer.*

Gas Lighting.

COMMON SENSE FOR GAS-USERS: A Catechism of Gas-Lighting for Householders, Gasfitters, Millowners, Architects, Engineers, etc. By ROBERT WILSON, C.E., Author of "A Treatise on Steam Boilers." Second Edition, with Folding Plates and Wood Engravings. Crown 8vo, price 1s. in wrapper.

"All gas-users will decidedly benefit, both in pocket and comfort, if they will avail themselves of Mr. Wilson's counsels."—*Engineering.*

Dynamo Construction.

HOW TO MAKE A DYNAMO: A Practical Treatise for Amateurs. Containing numerous Illustrations and Detailed Instructions for Constructing a Small Dynamo, to Produce the Electric Light. By ALFRED CROFTS. Second Edition, Revised and Enlarged. Crown 8vo, 2s. cloth. [*Just published.*

"The instructions given in this unpretentious little book are sufficiently clear and explicit to enable any amateur mechanic possessed of average skill and the usual tools to be found in an amateur's workshop, to build a practical dynamo machine."—*Electrician.*

CROSBY LOCKWOOD & SON'S CATALOGUE.

THE POPULAR WORKS OF MICHAEL REYNOLDS
("THE ENGINE DRIVER'S FRIEND").

Locomotive-Engine Driving.

LOCOMOTIVE-ENGINE DRIVING: A Practical Manual for Engineers in charge of Locomotive Engines. By MICHAEL REYNOLDS, Member of the Society of Engineers, formerly Locomotive Inspector L. B. and S. C. R. Eighth Edition. Including a KEY TO THE LOCOMOTIVE ENGINE. With Illustrations and Portrait of Author. Crown 8vo, 4s. 6d. cloth.

"Mr. Reynolds has supplied a want, and has supplied it well. We can confidently recommend the book, not only to the practical driver, but to everyone who takes an interest in the performance of locomotive engines."—*The Engineer.*

"Mr. Reynolds has opened a new chapter in the literature of the day. This admirable practical treatise, of the practical utility of which we have to speak in terms of warm commendation."—*Athenæum.*

"Evidently the work of one who knows his subject thoroughly."—*Railway Service Gazette.*

"Were the cautions and rules given in the book to become part of the every-day working of our engine-drivers, we might have fewer distressing accidents to deplore."—*Scotsman.*

Stationary Engine Driving.

STATIONARY ENGINE DRIVING: A Practical Manual for Engineers in charge of Stationary Engines. By MICHAEL REYNOLDS. Third Edition, Enlarged. With Plates and Woodcuts. Crown 8vo, 4s. 6d. cloth.

"The author is thoroughly acquainted with his subjects, and his advice on the various points treated is clear and practical. . . . He has produced a manual which is an exceedingly useful one for the class for whom it is specially intended."—*Engineering.*

"Our author leaves no stone unturned. He is determined that his readers shall not only know something about the stationary engine, but all about it."—*Engineer.*

"An engineman who has mastered the contents of Mr. Reynolds's book will require but little actual experience with boilers and engines before he can be trusted to look after them."—*English Mechanic.*

The Engineer, Fireman, and Engine-Boy.

THE MODEL LOCOMOTIVE ENGINEER, FIREMAN, and ENGINE-BOY. Comprising a Historical Notice of the Pioneer Locomotive Engines and their Inventors. By MICHAEL REYNOLDS. With numerous Illustrations and a fine Portrait of George Stephenson. Crown 8vo, 4s. 6d. cloth.

"From the technical knowledge of the author it will appeal to the railway man of to-day more forcibly than anything written by Dr. Smiles. . . . The volume contains information of a technical kind, and facts that every driver should be familiar with."—*English Mechanic.*

"We should be glad to see this book in the possession of everyone in the kingdom who has ever laid, or is to lay, hands on a locomotive engine."—*Iron.*

Continuous Railway Brakes.

CONTINUOUS RAILWAY BRAKES: A Practical Treatise on the several Systems in Use in the United Kingdom; their Construction and Performance. With copious Illustrations and numerous Tables. By MICHAEL REYNOLDS. Large crown 8vo, 9s. cloth.

"A popular explanation of the different brakes. It will be of great assistance in forming public opinion, and will be studied with benefit by those who take an interest in the brake."—*English Mechanic.*

"Written with sufficient technical detail to enable the principle and relative connection of the various parts of each particular brake to be readily grasped."—*Mechanical World.*

Engine-Driving Life.

ENGINE-DRIVING LIFE: Stirring Adventures and Incidents in the Lives of Locomotive-Engine Drivers. By MICHAEL REYNOLDS. Second Edition, with Additional Chapters. Crown 8vo, 2s. cloth. [*Just published.*

"From first to last perfectly fascinating. Wilkie Collins's most thrilling conceptions are thrown into the shade by true incidents, endless in their variety, related in every page."—*North British Mail.*

"Anyone who wishes to get a real insight into railway life cannot do better than read 'Engine-Driving Life' for himself; and if he once take it up he will find that the author's enthusiasm and real love of the engine-driving profession will carry him on till he has read every page."—*Saturday Review.*

Pocket Companion for Enginemen.

THE ENGINEMAN'S POCKET COMPANION AND PRACTICAL EDUCATOR FOR ENGINEMEN, BOILER ATTENDANTS, AND MECHANICS. By MICHAEL REYNOLDS. With Forty-five Illustrations and numerous Diagrams. Second Edition, Revised. Royal 18mo, 3s. 6d., strongly bound for pocket wear.

"This admirable work is well suited to accomplish its object, being the honest workmanship of a competent engineer."—*Glasgow Herald.*

"A most meritorious work, giving in a succinct and practical form all the information an engine-minder desirous of mastering the scientific principles of his daily calling would require."—*Miller.*

"A boon to those who are striving to become efficient mechanics."—*Daily Chronicle.*

French-English Glossary for Engineers, etc.

A POCKET GLOSSARY of TECHNICAL TERMS: ENGLISH-FRENCH, FRENCH-ENGLISH; with Tables suitable for the Architectural, Engineering, Manufacturing and Nautical Professions. By JOHN JAMES FLETCHER, Engineer and Surveyor; 200 pp. Waistcoat-pocket size, 1s. 6d., limp leather.

"It ought certainly to be in the waistcoat-pocket of every professional man. —*Iron.*
"It is a very great advantage for readers and correspondents in France and England to have so large a number of the words relating to engineering and manufacturers collected in a lilliputian volume. The little book will be useful both to students and travellers."—*Architect.*
"The glossary of terms is very complete, and many of the tables are new and well arranged. We cordially commend the book."—*Mechanical World.*

Portable Engines.

THE PORTABLE ENGINE; ITS CONSTRUCTION AND MANAGEMENT. A Practical Manual for Owners and Users of Steam Engines generally. By WILLIAM DYSON WANSBROUGH. With 90 Illustrations. Crown 8vo, 3s. 6d. cloth.

"This is a work of value to those who use steam machinery. . . . Should be read by everyone who has a steam engine, on a farm or elsewhere."—*Mark Lane Express.*
"We cordially commend this work to buyers and owners of steam engines, and to those who have to do with their construction or use."—*Timber Trades Journal.*
"Such a general knowledge of the steam engine as Mr. Wansbrough furnishes to the reader should be acquired by all intelligent owners and others who use the steam engine."—*Building News.*

CIVIL ENGINEERING, SURVEYING, etc.

MR. HUMBER'S IMPORTANT ENGINEERING BOOKS.

The Water Supply of Cities and Towns.

A COMPREHENSIVE TREATISE on the WATER-SUPPLY OF CITIES AND TOWNS. By WILLIAM HUMBER, A-M.Inst.C.E., and M. Inst. M.E., Author of "Cast and Wrought Iron Bridge Construction," &c. &c. Illustrated with 50 Double Plates, 1 Single Plate, Coloured Frontispiece, and upwards of 250 Woodcuts, and containing 400 pages of Text. Imp. 4to, £6 6s. elegantly and substantially half-bound in morocco.

List of Contents.

I. Historical Sketch of some of the means that have been adopted for the Supply of Water to Cities and Towns.—II. Water and the Foreign Matter usually associated with it.—III. Rainfall and Evaporation.—IV. Springs and the water-bearing formations of various districts.—V. Measurement and Estimation of the flow of Water.—VI. On the Selection of the Source of Supply.—VII. Wells.—VIII. Reservoirs.—IX. The Purification of Water.—X. Pumps.— XI. Pumping Machinery. — XII. Conduits.—XIII. Distribution of Water.—XIV. Meters, Service Pipes, and House Fittings.—XV. The Law and Economy of Water Works. XVI; Constant and Intermittent Supply.—XVII. Description of Plates. — Appendices, giving Tables of Rates of Supply, Velocities, &c. &c., together with Specifications of several Works illustrated, among which will be found: Aberdeen, Bideford, Canterbury, Dundee, Halifax, Lambeth, Rotherham, Dublin, and others.

"The most systematic and valuable work upon water supply hitherto produced in English, or in any other language. . . . Mr. Humber's work is characterised almost throughout by an exhaustiveness much more distinctive of French and German than of English technical treatises."—*Engineer.*
"We can congratulate Mr. Humber on having been able to give so large an amount of information on a subject so important as the water supply of cities and towns. The plates, fifty in number, are mostly drawings of executed works, and alone would have commanded the attention of every engineer whose practice may lie in this branch of the profession."—*Builder.*

Cast and Wrought Iron Bridge Construction.

A COMPLETE AND PRACTICAL TREATISE ON CAST AND WROUGHT IRON BRIDGE CONSTRUCTION, including Iron Foundations. In Three Parts—Theoretical, Practical, and Descriptive. By WILLIAM HUMBER, A.M.Inst.C.E., and M.Inst.M.E. Third Edition, Revised and much improved, with 115 Double Plates (20 of which now first appear in this edition), and numerous Additions to the Text. In Two Vols., imp. 4to, £6 16s. 6d. half-bound in morocco.

"A very valuable contribution to the standard literature of civil engineering. In addition to elevations, plans and sections, large scale details are given which very much enhance the instructive worth of those illustrations."—*Civil Engineer and Architect's Journal.*
"Mr. Humber's stately volumes, lately issued—in which the most important bridges erected during the last five years, under the direction of the late Mr. Brunel, Sir W. Cubitt, Mr. Hawkshaw, Mr. Page, Mr. Fowler, Mr. Hemans, and others among our most eminent engineers, are drawn and specified in great detail."—*Engineer*

MR. HUMBER'S GREAT WORK ON MODERN ENGINEERING.

Complete in Four Volumes, imperial 4to, price £12 12s., half-morocco. Each Volume sold separately as follows:—

A RECORD OF THE PROGRESS OF MODERN ENGINEERING. FIRST SERIES.
Comprising Civil, Mechanical, Marine, Hydraulic, Railway, Bridge, and other Engineering Works, &c. By WILLIAM HUMBER, A-M.Inst.C.E., &c. Imp. 4to, with 36 Double Plates, drawn to a large scale, Photographic Portrait of John Hawkshaw, C.E., F.R.S., &c., and copious descriptive Letterpress, Specifications, &c., £3 3s. half-morocco.

List of the Plates and Diagrams.

Victoria Station and Roof, L. B. & S. C. R. (8 plates); Southport Pier (2 plates); Victoria Station and Roof, L. C. & D. and G. W. R. (6 plates); Roof of Cremorne Music Hall; Bridge over G. N. Railway; Roof of Station, Dutch Rhenish Rail (2 plates); Bridge over the Thames, West London Extension Railway (5 plates); Armour Plates: Suspension Bridge, Thames (4 plates); The Allen Engine; Suspension Bridge, Avon (3 plates); Underground Railway (3 plates).

"Handsomely lithographed and printed. It will find favour with many who desire to preserve in a permanent form copies of the plans and specifications prepared for the guidance of the contractors for many important engineering works."—*Engineer.*

HUMBER'S RECORD OF MODERN ENGINEERING. SECOND SERIES.
Imp. 4to, with 36 Double Plates, Photographic Portrait of Robert Stephenson, C.E., M.P., F.R.S., &c., and copious descriptive Letterpress, Specifications, &c., £3 3s. half-morocco.

List of the Plates and Diagrams.

Birkenhead Docks, Low Water Basin (15 plates); Charing Cross Station Roof, C. C. Railway (3 plates); Digswell Viaduct, Great Northern Railway; Robbery Wood Viaduct, Great Northern Railway; Iron Permanent Way; Clydach Viaduct, Merthyr, Tredegar, and Abergavenny Railway; Ebbw Viaduct, Merthyr, Tredegar, and Abergavenny Railway; College Wood Viaduct, Cornwall Railway; Dublin Winter Palace Roof (3 plates); Bridge over the Thames, L. C. & D. Railway (6 plates); Albert Harbour, Greenock (4 plates).

"Mr. Humber has done the profession good and true service, by the fine collection of examples he has here brought before the profession and the public."—*Practical Mechanic's Journal.*

HUMBER'S RECORD OF MODERN ENGINEERING. THIRD SERIES.
Imp. 4to, with 40 Double Plates, Photographic Portrait of J. R. M'Clean, late Pres. Inst. C.E., and copious descriptive Letterpress, Specifications, &c., £3 3s. half-morocco.

List of the Plates and Diagrams.

MAIN DRAINAGE, METROPOLIS.—*North Side.*—Map showing Interception of Sewers; Middle Level Sewer (2 plates); Outfall Sewer, Bridge over River Lea (3 plates); Outfall Sewer, Bridge over Marsh Lane, North Woolwich Railway, and Bow and Barking Railway Junction; Outfall Sewer, Bridge over Bow and Barking Railway (3 plates); Outfall Sewer, Bridge over East London Waterworks' Feeder (2 plates); Outfall Sewer, Reservoir (2 plates); Outfall Sewer, Tumbling Bay and Outlet; Outfall Sewer, Penstocks. *South Side.*—Outfall Sewer, Bermondsey Branch (2 plates); Outfall Sewer, Reservoir and Outlet (4 plates); Outfall Sewer, Filth Hoist; Sections of Sewers (North and South Sides).
THAMES EMBANKMENT.—Section of River Wall; Steamboat Pier, Westminster (2 plates); Landing Stairs between Charing Cross and Waterloo Bridges; York Gate (2 plates); Overflow and Outlet at Savoy Street Sewer (3 plates); Steamboat Pier, Waterloo Bridge (3 plates); Junction of Sewers, Plans and Sections; Gullies, Plans and Sections; Rolling Stock; Granite and Iron Forts.

"The drawings have a constantly increasing value, and whoever desires to possess clear representations of the two great works carried out by our Metropolitan Board will obtain Mr. Humber's volume."—*Engineer.*

HUMBER'S RECORD OF MODERN ENGINEERING. FOURTH SERIES.
Imp. 4to, with 36 Double Plates, Photographic Portrait of John Fowler, late Pres. Inst. C.E., and copious descriptive Letterpress, Specifications, &c., £3 3s. half-morocco.

List of the Plates and Diagrams.

Abbey Mills Pumping Station, Main Drainage, Metropolis (4 plates); Barrow Docks (5 plates); Manquis Viaduct, Santiago and Valparaiso Railway (2 plates); Adam's Locomotive, St. Helen's Canal Railway (2 plates); Cannon Street Station Roof, Charing Cross Railway (3 plates); Road Bridge over the River Moka (2 plates); Telegraph Apparatus for Mesopotamia; Viaduct over the River Wye, Midland Railway (3 plates); St. Germans Viaduct, Cornwall Railway (2 plates); Wrought-Iron Cylinder for Diving Bell; Millwall Docks (6 plates); Milroy's Patent Excavator; Metropolitan District Railway (6 plates); Harbours, Ports, and Breakwaters (3 plates).

"We gladly welcome another year's issue of this valuable publication from the able pen of Mr. Humber. The accuracy and general excellence of this work are well known, while its usefulness in giving the measurements and details of some of the latest examples of engineering, as carried out by the most eminent men in the profession, cannot be too highly prized."—*Artisan.*

MR. HUMBER'S ENGINEERING BOOKS—continued.

Strains, Calculation of.

A HANDY BOOK FOR THE CALCULATION OF STRAINS IN GIRDERS AND SIMILAR STRUCTURES, AND THEIR STRENGTH. Consisting of Formulæ and Corresponding Diagrams, with numerous details for Practical Application, &c. By WILLIAM HUMBER, A-M.Inst.C.E., &c. Fourth Edition. Crown 8vo, nearly 100 Woodcuts and 3 Plates, 7s. 6d. cloth.

" The formulæ are neatly expressed, and the diagrams good."—*Athenæum.*
" We heartily commend this really *handy* book to our engineer and architect readers."—*English Mechanic.*

Barlow's Strength of Materials, enlarged by Humber

A TREATISE ON THE STRENGTH OF MATERIALS; with Rules for Application in Architecture, the Construction of Suspension Bridges, Railways, &c. By PETER BARLOW, F.R.S. A New Edition, revised by his Sons, P. W. BARLOW, F.R.S., and W. H. BARLOW, F.R.S.; to which are added, Experiments by HODGKINSON, FAIRBAIRN, and KIRKALDY; and Formulæ for Calculating Girders, &c. Arranged and Edited by W. HUMBER, A-M.Inst.C.E. Demy 8vo, 400 pp., with 19 large Plates and numerous Woodcuts, 18s. cloth.

" Valuable alike to the student, tyro, and the experienced practitioner, it will always rank in future, as it has hitherto done, as the standard treatise on that particular subject."—*Engineer.*
" There is no greater authority than Barlow."—*Building News.*
" As a scientific work of the first class, it deserves a foremost place on the bookshelves of every civil engineer and practical mechanic."—*English Mechanic.*

Trigonometrical Surveying.

AN OUTLINE OF THE METHOD OF CONDUCTING A TRIGONOMETRICAL SURVEY, *for the Formation of Geographical and Topographical Maps and Plans, Military Reconnaissance, Levelling, &c.,* with Useful Problems, Formulæ, and Tables. By Lieut.-General FROME, R.E. Fourth Edition, Revised and partly Re-written by Major General Sir CHARLES WARREN, G.C.M.G., R.E. With 19 Plates and 115 Woodcuts, royal 8vo, 16s. cloth.

" The simple fact that a fourth edition has been called for is the best testimony to its merits. No words of praise from us can strengthen the position so well and so steadily maintained by this work. Sir Charles Warren has revised the entire work, and made such additions as were necessary to bring every portion of the contents up to the present date."—*Broad Arrow.*

Oblique Bridges.

A PRACTICAL AND THEORETICAL ESSAY ON OBLIQUE BRIDGES. With 13 large Plates. By the late GEORGE WATSON BUCK, M.I.C.E. Third Edition, revised by his Son, J. H. WATSON BUCK, M.I.C.E.; and with the addition of Description to Diagrams for Facilitating the Construction of Oblique Bridges, by W. H. BARLOW, M.I.C.E. Royal 8vo, 12s. cloth.

" The standard text-book for all engineers regarding skew arches is Mr. Buck's treatise, and it would be impossible to consult a better."—*Engineer.*
" Mr. Buck's treatise is recognised as a standard text-book, and his treatment has divested the subject of many of the intricacies supposed to belong to it. As a guide to the engineer and architect, on a confessedly difficult subject, Mr. Buck's work is unsurpassed."—*Building News.*

Water Storage, Conveyance and Utilisation.

WATER ENGINEERING : A Practical Treatise on the Measurement, Storage, Conveyance and Utilisation of Water for the Supply of Towns, for Mill Power, and for other Purposes. By CHARLES SLAGG, Water and Drainage Engineer, A.M.Inst.C.E., Author of " Sanitary Work in the Smaller Towns, and in Villages," &c. With numerous Illustrations. Crown 8vo, 7s. 6d. cloth. [*Just published.*

" As a small practical treatise on the water supply of towns, and on some applications of water-power, the work is in many respects excellent."—*Engineering.*
" The author has collated the results deduced from the experiments of the most eminent authorities, and has presented them in a compact and practical form, accompanied by very clear and detailed explanations. . . . The application of water as a motive power is treated very carefully and exhaustively "—*Builder.*
" For anyone who desires to begin the study of hydraulics with a consideration of the practical applications of the science there is no better guide. '—*Architect.*

Statics, Graphic and Analytic.

GRAPHIC AND ANALYTIC STATICS, in their Practical Application to the Treatment of Stresses in Roofs, Solid Girders, Lattice, Bowstring and Suspension Bridges, Braced Iron Arches and Piers, and other Frameworks. By R. HUDSON GRAHAM, C.E. Containing Diagrams and Plates to Scale. With numerous Examples, many taken from existing Structures. Specially arranged for Class-work in Colleges and Universities. Second Edition, Revised and Enlarged. 8vo, 16s. cloth.

"Mr. Graham's book will find a place wherever graphic and analytic statics are used or studied."—*Engineer*.
"The work is excellent from a practical point of view, and has evidently been prepared with much care. The directions for working are ample, and are illustrated by an abundance of well-selected examples. It is an excellent text-book for the practical draughtsman."—*Athenæum*.

Student's Text-Book on Surveying.

PRACTICAL SURVEYING: A Text-Book for Students preparing for Examination or for Survey-work in the Colonies. By GEORGE W. USILL, A.M.I.C.E., Author of "The Statistics of the Water Supply of Great Britain." With Four Lithographic Plates and upwards of 330 Illustrations. Crown 8vo, 7s. 6d. cloth. [*Just published.*

"The best forms of instruments are described as to their construction, uses and modes of employment, and there are innumerable hints on work and equipment such as the author, in his experience as surveyor, draughtsman and teacher, has found necessary, and which the student in his inexperience will find most serviceable."—*Engineer*.
"We have no hesitation in saying that the student will find this treatise a better guide than any of its predecessors.... It deserves to be recognised as the first book which should be put in the hands of a pupil of Civil Engineering, and every gentleman of education who sets out for the Colonies would find it well to have a copy."—*Architect*.
"A very useful, practical handbook on field practice. Clear, accurate and not too condensed."—*Journal of Education*.

Survey Practice.

AID TO SURVEY PRACTICE, for Reference in Surveying, Levelling, Setting-out and in Route Surveys of Travellers by Land and Sea. With Tables, Illustrations, and Records. By LOWIS D'A. JACKSON, A.M.I.C.E., Author of "Hydraulic Manual," "Modern Metrology," &c. Second Edition, Enlarged. Large crown 8vo, 12s. 6d. cloth.

"Mr. Jackson has produced a valuable *vade-mecum* for the surveyor. We can recommend this book as containing an admirable supplement to the teaching of the accomplished surveyor."—*Athenæum*.
"As a text-book we should advise all surveyors to place it in their libraries, and study well the matured instructions afforded in its pages."—*Colliery Guardian*.
"The author brings to his work a fortunate union of theory and practical experience which, aided by a clear and lucid style of writing, renders the book a very useful one."—*Builder*.

Surveying, Land and Marine.

LAND AND MARINE SURVEYING, in Reference to the Preparation of Plans for Roads and Railways; Canals, Rivers, Towns' Water Supplies; Docks and Harbours. With Description and Use of Surveying Instruments. By W. D. HASKOLL, C.E., Author of "Bridge and Viaduct Construction," &c. Second Edition, with Additions. Large crown 8vo, 9s. cloth.

"This book must prove of great value to the student. We have no hesitation in recommending it, feeling assured that it will more than repay a careful study."—*Mechanical World*.
"We can strongly recommend it as a carefully-written and valuable text-book. It enjoys a well-deserved repute among surveyors."—*Builder*.
"This volume cannot fail to prove of the utmost practical utility. It may be safely recommended to all students who aspire to become clean and expert surveyors."—*Mining Journal*.

Tunnelling.

PRACTICAL TUNNELLING. Explaining in detail the Setting-out of the works, Shaft-sinking and Heading-driving, Ranging the Lines and Levelling underground, Sub-Excavating, Timbering, and the Construction of the Brickwork of Tunnels, with the amount of Labour required for, and the Cost of, the various portions of the work. By FREDERICK W. SIMMS, F.G.S., M.Inst.C.E. Third Edition, Revised and Extended by D. KINNEAR CLARK, M.Inst.C.E.; Imperial 8vo, with 21 Folding Plates and numerous Wood Engravings, 30s. cloth.

"The estimation in which Mr. Simms's book on tunnelling has been held for over thirty years cannot be more truly expressed than in the words of the late Prof. Rankine:—'The best source of information on the subject of tunnels is Mr. F. W. Simms's work on Practical Tunnelling.'"—*Architect*.
"It has been regarded from the first as a text book of the subject. ... Mr. Clarke has added immensely to the value of the book."—*Engineer*.

CIVIL ENGINEERING, SURVEYING, etc. 11

Levelling.
A TREATISE ON THE PRINCIPLES AND PRACTICE OF LEVELLING. Showing its Application to purposes of Railway and Civil Engineering, in the Construction of Roads; with Mr. TELFORD'S Rules for the same. By FREDERICK W. SIMMS, F.G.S., M.Inst.C.E. Seventh Edition, with the addition of LAW's Practical Examples for Setting-out Railway Curves, and TRAUTWINE'S Field Practice of Laying-out Circular Curves. With 7 Plates and numerous Woodcuts, 8vo, 8s. 6d. cloth. *⁎* TRAUTWINE on Curves may be had separate, 5s.

"The text-book on levelling in most of our engineering schools and colleges."—*Engineer.*
"The publishers have rendered a substantial service to the profession, especially to the younger members, by bringing out the present edition of Mr. Simms's useful work."—*Engineering.*

Heat, Expansion by.
EXPANSION OF STRUCTURES BY HEAT. By JOHN KEILY, C.E., late of the Indian Public Works and Victorian Railway Departments. Crown 8vo, 3s. 6d. cloth.

SUMMARY OF CONTENTS.
Section I. FORMULAS AND DATA.
Section II. METAL BARS.
Section III. SIMPLE FRAMES.
Section IV. COMPLEX FRAMES AND PLATES.
Section V. THERMAL CONDUCTIVITY.
Section VI. MECHANICAL FORCE OF HEAT.
Section VII. WORK OF EXPANSION AND CONTRACTION.
Section VIII. SUSPENSION BRIDGES.
Section IX. MASONRY STRUCTURES.

"The aim the author has set before him, viz., to show the effects of heat upon metallic and other structures, is a laudable one, for this is a branch of physics upon which the engineer or architect can find but little reliable and comprehensive data in books."—*Builder.*
"Whoever is concerned to know the effect of changes of temperature on such structures as suspension bridges and the like, could not do better than consult Mr. Keily's valuable and handy exposition of the geometrical principles involved in these changes."—*Scotsman.*

Practical Mathematics.
MATHEMATICS FOR PRACTICAL MEN: Being a Commonplace Book of Pure and Mixed Mathematics. Designed chiefly for the use of Civil Engineers, Architects and Surveyors. By OLINTHUS GREGORY, LL.D., F.R.A.S., Enlarged by HENRY LAW, C.E. 4th Edition, carefully Revised by J. R. YOUNG, formerly Professor of Mathematics, Belfast College. With 13 Plates, 8vo, £1 1s. cloth.

"The engineer or architect will here find ready to his hand rules for solving nearly every mathematical difficulty that may arise in his practice The rules are in all cases explained by means of examples, in which every step of the process is clearly worked out."—*Builder.*
"It is an instructive book for the student, and a text-book for him who, having once mastered the subjects it treats of, needs occasionally to refresh his memory upon them."—*Building News.*

Hydraulic Tables.
HYDRAULIC TABLES, CO-EFFICIENTS, and FORMULÆ *for finding the Discharge of Water from Orifices, Notches, Weirs, Pipes, and Rivers.* With New Formulæ, Tables, and General Information on Rainfall, Catchment-Basins, Drainage, Sewerage, Water Supply for Towns and Mill Power. By JOHN NEVILLE, Civil Engineer, M.R.I.A. Third Edition, carefully Revised, with Additions. Numerous Illustrations. Cr. 8vo, 14s. cloth.

"Alike valuable to students and engineers in practice, its study will prevent the annoyance of avoidable failures, and assist them to select the readiest means of successfully carrying out any given work connected with hydraulic engineering."—*Mining Journal.*
"It is, of all English books on the subject, the one nearest to completeness. . . . From the good arrangement of the matter, the clear explanations, and abundance of formulæ, the carefully calculated tables, and, above all, the thorough acquaintance with both theory and construction, which is displayed from first to last, the book will be found to be an acquisition."—*Architect.*

Hydraulics.
HYDRAULIC MANUAL. Consisting of Working Tables and Explanatory Text. Intended as a Guide in Hydraulic Calculations and Field Operations. By LOWIS D'A. JACKSON, Author of "Aid to Survey Practice," "Modern Metrology," &c. Fourth Edition, Enlarged. Large cr. 8vo, 16s. cl.

"The author has had a wide experience in hydraulic engineering and has been a careful observer of the facts which have come under his notice, and from the great mass of material at his command he has constructed a manual which may be accepted as a trustworthy guide to this branch of the engineer's profession. We can heartily recommend this volume to all who desire to be acquainted with the latest development of this important subject."—*Engineering.*
"The most useful feature of this work is its freedom from what is superannuated, and its thorough adoption of recent experiments; the text is, in fact, in great part a short account of the great modern experiments."—*Nature.*

Drainage.

ON THE DRAINAGE OF LANDS, TOWNS AND BUILDINGS. By G. D. DEMPSEY, C.E., Author of "The Practical Railway Engineer," &c. Revised, with large Additions on RECENT PRACTICE IN DRAINAGE ENGINEERING, by D. KINNEAR CLARK, M.Inst.C.E. Author of "Tramways: Their Construction and Working," "A Manual of Rules, Tables, and Data for Mechanical Engineers," &c. &c. Crown 8vo, 7s. 6d. cloth.

"The new matter added to Mr. Dempsey's excellent work is characterised by the comprehensive grasp and accuracy of detail for which the name of Mr. D. K. Clark is a sufficient voucher."—*Athenæum.*

"As a work on recent practice in drainage engineering, the book is to be commended to all who are making that branch of engineering science their special study."—*Iron.*

"A comprehensive manual on drainage engineering, and a useful introduction to the student." *Building News.*

Tramways and their Working.

TRAMWAYS: THEIR CONSTRUCTION AND WORKING. Embracing a Comprehensive History of the System; with an exhaustive Analysis of the various Modes of Traction, including Horse-Power, Steam, Heated Water, and Compressed Air; a Description of the Varieties of Rolling Stock; and ample Details of Cost and Working Expenses: the Progress recently made in Tramway Construction, &c. &c. By D. KINNEAR CLARK, M.Inst.C.E. With over 200 Wood Engravings, and 13 Folding Plates. Two Vols., large crown 8vo, 30s. cloth.

"All interested in tramways must refer to it, as all railway engineers have turned to the author's work 'Railway Machinery.'"—*Engineer.*

"An exhaustive and practical work on tramways, in which the history of this kind of locomotion, and a description and cost of the various modes of laying tramways, are to be found."—*Building News.*

"The best form of rails, the best mode of construction, and the best mechanical appliances are so fairly indicated in the work under review, that any engineer about to construct a tramway will be enabled at once to obtain the practical information which will be of most service to him."—*Athenæum.*

Oblique Arches.

A PRACTICAL TREATISE ON THE CONSTRUCTION OF OBLIQUE ARCHES. By JOHN HART. Third Edition, with Plates. Imperial 8vo, 8s. cloth.

Curves, Tables for Setting-out.

TABLES OF TANGENTIAL ANGLES AND MULTIPLES *for Setting-out Curves from* 5 *to* 200 *Radius.* By ALEXANDER BEAZELEY, M.Inst.C.E. Third Edition. Printed on 48 Cards, and sold in a cloth box, waistcoat-pocket size, 3s. 6d.

"Each table is printed on a small card, which, being placed on the theodolite, leaves the hands free to manipulate the instrument—no small advantage as regards the rapidity of work."—*Engineer.*

"Very handy; a man may know that all his day's work must fall on two of these cards, which he puts into his own card-case, and leaves the rest behind."—*Athenæum.*

Earthwork.

EARTHWORK TABLES. Showing the Contents in Cubic Yards of Embankments, Cuttings, &c., of Heights or Depths up to an average of 80 feet. By JOSEPH BROADBENT, C.E., and FRANCIS CAMPIN, C.E. Crown 8vo, 5s. cloth.

"The way in which accuracy is attained, by a simple division of each cross section into three elements, two in which are constant and one variable, is ingenious."—*Athenæum.*

Tunnel Shafts.

THE CONSTRUCTION OF LARGE TUNNEL SHAFTS: *A Practical and Theoretical Essay.* By J. H. WATSON BUCK, M.Inst.C.E., Resident Engineer, London and North-Western Railway. Illustrated with Folding Plates, royal 8vo, 12s. cloth.

"Many of the methods given are of extreme practical value to the mason; and the observations on the form of arch, the rules for ordering the stone, and the construction of the templates will be found of considerable use. We commend the book to the engineering profession."—*Building News.*

"Will be regarded by civil engineers as of the utmost value, and calculated to save much time and obviate many mistakes."—*Colliery Guardian.*

Girders, Strength of.

GRAPHIC TABLE FOR FACILITATING THE COMPUTATION OF THE WEIGHTS OF WROUGHT IRON AND STEEL GIRDERS, *etc.,* for Parliamentary and other Estimates. By J. H. WATSON BUCK, M.Inst.C.E. On a Sheet, 2s. 6d.

CIVIL ENGINEERING, SURVEYING, etc.   13

River Engineering.

RIVER BARS: The Causes of their Formation, and their Treatment by " Induced Tidal Scour;" with a Description of the Successful Reduction by this Method of the Bar at Dublin. By A. J. MANN, Assist. Eng. to the Dublin Port and Docks Board. Royal 8vo, 7s. 6d. cloth.

"We recommend all interested in harbour works—and, indeed, those concerned in the improvements of rivers generally—to read Mr. Mann's interesting work on the treatment of river bars."—*Engineer*.

Trusses.

TRUSSES OF WOOD AND IRON. Practical Applications of Science in Determining the Stresses, Breaking Weights, Safe Loads, Scantlings, and Details of Construction, with Complete Working Drawings. By WILLIAM GRIFFITHS, Surveyor, Assistant Master, Tranmere School of Science and Art. Oblong 8vo, 4s. 6d. cloth.

"This handy little book enters so minutely into every detail connected with the construction of roof trusses, that no student need be ignorant of these matters."—*Practical Engineer*.

Railway Working.

SAFE RAILWAY WORKING. A Treatise on Railway Accidents: Their Cause and Prevention; with a Description of Modern Appliances and Systems. By CLEMENT E. STRETTON, C.E., Vice-President and Consulting Engineer, Amalgamated Society of Railway Servants. With Illustrations and Coloured Plates, crown 8vo, 4s. 6d. strongly bound.

"A book for the engineer, the directors, the managers; and, in short, all who wish for information on railway matters will find a perfect encyclopædia in 'Safe Railway Working.' "—*Railway Review*.

"We commend the remarks on railway signalling to all railway managers, especially where a uniform code and practice is advocated."—*Herepath's Railway Journal*.

"The author may be congratulated on having collected, in a very convenient form, much valuable information on the principal questions affecting the safe working of railways."—*Railway Engineer*.

Field-Book for Engineers.

THE ENGINEER'S, MINING SURVEYOR'S, AND CONTRACTOR'S FIELD-BOOK. Consisting of a Series of Tables, with Rules, Explanations of Systems, and use of Theodolite for Traverse Surveying and Plotting the Work with minute accuracy by means of Straight Edge and Set Square only; Levelling with the Theodolite, Casting-out and Reducing Levels to Datum, and Plotting Sections in the ordinary manner; setting-out Curves with the Theodolite by Tangential Angles and Multiples, with Right and Left-hand Readings of the Instrument: Setting-out Curves without Theodolite, on the System of Tangential Angles by sets of Tangents and Offsets: and Earthwork Tables to 80 feet deep, calculated for every 6 inches in depth. By W. DAVIS HASKOLL, C.E. With numerous Woodcuts. Fourth Edition, Enlarged. Crown 8vo, 12s. cloth.

"The book is very handy; the separate tables of sines and tangents to every minute will make it useful for many other purposes, the genuine traverse tables existing all the same."—*Athenæum*.

"Every person engaged in engineering field operations will estimate the importance of such a work and the amount of valuable time which will be saved by reference to a set of reliable tables prepared with the accuracy and fulness of those given in this volume."—*Railway News*.

Earthwork, Measurement of.

A MANUAL ON EARTHWORK. By ALEX. J. S. GRAHAM, C.E. With numerous Diagrams. 18mo, 2s. 6d. cloth.

"A great amount of practical information, very admirably arranged, and available for rough estimates, as well as for the more exact calculations required in the engineer's and contractor's offices."—*Artisan*.

Strains in Ironwork.

THE STRAINS ON STRUCTURES OF IRONWORK; with Practical Remarks on Iron Construction. By F. W. SHEILDS, M.Inst.C.E. Second Edition, with 5 Plates. Royal 8vo, 5s. cloth.

"The student cannot find a better little book on this subject."—*Engineer*.

Cast Iron and other Metals, Strength of.

A PRACTICAL ESSAY ON THE STRENGTH OF CAST IRON AND OTHER METALS. By THOMAS TREDGOLD, C.E. Fifth Edition, including HODGKINSON'S Experimental Researches. 8vo, 12s. cloth.

ARCHITECTURE, BUILDING, etc.

Construction.
THE SCIENCE OF BUILDING: An Elementary Treatise on the Principles of Construction. By E. WYNDHAM TARN, M.A., Architect. Second Edition, Revised, with 58 Engravings. Crown 8vo, 7s. 6d. cloth.
"A very valuable book, which we strongly recommend to all students."—*Builder.*
"No architectural student should be without this handbook of constructional knowledge."—*Architect.*

Villa Architecture.
A HANDY BOOK OF VILLA ARCHITECTURE: Being a Series of Designs for Villa Residences in various Styles. With Outline Specifications and Estimates. By C. WICKES, Architect, Author of "The Spires and Towers of England," &c. 61 Plates, 4to, £1 11s. 6d. half-morocco, gilt edges.
"The whole of the designs bear evidence of their being the work of an artistic architect, and they will prove very valuable and suggestive."—*Building News.*

Text-Book for Architects.
THE ARCHITECT'S GUIDE: Being a Text-Book of Useful Information for Architects, Engineers, Surveyors, Contractors, Clerks of Works, &c. &c. By FREDERICK ROGERS, Architect, Author of "Specifications for Practical Architecture," &c. Second Edition, Revised and Enlarged. With numerous Illustrations. Crown 8vo, 6s. cloth.
"As a text-book of useful information for architects, engineers, surveyors, &c., it would be hard to find a handier or more complete little volume."—*Standard.*
"A young architect could hardly have a better guide-book."—*Timber Trades Journal.*

Taylor and Cresy's Rome.
THE ARCHITECTURAL ANTIQUITIES OF ROME. By the late G. L. TAYLOR, Esq., F.R.I.B.A., and EDWARD CRESY, Esq. New Edition, thoroughly Revised by the Rev. ALEXANDER TAYLOR, M.A. (son of the late G. L. Taylor, Esq.), Fellow of Queen's College, Oxford, and Chaplain of Gray's Inn. Large folio, with 130 Plates, half-bound, £3 3s.
N.B.—*This is the only book which gives on a large scale, and with the precision of architectural measurement, the principal Monuments of Ancient Rome in plan, elevation, and detail.*
"Taylor and Cresy's work has from its first publication been ranked among those professional books which cannot be bettered. . . . It would be difficult to find examples of drawings, even among those of the most painstaking students of Gothic, more thoroughly worked out than are the one hundred and thirty plates in this volume."—*Architect.*

Architectural Drawing.
PRACTICAL RULES ON DRAWING, for the Operative Builder and Young Student in Architecture. By GEORGE PYNE. With 14 Plates, 4to, 7s. 6d. boards.

Civil Architecture.
THE DECORATIVE PART OF CIVIL ARCHITECTURE. By Sir WILLIAM CHAMBERS, F.R.S. With Illustrations, Notes, and an Examination of Grecian Architecture, by JOSEPH GWILT, F.S.A. Edited by W. H. LEEDS. 66 Plates, 4to, 21s. cloth.

House Building and Repairing.
THE HOUSE-OWNER'S ESTIMATOR; or, What will it Cost to Build, Alter, or Repair? A Price Book adapted to the Use of Unprofessional People, as well as for the Architectural Surveyor and Builder. By JAMES D. SIMON, A.R.I.B.A. Edited and Revised by FRANCIS T. W. MILLER, A.R.I.B.A. With numerous Illustrations. Fourth Edition, Revised. Crown 8vo, 3s. 6d. cloth. [*Just published.*
"In two years it will repay its cost a hundred times over"—*Field.*
"A very handy book."—*English Mechanic.*

Designing, Measuring, and Valuing.

THE STUDENT'S GUIDE to the PRACTICE of MEASURING AND VALUING ARTIFICERS' WORKS. Containing Directions for taking Dimensions, Abstracting the same, and bringing the Quantities into Bill, with Tables of Constants for Valuation of Labour, and for the Calculation of Areas and Solidities. Originally edited by EDWARD DOBSON, Architect. Revised, with considerable Additions on Mensuration and Construction, and a New Chapter on Dilapidations, Repairs, and Contracts, by E. WYNDHAM TARN, M.A. Sixth Edition, including a Complete Form of a Bill of Quantities. With 8 Plates and 63 Woodcuts. Crown 8vo, 7s. 6d. clo [*Just published.*

"Well fulfils the promise of its title-page, and we can thoroughly recommend it to the class for whose use it has been compiled. Mr. Tarn's additions and revisions have much increased the usefulness of the work, and have especially augmented its value to students."—*Engineering.*
"This edition will be found the most complete treatise on the principles of measuring and valuing artificers' work that has yet been published."—*Building News.*

Pocket Estimator and Technical Guide.

THE POCKET TECHNICAL GUIDE, MEASURER AND ESTIMATOR FOR BUILDERS AND SURVEYORS. Containing Technical Directions for Measuring Work in all the Building Trades, with a Treatise on the Measurement of Timber and Complete Specifications for Houses, Roads, and Drains, and an easy Method of Estimating the various parts of a Building collectively. By A. C. BEATON, Author of "Quantities and Measurements," &c. Fifth Edition, carefully Revised and Priced according to the Present Value of Materials and Labour, with 53 Woodcuts, leather, waistcoat-pocket size, 1s. 6d. gilt edges. [*Just published.*

"No builder, architect, surveyor, or valuer should be without his 'Beaton.'"—*Building News.*
"Contains an extraordinary amount of information in daily requisition in measuring and estimating. Its presence in the pocket will save valuable time and trouble."—*Building World.*

Donaldson on Specifications.

THE HANDBOOK OF SPECIFICATIONS; or, Practical Guide to the Architect, Engineer, Surveyor, and Builder, in drawing up Specifications and Contracts for Works and Constructions. Illustrated by Precedents of Buildings actually executed by eminent Architects and Engineers. By Professor T. L. DONALDSON, P.R.I.B.A., &c. New Edition, in One large Vol., 8vo, with upwards of 1,000 pages of Text, and 33 Plates, £1 11s. 6d. cloth.

"In this work forty-four specifications of executed works are given, including the specifications for parts of the new Houses of Parliament, by Sir Charles Barry, and for the new Royal Exchange, by Mr. Tite, M.P. The latter, in particular, is a very complete and remarkable document. It embodies, as Mr. Donaldson mentions, 'the bill of quantities with the description of the works.' . . . It is valuable as a record, and more valuable still as a book of precedents. . . . Suffice it to say that Donaldson's 'Handbook of Specifications' must be bought by all architects."—*Builder.*

Bartholomew and Rogers' Specifications.

SPECIFICATIONS FOR PRACTICAL ARCHITECTURE. A Guide to the Architect, Engineer, Surveyor, and Builder. With an Essay on the Structure and Science of Modern Buildings. Upon the Basis of the Work by ALFRED BARTHOLOMEW, thoroughly Revised, Corrected, and greatly added to by FREDERICK ROGERS, Architect. Second Edition, Revised, with Additions. With numerous Illustrations, medium 8vo, 15s. cloth.

"The collection of specifications prepared by Mr. Rogers on the basis of Bartholomew's work is too well known to need any recommendation from us. It is one of the books with which every young architect must be equipped; for time has shown that the specifications cannot be set aside through any defect in them."—*Architect.*
"Good forms for specifications are of considerable value, and it was an excellent idea to compile a work on the subject upon the basis of the late Alfred Bartholomew's valuable work. The second edition of Mr. Rogers's book is evidence of the want of a book dealing with modern requirements and materials."—*Building News.*

Building; Civil and Ecclesiastical.

A BOOK ON BUILDING, Civil and Ecclesiastical, including Church Restoration; with the Theory of Domes and the Great Pyramid, &c. By Sir EDMUND BECKETT, Bart., LL.D., F.R.A.S., Author of "Clocks and Watches, and Bells," &c. Second Edition, Enlarged. Fcap. 8vo, 5s. cloth.

"A book which is always amusing and nearly always instructive. The style throughout is in the highest degree condensed and epigrammatic."—*Times.*

16   CROSBY LOCKWOOD & SON'S CATALOGUE.

Geometry for the Architect, Engineer, etc.

PRACTICAL GEOMETRY, for the Architect, Engineer and Mechanic. Giving Rules for the Delineation and Application of various Geometrical Lines, Figures and Curves. By E. W. TARN, M.A., Architect, Author of "The Science of Building," &c. Second Edition. With Appendices on Diagrams of Strains and Isometrical Projection. With 172 Illustrations, demy 8vo, 9s. cloth.

"No book with the same objects in view has ever been published in which the clearness of the rules laid down and the illustrative diagrams have been so satisfactory."—*Scotsman*.
"This is a manual for the practical man, whether architect, engineer, or mechanic. . . . The object of the author being to avoid all abstruse formulæ or complicated methods, and to enable persons with but a moderate knowledge of geometry to work out the problems required."—*English Mechanic*.

The Science of Geometry.

THE GEOMETRY OF COMPASSES; or, Problems Resolved by the mere Description of Circles, and the use of Coloured Diagrams and Symbols. By OLIVER BYRNE. Coloured Plates. Crown 8vo, 3s. 6d. cloth.

"The treatise is a good one, and remarkable—like all Mr. Byrne's contributions to the science of geometry—for the lucid character of its teaching."—*Building News*.

DECORATIVE ARTS, etc.

Woods and Marbles (Imitation of).

SCHOOL OF PAINTING FOR THE IMITATION OF WOODS AND MARBLES, as Taught and Practised by A. R. VAN DER BURG and P. VAN DER BURG, Directors of the Rotterdam Painting Institution. Royal folio, 18¾ by 12½ in., Illustrated with 24 full-size Coloured Plates; also 12 plain Plates, comprising 154 Figures. Second and Cheaper Edition. Price £1 11s. 6d.

List of Plates.

1. Various Tools required for Wood Painting—2, 3. Walnut: Preliminary Stages of Graining and Finished Specimen—4. Tools used for Marble Painting and Method of Manipulation—5, 6. St. Remi Marble: Earlier Operations and Finished Specimen—7. Methods of Sketching different Grains, Knots, &c.—8, 9. Ash: Preliminary Stages and Finished Specimen—10. Methods of Sketching Marble Grains—11, 12. Breche Marble: Preliminary Stages of Working and Finished Specimen—13. Maple: Methods of Producing the different Grains—14, 15. Bird's-eye Maple: Preliminary Stages and Finished Specimen—16. Methods of Sketching the different Species of White Marble—17, 18. White Marble: Preliminary Stages of Process and Finished Specimen—19. Mahogany: Specimens of various Grains and Methods of Manipulation—20, 21. Mahogany: Earlier Stages and Finished Specimen—22, 23, 24. Sienna Marble: Varieties of Grain, Preliminary Stages and Finished Specimen—25, 26, 27. Juniper Wood: Methods of producing Grain, &c.: Preliminary Stages and Finished Specimen—28, 29, 30. Vert de Mer Marble: Varieties of Grain and Methods of Working Unfinished and Finished Specimens—31, 32, 33. Oak: Varieties of Grain, Tools Employed, and Methods of Manipulation, Preliminary Stages and Finished Specimen—34, 35, 36. Waulsort Marble: Varieties of Grain, Unfinished and Finished Specimens.

*** OPINIONS OF THE PRESS.

"Those who desire to attain skill in the art of painting woods and marbles will find advantage in consulting this book. . . . Some of the Working Men's Clubs should give their young men the opportunity to study it."—*Builder*.
"A comprehensive guide to the art. The explanations of the processes, the manipulation and management of the colours, and the beautifully executed plates will not be the least valuable to the student who aims at making his work a faithful transcript of nature."—*Building News*.
"Students and novices are fortunate who are able to become the possessors of so noble a work."—*Architect*.

House Decoration.

ELEMENTARY DECORATION. A Guide to the Simpler Forms of Everyday Art, as applied to the Interior and Exterior Decoration of Dwelling Houses, &c. By JAMES W. FACEY, Jun. With 68 Cuts. 12mo, 2s. cloth limp.

"As a technical guide-book to the decorative painter it will be found reliable."—*Building News*.

PRACTICAL HOUSE DECORATION : A Guide to the Art of Ornamental Painting, the Arrangement of Colours in Apartments, and the principles of Decorative Design. With some Remarks upon the Nature and Properties of Pigments. By JAMES WILLIAM FACEY, Author of "Elementary Decoration," &c. With numerous Illustrations. 12mo, 2s. 6d. cloth limp.

N.B.—*The above Two Works together in One Vol., strongly half-bound, 5s.*

Colour.

A GRAMMAR OF COLOURING. Applied to Decorative Painting and the Arts. By GEORGE FIELD. New Edition, Revised, Enlarged, and adapted to the use of the Ornamental Painter and Designer. By ELLIS A. DAVIDSON. With New Coloured Diagrams and Engravings. 12mo, 3s. 6d. cloth boards.

"The book is a most useful resume of the properties of pigments."—*Builder.*

House Painting, Graining, etc.

HOUSE PAINTING, GRAINING, MARBLING, AND SIGN WRITING, A Practical Manual of. By ELLIS A. DAVIDSON. Fifth Edition. With Coloured Plates and Wood Engravings. 12mo, 6s. cloth boards.

"A mass of information, of use to the amateur and of value to the practical man."—*English Mechanic.*
"Simply invaluable to the youngster entering upon this particular calling, and highly serviceable to the man who is practising it."—*Furniture Gazette.*

Decorators, Receipts for.

THE DECORATOR'S ASSISTANT : A Modern Guide to Decorative Artists and Amateurs, Painters, Writers, Gilders, &c. Containing upwards of 600 Receipts, Rules and Instructions ; with a variety of Information for General Work connected with every Class of Interior and Exterior Decorations, &c. Third Edition, Revised. 152 pp., crown 8vo, 1s. in wrapper.

"Full of receipts of value to decorators, painters, gilders, &c. The book contains the gist of larger treatises on colour and technical processes. It would be difficult to meet with a work so full of varied information on the painter's art."—*Building News.*
"We recommend the work to all who, whether for pleasure or profit, require a guide to decoration."—*Plumber and Decorator.*

Moyr Smith on Interior Decoration.

ORNAMENTAL INTERIORS, ANCIENT AND MODERN. By J. MOYR SMITH. Super-royal 8vo, with 32 full-page Plates and numerous smaller Illustrations, handsomely bound in cloth, gilt top, price 18s.

☞ *In* "ORNAMENTAL INTERIORS" *the designs of more than thirty artist-decorators and architects of high standing have been illustrated. The book may therefore fairly claim to give a good general view of the works of the modern school of decoration, besides giving characteristic examples of earlier decorative arrangements.*

"ORNAMENTAL INTERIORS" *gives a short account of the styles of Interior Decoration as practised by the Ancients in Egypt, Greece, Assyria, Rome and Byzantium. This part is illustrated by characteristic designs.*

*** OPINIONS OF THE PRESS.
"The book is well illustrated and handsomely got up, and contains some true criticism and a good many good examples of decorative treatment."—*The Builder.*
"Well fitted for the dilettante, amateur, and professional designer."—*Decoration.*
"This is the most elaborate, and beautiful work on the artistic decoration of interiors that we have seen. . . . The scrolls, panels and other designs from the author's own pen are very beautiful and chaste ; but he takes care that the designs of other men shall figure even more than his own."—*Liverpool Albion.*
"To all who take an interest in elaborate domestic ornament this handsome volume will be welcome."—*Graphic.*
"Mr. Moyr Smith deserves the thanks of art workers for having placed within their reach a book that seems eminently adapted to afford, by example and precept, that guidance of which most craftsmen stand in need."—*Furniture Gazette.*

British and Foreign Marbles.

MARBLE DECORATION and the Terminology of British and Foreign Marbles. A Handbook for Students. By GEORGE H. BLAGROVE, Author of " Shoring and its Application," &c. With 28 Illustrations. Crown 8vo, 3s. 6d. cloth.

"This most useful and much wanted handbook should be in the hands of every architect and builder."—*Building World.*
"It is an excellent manual for students, and interesting to artistic readers generally."—*Saturday Review.*
"A carefully and usefully written eatise ; the work is essentially practical."—*Scotsman.*

Marble Working, etc.

MARBLE AND MARBLE WORKERS : A Handbook for Architects, Artists, Masons and Students. By ARTHUR LEE, Author of " A Visit to Carrara," " The Working of Marble," &c. Small crown 8vo, 2s. cloth .

"A really valuable addition to the technical literature of architects and masons."—*Building News.*

c

DELAMOTTE'S WORKS ON ILLUMINATION AND ALPHABETS.

A PRIMER OF THE ART OF ILLUMINATION, for the Use of Beginners: with a Rudimentary Treatise on the Art, Practical Directions for its exercise, and Examples taken from Illuminated MSS., printed in Gold and Colours. By F. DELAMOTTE. New and Cheaper Edition. Small 4to, 6s. ornamental boards.

"The examples of ancient MSS. recommended to the student, which, with much good sense, the author chooses from collections accessible to all, are selected with judgment and knowledge, as well as taste."—*Athenæum.*

ORNAMENTAL ALPHABETS, Ancient and Mediæval, from the Eighth Century, with Numerals; including Gothic, Church-Text, large and small, German, Italian, Arabesque, Initials for Illumination, Monograms, Crosses, &c. &c., for the use of Architectural and Engineering Draughtsmen, Missal Painters, Masons, Decorative Painters, Lithographers, Engravers, Carvers, &c. &c. Collected and Engraved by F. DELAMOTTE, and printed in Colours. New and Cheaper Edition. Royal 8vo, oblong, 2s. 6d. ornamental boards.

"For those who insert enamelled sentences round gilded chalices, who blazon shop legends over shop-doors, who letter church walls with pithy sentences from the Decalogue, this book will be useful."—*Athenæum.*

EXAMPLES OF MODERN ALPHABETS, Plain and Ornamental; including German, Old English, Saxon, Italic, Perspective, Greek, Hebrew, Court Hand, Engrossing, Tuscan, Riband, Gothic, Rustic, and Arabesque; with several Original Designs, and an Analysis of the Roman and Old English Alphabets, large and small, and Numerals, for the use of Draughtsmen, Surveyors, Masons, Decorative Painters, Lithographers, Engravers, Carvers, &c. Collected and Engraved by F. DELAMOTTE, and printed in Colours. New and Cheaper Edition. Royal 8vo, oblong, 2s. 6d. ornamental boards.

"There is comprised in it every possible shape into which the letters of the alphabet and numerals can be formed, and the talent which has been expended in the conception of the various plain and ornamental letters is wonderful."—*Standard.*

MEDIÆVAL ALPHABETS AND INITIALS FOR ILLUMINATORS. By G. DELAMOTTE. Containing 21 Plates and Illuminated Title, printed in Gold and Colours. With an Introduction by J. WILLIS BROOKS. Fourth and Cheaper Edition. Small 4to, 4s. ornamental boards.

"A volume in which the letters of the alphabet come forth glorified in gilding and all the colours of the prism interwoven and intertwined and intermingled."—*Sun.*

THE EMBROIDERER'S BOOK OF DESIGN. Containing Initials, Emblems, Cyphers, Monograms, Ornamental Borders, Ecclesiastical Devices, Mediæval and Modern Alphabets, and National Emblems. Collected by F. DELAMOTTE, and printed in Colours. Oblong royal 8vo, 1s. 6d. ornamental wrapper.

"The book will be of great assistance to ladies and young children who are endowed with the art of plying the needle in this most ornamental and useful pretty work."—*East Anglian Times.*

Wood Carving.

INSTRUCTIONS IN WOOD-CARVING, for Amateurs; with Hints on Design. By A LADY. With Ten large Plates, 2s. 6d. in emblematic wrapper.

"The handicraft of the wood-carver, so well as a book can impart it, may be learnt from 'A Lady's' publication."—*Athenæum.*
"The directions given are plain and easily understood."—*English Mechanic.*

Glass Painting.

GLASS STAINING AND THE ART OF PAINTING ON GLASS. From the German of Dr. GESSERT and EMANUEL OTTO FROMBERG. With an Appendix on THE ART OF ENAMELLING. 12mo, 2s. 6d. cloth limp.

Letter Painting.

THE ART OF LETTER PAINTING MADE EASY. By JAMES GREIG BADENOCH. With 12 full-page Engravings of Examples, 1s. 6d. cloth limp.

"The system is a simple one, but quite original, and well worth the careful attention of letter painters. It can be easily mastered and remembered."—*Building News.*

CARPENTRY, TIMBER, etc.

Tredgold's Carpentry, Enlarged by Tarn.
THE ELEMENTARY PRINCIPLES OF CARPENTRY. A Treatise on the Pressure and Equilibrium of Timber Framing, the Resistance of Timber, and the Construction of Floors, Arches, Bridges, Roofs, Uniting Iron and Stone with Timber, &c. To which is added an Essay on the Nature and Properties of Timber, &c., with Descriptions of the kinds of Wood used in Building; also numerous Tables of the Scantlings of Timber for different purposes, the Specific Gravities of Materials, &c. By THOMAS TREDGOLD, C.E. With an Appendix of Specimens of Various Roofs of Iron and Stone, Illustrated. Seventh Edition, thoroughly revised and considerably enlarged by E. WYNDHAM TARN, M.A., Author of "The Science of Building," &c. With 61 Plates, Portrait of the Author, and several Woodcuts. In one large vol., 4to, price £1 5s. cloth.
"Ought to be in every architect's and every builder's library."—*Builder*.
"A work whose monumental excellence must commend it wherever skilful carpentry is concerned. The author's principles are rather confirmed than impaired by time. The additional plates are of great intrinsic value."—*Building News*.

Woodworking Machinery.
WOODWORKING MACHINERY: Its Rise, Progress, and Construction. With Hints on the Management of Saw Mills and the Economical Conversion of Timber. Illustrated with Examples of Recent Designs by leading English, French, and American Engineers. By M. POWIS BALE, A.M.Inst.C.E.,M.I.M.E. Large crown 8vo, 12s. 6d. cloth.
"Mr. Bale is evidently an expert on the subject and he has collected so much information that his book is all-sufficient for builders and others engaged in the conversion of timber."—*Architect*.
"The most comprehensive compendium of wood-working machinery we have seen. The author is a thorough master of his subject."—*Building News*.
"The appearance of this book at the present time will, we should think, give a considerable impetus to the onward march of the machinist engaged in the designing and manufacture of wood-working machines. It should be in the office of every wood-working factory."—*English Mechanic*.

Saw Mills.
SAW MILLS: Their Arrangement and Management, and the Economical Conversion of Timber. (A Companion Volume to "Woodworking Machinery.") By M. POWIS BALE. With numerous Illustrations. Crown 8vo, 10s. 6d. cloth.
"The *administration* of a large sawing establishment is discussed, and the subject examined from a financial standpoint. Hence the size, shape, order, and disposition of saw-mills and the like are gone into in detail, and the course of the timber is traced from its reception to its delivery in its converted state. We could not desire a more complete or practical treatise."—*Builder*.
"We highly recommend Mr. Bale's work to the attention and perusal of all those who are engaged in the art of wood conversion, or who are about building or remodelling saw-mills on improved principles."—*Building News*.

Carpentering.
THE CARPENTER'S NEW GUIDE; or, Book of Lines for Carpenters; comprising all the Elementary Principles essential for acquiring a knowledge of Carpentry. Founded on the late PETER NICHOLSON's Standard Work. A New Edition, Revised by ARTHUR ASHPITEL, F.S.A. Together with Practical Rules on Drawing, by GEORGE PYNE. With 74 Plates, 4to, £1 1s. cloth.

Handrailing.
A PRACTICAL TREATISE ON HANDRAILING: Showing New and Simple Methods for Finding the Pitch of the Plank, Drawing the Moulds, Bevelling, Jointing-up, and Squaring the Wreath. By GEORGE COLLINGS. Illustrated with Plates and Diagrams. 12mo, 1s. 6d. cloth limp.
"Will be found of practical utility in the execution of this difficult branch of joinery."—*Builder*.
"Almost every difficult phase of this somewhat intricate branch of joinery is elucidated by the aid of plates and explanatory letterpress."—*Furniture Gazette*.

Circular Work.
CIRCULAR WORK IN CARPENTRY AND JOINERY: A Practical Treatise on Circular Work of Single and Double Curvature. By GEORGE COLLINGS, Author of "A Practical Treatise on Handrailing." Illustrated with numerous Diagrams. 12mo, 2s. 6d. cloth limp.
"An excellent example of what a book of this kind should be. Cheap in price, clear in definition and practical in the examples selected."—*Builder*.

Timber Merchant's Companion.

THE TIMBER MERCHANT'S AND BUILDER'S COMPANION. Containing New and Copious Tables of the Reduced Weight and Measurement of Deals and Battens, of all sizes, from One to a Thousand Pieces, and the relative Price that each size bears per Lineal Foot to any given Price per Petersburg Standard Hundred; the Price per Cube Foot of Square Timber to any given Price per Load of 50 Feet; the proportionate Value of Deals and Battens by the Standard, to Square Timber by the Load of 50 Feet; the readiest mode of ascertaining the Price of Scantling per Lineal Foot of any size, to any given Figure per Cube Foot, &c. &c. By WILLIAM DOWSING. Fourth Edition, Revised and Corrected. Cr. 8vo, 3s. cl.

"Everything is as concise and clear as it can possibly be made. There can be no doubt that every timber merchant and builder ought to possess it."—*Hull Advertiser.*
"We are glad to see a fourth edition of these admirable tables, which for correctness and simplicity of arrangement leave nothing to be desired."—*Timber Trades Journal.*
"An exceedingly well-arranged, clear, and concise manual of tables for the use of all who buy or sell timber."—*Journal of Forestry.*

Practical Timber Merchant.

THE PRACTICAL TIMBER MERCHANT. Being a Guide for the use of Building Contractors, Surveyors, Builders, &c., comprising useful Tables for all purposes connected with the Timber Trade, Marks of Wood, Essay on the Strength of Timber, Remarks on the Growth of Timber, &c. By W. RICHARDSON. Fcap. 8vo, 3s. 6d. cloth.

"Contains much valuable information for the use of timber merchants, builders, foresters, and all others connected with the growth, sale, and manufacture of timber."—*Journal of Forestry.*

Timber Freight Book.

THE TIMBER MERCHANT'S, SAW MILLER'S, AND IMPORTER'S FREIGHT BOOK AND ASSISTANT. Comprising Rules, Tables, and Memoranda relating to the Timber Trade. By WILLIAM RICHARDSON, Timber Broker; together with a Chapter on "SPEEDS OF SAW MILL MACHINERY," by M. POWIS BALE, M.I.M.E., &c. 12mo, 3s. 6d. cl. boards.

"A very useful manual of rules, tables, and memoranda relating to the timber trade. We recommend it as a compendium of calculation to all timber measurers and merchants, and as supplying a real want in the trade."—*Building News.*

Packing-Case Makers, Tables for.

PACKING-CASE TABLES; showing the number of Superficial Feet in Boxes or Packing-Cases, from six inches square and upwards. By W. RICHARDSON, Timber Broker. Second Edition. Oblong 4to, 3s. 6d. cl.

"Invaluable labour-saving tables."—*Ironmonger.*
"Will save much labour and calculation."—*Grocer.*

Superficial Measurement.

THE TRADESMAN'S GUIDE TO SUPERFICIAL MEASUREMENT. Tables calculated from 1 to 200 inches in length, by 1 to 108 inches in breadth. For the use of Architects, Surveyors, Engineers, Timber Merchants, Builders, &c. By JAMES HAWKINGS. Third Edition. Fcap., 5s. 6d. cloth.

"A useful collection of tables to facilitate rapid calculation of surfaces. The exact area of any surface of which the limits have been ascertained can be instantly determined. The book will be found of the greatest utility to all engaged in building operations."—*Scotsman.*
"These tables will be found of great assistance to all who require to make calculations in superficial measurement."—*English Mechanic.*

Forestry.

THE ELEMENTS OF FORESTRY. Designed to afford Information concerning the Planting and Care of Forest Trees for Ornament or Profit, with Suggestions upon the Creation and Care of Woodlands. By F. B. HOUGH. Large crown 8vo, 10s. cloth.

Timber Importer's Guide.

THE TIMBER IMPORTER'S, TIMBER MERCHANT'S AND BUILDER'S STANDARD GUIDE. By RICHARD E. GRANDY. Comprising an Analysis of Deal Standards, Home and Foreign with Comparative Values and Tabular Arrangements for fixing Nett Landed Cost on Baltic and North American Deals, including all intermediate Expenses, Freight, Insurance, &c. &c. Together with copious Information for the Retailer and Builder. Third Edition, Revised. 12mo, 2s. cloth limp.

"Everything it pretends to be: built up gradually, it leads one from a forest to a treenail, and throws in, as a makeweight, a host of material concerning bricks, columns, cisterns, &c."—*English Mechanic.*

MARINE ENGINEERING, NAVIGATION, etc.

Chain Cables.

CHAIN CABLES AND CHAINS. Comprising Sizes and Curves of Links, Studs, &c., Iron for Cables and Chains, Chain Cable and Chain Making, Forming and Welding Links, Strength of Cables and Chains, Certificates for Cables, Marking Cables, Prices of Chain Cables and Chains, Historical Notes, Acts of Parliament, Statutory Tests, Charges for Testing, List of Manufacturers of Cables, &c. &c. By THOMAS W. TRAILL, F.E.R.N., M.Inst.C.E., Engineer Surveyor in Chief, Board of Trade, Inspector of Chain Cable and Anchor Proving Establishments, and General Superintendent, Lloyd's Committee on Proving Establishments. With numerous Tables, Illustrations and Lithographic Drawings. Folio, £2 2s. cloth.
"It contains a vast amount of valuable information. Nothing seems to be wanting to make it a complete and standard work of reference on the subject."—*Nautical Magazine.*

Marine Engineering.

MARINE ENGINES AND STEAM VESSELS (A Treatise on). By ROBERT MURRAY, C.E. Eighth Edition, thoroughly Revised, with considerable Additions by the Author and by GEORGE CARLISLE, C.E., Senior Surveyor to the Board of Trade at Liverpool. 12mo, 5s. cloth boards.
"Well adapted to give the young steamship engineer or marine engine and boiler maker a general introduction into his practical work."—*Mechanical World.*
"We feel sure that this thoroughly revised edition will continue to be as popular in the future as it has been in the past, as for its size, it contains more useful information than any similar treatise."—*Industries.*
"The information given is both sound and sensible, and well qualified to direct young sea-going hands on the straight road to the extra chief's certificate."—*Glasgow Herald.*
"An indispensable manual for the student of marine engineering."—*Liverpool Mercury.*

Pocket-Book for Naval Architects and Shipbuilders.

THE NAVAL ARCHITECT'S AND SHIPBUILDER'S POCKET-BOOK of Formulæ, Rules, and Tables, and MARINE ENGINEER'S AND SURVEYOR'S Handy Book of Reference. By CLEMENT MACKROW, Member of the Institution of Naval Architects, Naval Draughtsman. Third Edition, Revised. With numerous Diagrams, &c. Fcap., 12s. 6d. leather.
"Should be used by all who are engaged in the construction or design of vessels. . . . Will be found to contain the most useful tables and formulæ required by shipbuilders, carefully collected from the best authorities, and put together in a popular and simple form."—*Engineer.*
"The professional shipbuilder has now, in a convenient and accessible form, reliable data for solving many of the numerous problems that present themselves in the course of his work."—*Iron.*
"There is scarcely a subject on which a naval architect or shipbuilder can require to refresh his memory which will not be found within the covers of Mr. Mackrow's book."—*English Mechanic.*

Pocket-Book for Marine Engineers.

A POCKET-BOOK OF USEFUL TABLES AND FORMULÆ FOR MARINE ENGINEERS. By FRANK PROCTOR, A.I.N.A. Third Edition. Royal 32mo, leather, gilt edges, with strap, 4s.
"We recommend it to our readers as going far to supply a long-felt want."—*Naval Science.*
"A most useful companion to all marine engineers."—*United Service Gazette.*

Introduction to Marine Engineering.

ELEMENTARY ENGINEERING: A Manual for Young Marine Engineers and Apprentices. In the Form of Questions and Answers on Metals, Alloys, Strength of Materials, Construction and Management of Marine Engines, &c. &c. With an Appendix of Useful Tables. By J. S. BREWER, Government Marine Surveyor, Hongkong. Small crown 8vo, 2s. 6d. cloth. [*Just published.*
"Contains much valuable information for the class for whom it is intended, especially in the chapters on the management of boilers and engines."—*Nautical Magazine.*
"A useful introduction to the more elaborate text books."—*Scotsman.*
"To a student who has the requisite desire and resolve to attain a thorough knowledge, Mr. Brewer offers decidedly useful help."—*Athenæum.*

Navigation.

PRACTICAL NAVIGATION. Consisting of THE SAILOR'S SEA-BOOK, by JAMES GREENWOOD and W. H. ROSSER; together with the requisite Mathematical and Nautical Tables for the Working of the Problems. By HENRY LAW, C.E., and Professor J. R. YOUNG. Illustrated. 12mo, 7s. strongly half-bound.

MINING AND MINING INDUSTRIES.

Metalliferous Mining.

BRITISH MINING: A Treatise on the History, Discovery, Practical Development, and Future Prospects of Metalliferous Mines in the United Kingdom. By ROBERT HUNT, F.R.S., Keeper of Mining Records; Editor of "Ure's Dictionary of Arts, Manufactures, and Mines," &c. Upwards of 950 pp., with 230 Illustrations. Second Edition, Revised. Super-royal 8vo, £3 2s. cloth.

"One of the most valuable works of reference of modern times. Mr. Hunt, as keeper of mining records of the United Kingdom, has had opportunities for such a task not enjoyed by anyone else, and has evidently made the most of them. . . . The language and style adopted are good, and the treatment of the various subjects laborious, conscientious, and scientific."—*Engineering*.

"The book is, in fact, a treasure-house of statistical information on mining subjects, and we know of no other work embodying so great a mass of matter of this kind. Were this the only merit of Mr. Hunt's volume, it would be sufficient to render it indispensable in the library of everyone interested in the development of the mining and metallurgical industries of the country."—*Athenæum*.

"A mass of information not elsewhere available, and of the greatest value to those who may be interested in our great mineral industries."—*Engineer*.

"A sound, business-like collection of interesting facts. . . . The amount of information Mr. Hunt has brought together is enormous. . . . The volume appears likely to convey more instruction upon the subject than any work hitherto published."—*Mining Journal*.

"The work will be for the mining industry what Dr. Percy's celebrated treatise has been for the metallurgical—a book that cannot with advantage be omitted from the library."—*Iron and Coal Trades Review*.

"The volume is massive and exhaustive, and the high intellectual powers and patient, persistent application which characterise the author have evidently been brought into play in its production. Its contents are invaluable."—*Colliery Guardian*.

Coal and Iron.

THE COAL AND IRON INDUSTRIES OF THE UNITED KINGDOM. Comprising a Description of the Coal Fields, with Returns ot their Produce and its Distribution, and Analyses of Special Varieties. Also an Account of the occurrence of Iron Ores in Veins or Seams; Analyses of each Variety; and a History of the Rise and Progress of Pig Iron Manufacture since the year 1740. By RICHARD MEADE, Assistant Keeper of Mining Records. With Maps of the Coal Fields and Ironstone Deposits of the United Kingdom. 8vo, £1 8s. cloth.

"The book is one which must find a place on the shelves of all interested in coal and iron production, and in the iron, steel, and other metallurgical industries."—*Engineer*.

"Of this book we may unreservedly say that it is the best of its class which we have ever met. . . . A book of reference which no one engaged in the iron or coal trades should omit from his library."—*Iron and Coal Trades Review*.

"An exhaustive treatise and a valuable work of reference."—*Mining Journal*.

Prospecting for Gold and other Metals.

THE PROSPECTOR'S HANDBOOK: A Guide for the Prospector and Traveller in Search of Metal-Bearing or other Valuable Minerals. By J. W. ANDERSON, M.A. (Camb.), F.R.G.S., Author of "Fiji and New Caledonia." Fourth Edition, thoroughly Revised and Enlarged. Small crown 8vo, 3s. 6d. cloth. [*Just published*.

"Will supply a much felt want, especially among Colonists, n whose way are so often thrown many mineralogical specimens the value of which it is difficult for anyone, not a specialist, to determine. The author has placed his instructions before his readers in the plainest possible terms, and his book is the best of its kind."—*Engineer*.

"How to find commercial minerals, and how to identify them when they are found, are the leading points to which attention is directed. The author has managed to pack as much practical detail into his pages as would supply material for a book three times its size."—*Mining Journal*.

"Those toilers who explore the trodden or untrodden tracks on the face of the globe will find much that is useful to them in this book."—*Athenæum*.

Mining Notes and Formulæ.

NOTES AND FORMULÆ FOR MINING STUDENTS. By JOHN HERMAN MERIVALE, M.A., Certificated Colliery Manager, Professor of Mining in the Durham College of Science, Newcastle-upon-Tyne. Second Edition, carefully Revised. Small crown 8vo, cloth, price 2s. 6d.

"Invaluable to anyone who is working up for an examination on mining subjects."—*Coal and Iron Trades Review*.

"The author has done his work in an exceedingly creditable manner, and has produced a book that will be of service to students, and those who are practically engaged in mining operations."—*Engineer*

"A vast amount of technical matter of the utmost value to mining engineers, and of considerable interest to students."—*Schoolmaster*.

Gold, Metallurgy of.

THE METALLURGY OF GOLD: A Practical Treatise on the Metallurgical Treatment of Gold-bearing Ores. Including the Processes of Concentration and Chlorination, and the Assaying, Melting and Refining of Gold. By M. EISSLER, Mining Engineer and Metallurgical Chemist, formerly Assistant Assayer of the U. S. Mint, San Francisco. Second Edition, Revised and much Enlarged. With 132 Illustrations. Crown 8vo, 9s. cloth.
[*Just published.*
"This book thoroughly deserves its title of a 'Practical Treatise.' The whole process of gold milling, from the breaking of the quartz to the assay of the bullion, is described in clear and orderly narrative and with much, but not too much, fulness of detail."—*Saturday Review*.
"The work is a storehouse of information and valuable data, and we strongly recommend it to all professional men engaged in the gold-mining industry."—*Mining Journal*.
"Anyone who wishes to have an intelligent acquaintance with the characteristics of gold and gold ores, the methods of extracting the metal, concentrating and chlorinating it, and further on of refining and assaying it, will find all he wants in Mr. Eissler's book."—*Financial News*.

Silver, Metallurgy of.

THE METALLURGY OF SILVER: A Practical Treatise on the Amalgamation, Roasting and Lixiviation of Silver Ores. Including the Assaying, Melting and Refining of Silver Bullion. By M. EISSLER, Author of "The Metallurgy of Gold." With 124 Illustrations. Crown 8vo, 10s. 6d. cloth.
[*Just published.*
"A practical treatise, and a technical work which we are convinced will supply a long-felt want amongst practical men, and at the same time be of value to students and others indirectly connected with the industries."—*Mining Journal*.
"From first to last the book is thoroughly sound and reliable."—*Colliery Guardian*.
"For chemists, practical miners, assayers and investors alike, we do not know of any work on the subject so handy and yet so comprehensive."—*Glasgow Herald*.

Mineral Surveying and Valuing.

THE MINERAL SURVEYOR AND VALUER'S COMPLETE GUIDE, comprising a Treatise on Improved Mining Surveying and the Valuation of Mining Properties, with New Traverse Tables. By WM. LINTERN, Mining and Civil Engineer. Third Edition, with an Appendix on "Magnetic and Angular Surveying," With Four Plates. 12mo, 4s. cloth.
"An enormous fund of information of great value."—*Mining Journal*.
"Mr. Lintern's book forms a valuable and thoroughly trustworthy guide."—*Iron and Coal Trades Review*.
"This new edition must be of the highest value to colliery surveyors, proprietors and managers."—*Colliery Guardian*.

Metalliferous Minerals and Mining.

TREATISE ON METALLIFEROUS MINERALS AND MINING. By D. C. DAVIES, F.G.S., Mining Engineer, &c., Author of "A Treatise on Slate and Slate Quarrying." Illustrated with numerous Wood Engravings. Fourth Edition, carefully Revised. Crown 8vo, 12s. 6d. cloth.
"Neither the practical miner nor the general reader interested in mines can have a better book for his companion and his guide."—*Mining Journal*.
"The volume is one which no student of mineralogy should be without."—*Colliery Guardian*.
"A book that will not only be useful to the geologist, the practical miner, and the metallurgist, but also very interesting to the general public."—*Iron*.
"As a history of the present state of mining throughout the world this book has a real value, and it supplies an actual want, for no such information has hitherto been brought together within such limited space."—*Athenæum*.

Earthy Minerals and Mining.

TREATISE ON EARTHY AND OTHER MINERALS AND MINING. By D. C. DAVIES, F.G.S. Uniform with, and forming a Companion Volume to, the same Author's "Metalliferous Minerals and Mining." With 76 Wood Engravings. Second Edition. Crown 8vo, 12s. 6d. cloth.
"It is essentially a practical work, intended primarily for the use of practical men. . . . We do not remember to have met with any English work on mining matters that contains the same amount of information packed in equally convenient form."—*Academy*.
"The book is clearly the result of many years' careful work and thought, and we should be inclined to rank it as among the very best of the handy technical and trades manuals which have recently appeared."—*British Quarterly Review*.
"The volume contains a great mass of practical information carefully methodised and presented in a very intelligible shape."—*Scotsman*.
"The subject matter of the volume will be found of high value by all—and they are a numerous class—who trade in earthy minerals."—*Athenæum*.

Underground Pumping Machinery.

MINE DRAINAGE. Being a Complete and Practical Treatise on Direct-Acting Underground Steam Pumping Machinery, with a Description of a large number of the best known Engines, their General Utility and the Special Sphere of their Action, the Mode of their Application, and their merits compared with other forms of Pumping Machinery. By STEPHEN MICHELL. 8vo, 15s. cloth.

"Will be highly esteemed by colliery owners and lessees, mining engineers, and students generally who require to be acquainted with the best means of securing the drainage of mines. It is a most valuable work, and stands almost alone in the literature of steam pumping machinery."—*Colliery Guardian.*

"Much valuable information is given, so that the book is thoroughly worthy of an extensive circulation amongst practical men and purchasers of machinery."—*Mining Journal.*

Mining Tools.

A MANUAL OF MINING TOOLS. For the Use of Mine Managers, Agents, Students, &c. By WILLIAM MORGANS, Lecturer on Practical Mining at the Bristol School of Mines. 12mo, 2s. 6d. cloth limp.

ATLAS OF ENGRAVINGS to Illustrate the above, containing 235 Illustrations of Mining Tools, drawn to scale. 4to, 4s. 6d. cloth.

"Students in the science of mining, and overmen, captains, managers, and viewers may gain practical knowledge and useful hints by the study of Mr. Morgans' manual."—*Colliery Guardian.*
"A valuable work, which will tend materially to improve our mining literature."—*Mining Journal.*

Coal Mining.

COAL AND COAL MINING: A Rudimentary Treatise on. By Sir WARINGTON W. SMYTH, M.A., F.R.S., &c., Chief Inspector of the Mines of the Crown. New Edition, Revised and Corrected. With numerous Illustrations. 12mo, 4s. cloth boards.

"As an outline is given of every known coal-field in this and other countries, as well as of the principal methods of working, the book will doubtless interest a very large number of readers."—*Mining Journal.*

Granite Quarrying.

GRANITES AND OUR GRANITE INDUSTRIES. By GEORGE F. HARRIS, F.G.S., Membre de la Société Belge de Géologie, Lecturer on Economic Geology at the Birkbeck Institution, &c. With Illustrations. Crown 8vo, 2s. 6d. cloth.

"A clearly and well-written manual for persons engaged or interested in the granite industry."—*Scotsman.*
"An interesting work, which will be deservedly esteemed. We advise the author to write again."—*Colliery Guardian.*
"An exceedingly interesting and valuable monograph, on a subject which has hitherto received unaccountably little attention in the shape of systematic literary treatment."—*Scottish Leader.*

NATURAL AND APPLIED SCIENCE.

Text Book of Electricity.

THE STUDENT'S TEXT-BOOK OF ELECTRICITY. By HENRY M. NOAD, Ph.D., F.R.S., F.C.S. New Edition, carefully Revised. With an Introduction and Additional Chapters, by W. H. PREECE, M.I.C.E., Vice-President of the Society of Telegraph Engineers, &c. With 470 Illustrations. Crown 8vo, 12s. 6d. cloth.

"The original plan of this book has been carefully adhered to so as to make it a reflex of the existing state of electrical science, adapted for students. . . . Discovery seems to have progressed with marvellous strides; nevertheless it has now apparently ceased, and practical applications have commenced their career; and it is to give a faithful account of these that this fresh edition of Dr. Noad's valuable text-book is launched forth."—*Extract from Introduction by W. H. Preece, Esq.*

"We can recommend Dr. Noad's book for clear style, great range of subject, a good index and a plethora of woodcuts. Such collections as the present are indispensable."—*Athenæum.*
"An admirable text book for every student — beginner or advanced — of electricity."—*Engineering.*

Electricity.

A MANUAL OF ELECTRICITY: Including Galvanism, Magnetism, Dia-Magnetism, Electro-Dynamics, Magno-Electricity, and the Electric Telegraph. By HENRY M. NOAD, Ph.D., F.R.S., F.C.S. Fourth Edition. With 500 Woodcuts. 8vo, £1 4s. cloth.

"It is worthy of a place in the library of every public institution."—*Mining Journal.*

NATURAL AND APPLIED SCIENCE. 25

Electric Light.
ELECTRIC LIGHT : Its Production and Use. Embodying Plain Directions for the Treatment of Voltaic Batteries, Electric Lamps, and Dynamo-Electric Machines. By J. W. URQUHART, C.E., Author of "Electroplating: A Practical Handbook." Edited by F. C. WEBB, M.I.C.E., M.S.T.E. Second Edition, Revised, with large Additions and 128 Illusts. 7s. 6d. cloth.
"The book is by far the best that we have yet met with on the subject."—*Athenæum.*
"It is the only work at present available which gives, in language intelligible for the most part to the ordinary reader, a general but concise history of the means which have been adopted up to the present time in producing the electric light."—*Metropolitan.*
"The book contains a general account of the means adopted in producing the electric light, not only as obtained from voltaic or galvanic batteries, but treats at length of the dynamo-electric machine in several of its forms."—*Colliery Guardian.*

Electric Lighting.
THE ELEMENTARY PRINCIPLES OF ELECTRIC LIGHTING. By ALAN A. CAMPBELL SWINTON, Associate I.E.E. Second Edition, Enlarged and Revised. With 16 Illustrations. Crown 8vo, 1s. 6d. cloth.
"Anyone who desires a short and thoroughly clear exposition of the elementary principles of electric-lighting cannot do better than read this little work."—*Bradford Observer.*

Dr. Lardner's School Handbooks.
NATURAL PHILOSOPHY FOR SCHOOLS. By Dr. LARDNER. 328 Illustrations. Sixth Edition. One Vol., 3s. 6d. cloth.
"A very convenient class-book for junior students in private schools. It is intended to convey, in clear and precise terms, general notions of all the principal divisions of Physical Science."—*British Quarterly Review.*

ANIMAL PHYSIOLOGY FOR SCHOOLS. By Dr. LARDNER. With 190 Illustrations. Second Edition. One Vol., 3s. 6d. cloth.
"Clearly written, well arranged, and excellently illustrated."—*Gardener's Chronicle.*

Dr. Lardner's Electric Telegraph.
THE ELECTRIC TELEGRAPH. By Dr. LARDNER. Revised and Re-written by E. B. BRIGHT, F.R.A.S. 140 Illustrations. Small 8vo, 2s. 6d. cloth.
"One of the most readable books extant on the Electric Telegraph."—*English Mechanic.*

Astronomy.
ASTRONOMY. By the late Rev. ROBERT MAIN, M.A., F.R.S., formerly Radcliffe Observer at Oxford. Third Edition, Revised and Corrected to the present time, by WILLIAM THYNNE LYNN, B.A., F.R.A.S., formerly of the Royal Observatory, Greenwich. 12mo, 2s. cloth limp.
"A sound and simple treatise, very carefully edited, and a capital book for beginners."—*Knowledge.*
"Accurately brought down to the requirements of the present time by Mr. Lynn."—*Educational Times.*

The Blowpipe.
THE BLOWPIPE IN CHEMISTRY, MINERALOGY, AND GEOLOGY. Containing all known Methods of Anhydrous Analysis, many Working Examples, and Instructions for Making Apparatus. By Lieut.-Colonel W. A. Ross, R.A., F.G.S. With 120 Illustrations. Second Edition, Revised and Enlarged. Crown 8vo, 5s. cloth. [*Just published.*
"The student who goes conscientiously through the course of experimentation here laid down will gain a better insight into inorganic chemistry and mineralogy than if he had 'got up' any of the best text-books of the day, and passed any number of examinations in their contents."—*Chemical News.*

The Military Sciences.
AIDE-MEMOIRE TO THE MILITARY SCIENCES. Framed from Contributions of Officers and others connected with the different Services. Originally edited by a Committee of the Corps of Royal Engineers. Second Edition, most carefully Revised by an Officer of the Corps, with many Additions; containing nearly 350 Engravings and many hundred Woodcuts. Three Vols., royal 8vo, extra cloth boards, and lettered, £4 10s.
"A compendious encyclopædia of military knowledge, to which we are greatly indebted."—*Edinburgh Review.*
"The most comprehensive book of reference to the military and collateral sciences."—*Volunteer Service Gazette.*

Field Fortification.
A TREATISE ON FIELD FORTIFICATION, THE ATTACK OF FORTRESSES, MILITARY MINING, AND RECONNOITRING. By Colonel I. S. MACAULAY, late Professor of Fortification in the R.M.A., Woolwich. Sixth Edition, crown 8vo, cloth, with separate Atlas of 12 Plates, 12s.

Temperaments.

OUR TEMPERAMENTS, THEIR STUDY AND THEIR TEACHING. A Popular Outline. By ALEXANDER STEWART, F.R.C.S. Edin. In one large 8vo volume, with 30 Illustrations, including A Selection from Lodge's "Historical Portraits," showing the Chief Forms of Faces. Price 15s. cloth, gilt top.

"The book is exceedingly interesting, even for those who are not systematic students of anthropology. . . . To those who think the proper study of mankind is man, it will be full of attraction."—*Daily Telegraph.*

"The author's object is to enable a student to read a man's temperament in his aspect. The work is well adapted to its end. It is worthy of the attention of students of human nature."—*Scotsman.*

"The volume is heavy to hold, but light to read. Though the author has treated his subject exhaustively, he writes in a popular and pleasant manner that renders it attractive to the general reader."—*Punch.*

Antiseptic Nursing.

ANTISEPTICS: A Handbook for Nurses. Being an Epitome of Antiseptic Treatment. With Notes on Antiseptic Substances, Disinfection, Monthly Nursing, &c. By Mrs. ANNIE HEWER, late Hospital Sister, Diplomée Obs. Soc. Lond. Crown 8vo, 1s. 6d. cloth. [*Just published.*

"This excellent little work . . . is very readable and contains much information. We can strongly recommend it to those who are undergoing training at the various hospitals, and also to those who are engaged in the practice of nursing, as they cannot fail to obtain practical hints from its perusal."—*Lancet.*

"The student or the busy practitioner would do well to look through its pages, offering as they do a suggestive and faithful picture of antiseptic methods."—*Hospital Gazette.*

"A clear, concise, and excellent little handbook."—*The Hospital.*

Pneumatics and Acoustics.

PNEUMATICS: including Acoustics and the Phenomena of Wind Currents, for the Use of Beginners. By CHARLES TOMLINSON, F.R.S., F.C.S., &c. Fourth Edition, Enlarged. With numerous Illustrations. 12mo, 1s. 6d. cloth.

"Beginners in the study of this important application of science could not have a better manual."—*Scotsman.*

"A valuable and suitable text-book for students of Acoustics and the Phenomena of Wind Currents."—*Schoolmaster.*

Conchology.

A MANUAL OF THE MOLLUSCA: Being a Treatise on Recent and Fossil Shells. By S. P. WOODWARD, A.L.S., F.G.S., late Assistant Palæontologist in the British Museum. Fifth Edition. With an Appendix on *Recent and Fossil Conchological Discoveries,* by RALPH TATE, A.L.S., F.G.S. Illustrated by A. N. WATERHOUSE and JOSEPH WILSON LOWRY. With 23 Plates and upwards of 300 Woodcuts. Crown 8vo, 7s. 6d. cloth boards.

"A most valuable storehouse of conchological and geological information."—*Science Gossip.*

Geology.

RUDIMENTARY TREATISE ON GEOLOGY, PHYSICAL AND HISTORICAL. Consisting of "Physical Geology," which sets forth the leading Principles of the Science; and "Historical Geology," which treats of the Mineral and Organic Conditions of the Earth at each successive epoch, especial reference being made to the British Series of Rocks. By RALPH TATE, A.L.S., F.G.S., &c., &c. With 250 Illustrations. 12mo, 5s. cloth boards.

"The fulness of the matter has elevated the book into a manual. Its information is exhaustive and well arranged."—*School Board Chronicle.*

Geology and Genesis.

THE TWIN RECORDS OF CREATION; or, Geology and Genesis: their Perfect Harmony and Wonderful Concord. By GEORGE W. VICTOR LE VAUX. Numerous Illustrations. Fcap. 8vo, 5s. cloth.

"A valuable contribution to the evidences of Revelation, and disposes very conclusively of the arguments of those who would set God's Works against God's Word. No real difficulty is shirked, and no sophistry is left unexposed."—*The Rock.*

"The remarkable peculiarity of this author is that he combines an unbounded admiration of science with an unbounded admiration of the Written record. The two impulses are balanced to a nicety; and the consequence is that difficulties, which to minds less evenly poised would be serious, find immediate solutions of the happiest kinds."—*London Review.*

DR. LARDNER'S HANDBOOKS OF NATURAL PHILOSOPHY.

THE HANDBOOK OF MECHANICS. Enlarged and almost rewritten by BENJAMIN LOEWY, F.R.A.S. With 378 Illustrations. Post 8vo, 6s. cloth.

"The perspicuity of the original has been retained, and chapters which had become obsolete have been replaced by others of more modern character. The explanations throughout are studiously popular, and care has been taken to show the application of the various branches of physics to the industrial arts, and to the practical business of life."—*Mining Journal.*

"Mr. Loewy has carefully revised the book, and brought it up to modern requirements."—*Nature.*

"Natural philosophy has had few exponents more able or better skilled in the art of popularising the subject than Dr. Lardner; and Mr. Loewy is doing good service in fitting this treatise, and the others of the series, for use at the present time."—*Scotsman.*

THE HANDBOOK OF HYDROSTATICS AND PNEUMATICS. New Edition, Revised and Enlarged, by BENJAMIN LOEWY, F.R.A.S. With 236 Illustrations. Post 8vo, 5s. cloth.

"For those 'who desire to attain an accurate knowledge of physical science without the profound methods of mathematical investigation,' this work is not merely intended, but well adapted."—*Chemical News.*

"The volume before us has been carefully edited, augmented to nearly twice the bulk of the former edition, and all the most recent matter has been added. . . . It is a valuable text-book."—*Nature.*

"Candidates for pass examinations will find it, we think, specially suited to their requirements.'
English Mechanic.

THE HANDBOOK OF HEAT. Edited and almost entirely rewritten by BENJAMIN LOEWY, F.R.A.S., &c. 117 Illustrations. Post 8vo, 6s. cloth.

"The style is always clear and precise, and conveys instruction without leaving any cloudiness or lurking doubts behind."—*Engineering.*

"A most exhaustive book on the subject on which it treats, and is so arranged that it can be understood by all who desire to attain an accurate knowledge of physical science. . . . . Mr. Loewy has included all the latest discoveries in the varied laws and effects of heat."—*Standard.*

"A complete and handy text-book for the use of students and general readers."—*English Mechanic.*

THE HANDBOOK OF OPTICS. By DIONYSIUS LARDNER, D.C.L., formerly Professor of Natural Philosophy and Astronomy in University College, London. New Edition. Edited by T. OLVER HARDING, B.A. Lond., of University College, London. With 298 Illustrations. Small 8vo, 448 pages, 5s. cloth.

"Written by one of the ablest English scientific writers, beautifully and elaborately illustrated."
Mechanic's Magazine.

THE HANDBOOK OF ELECTRICITY, MAGNETISM, AND ACOUSTICS. By Dr. LARDNER. Ninth Thousand. Edit. by GEORGE CAREY FOSTER, B.A., F.C.S. With 400 Illustrations. Small 8vo, 5s. cloth.

"The book could not have been entrusted to anyone better calculated to preserve the terse and lucid style of Lardner, while correcting his errors and bringing up his work to the present state of scientific knowledge."—*Popular Science Review.*

⁎⁎* *The above Five Volumes, though each is Complete in itself, form* A COMPLETE COURSE OF NATURAL PHILOSOPHY.

Dr. Lardner's Handbook of Astronomy.

THE HANDBOOK OF ASTRONOMY. Forming a Companion to the "Handbook of Natural Philosophy." By DIONYSIUS LARDNER, D.C.L., formerly Professor of Natural Philosophy and Astronomy in University College, London. Fourth Edition. Revised and Edited by EDWIN DUNKIN, F.R.A.S., Royal Observatory, Greenwich. With 38 Plates and upwards of 100 Woodcuts. In One Vol., small 8vo, 550 pages, 9s. 6d. cloth.

"Probably no other book contains the same amount of information in so compendious and well-arranged a form—certainly none at the price at which this is offered to the public."—*Athenæum.*

"We can do no other than pronounce this work a most valuable manual of astronomy, and we strongly recommend it to all who wish to acquire a general—but at the same time correct—acquaintance with this sublime science."—*Quarterly Journal of Science.*

"One of the most deservedly popular books on the subject . . . We would recommend not only the student of the elementary principles of the science, but he who aims at mastering the higher and mathematical branches of astronomy, not to be without this work beside him."—*Practical Magazine.*

DR. LARDNER'S MUSEUM OF SCIENCE AND ART.

THE MUSEUM OF SCIENCE AND ART. Edited by DIONYSIUS LARDNER, D.C.L., formerly Professor of Natural Philosophy and Astronomy in University College, London. With upwards of 1,200 Engravings on Wood. In 6 Double Volumes, £1 1s., in a new and elegant cloth binding; or handsomely bound in half-morocco, 31s. 6d.

Contents:

The Planets: Are they Inhabited Worlds?—Weather Prognostics—Popular Fallacies in Questions of Physical Science—Latitudes and Longitudes—Lunar Influences—Meteoric Stones and Shooting Stars—Railway Accidents—Light—Common Things: Air—Locomotion in the United States—Cometary Influences—Common Things: Water—The Potter's Art—Common Things: Fire—Locomotion and Transport, their Influence and Progress—The Moon—Common Things: The Earth—The Electric Telegraph—Terrestrial Heat—The Sun—Earthquakes and Volcanoes—Barometer, Safety Lamp, and Whitworth's Micrometric Apparatus—Steam—The Steam Engine—The Eye—The Atmosphere—Time—Common Things: Pumps—Common Things: Spectacles, the Kaleidoscope—Clocks and Watches—Microscopic Drawing and Engraving—Locomotive—Thermometer—New Planets: Leverrier and Adams's Planet—Magnitude and Minuteness—Common Things: The Almanack—Optical Images—How to observe the Heavens—Common Things: The Looking-glass—Stellar Universe—The Tides—Colour—Common Things: Man—Magnifying Glasses—Instinct and Intelligence—The Solar Microscope—The Camera Lucida—The Magic Lantern—The Camera Obscura—The Microscope—The White Ants: Their Manners and Habits—The Surface of the Earth, or First Notions of Geography—Science and Poetry—The Bee—Steam Navigation—Electro-Motive Power—Thunder, Lightning, and the Aurora Borealis—The Printing Press—The Crust of the Earth—Comets—The Stereoscope—The Pre-Adamite Earth—Eclipses—Sound.

⁎⁎ OPINIONS OF THE PRESS.

"This series, besides affording popular but sound instruction on scientific subjects, with which the humblest man in the country ought to be acquainted, also undertakes that teaching of 'Common Things' which every well-wisher of his kind is anxious to promote. Many thousand copies of this serviceable publication have been printed, in the belief and hope that the desire for instruction and improvement widely prevails; and we have no fear that such enlightened faith will meet with disappointment."—*Times.*

"A cheap and interesting publication, alike informing and attractive. The papers combine subjects of importance and great scientific knowledge, considerable inductive powers, and a popular style of treatment."—*Spectator.*

"The 'Museum of Science and Art' is the most valuable contribution that has ever been made to the Scientific Instruction of every class of society."—Sir DAVID BREWSTER, in the *North British Review.*

"Whether we consider the liberality and beauty of the illustrations, the charm of the writing, or the durable interest of the matter, we must express our belief that there is hardly to be found among the new books one that would be welcomed by people of so many ages and classes as a valuable present."—*Examiner.*

⁎⁎ *Separate books formed from the above, suitable for Workmen's Libraries, Science Classes, etc.*

Common Things Explained. Containing Air, Earth, Fire, Water, Time, Man, the Eye, Locomotion, Colour, Clocks and Watches, &c. 233 Illustrations, cloth gilt, 5s.

The Microscope. Containing Optical Images, Magnifying Glasses, Origin and Description of the Microscope, Microscopic Objects, the Solar Microscope, Microscopic Drawing and Engraving, &c. 147 Illustrations, cloth gilt, 2s.

Popular Geology. Containing Earthquakes and Volcanoes, the Crust of the Earth, &c. 201 Illustrations, cloth gilt, 2s. 6d.

Popular Physics. Containing Magnitude and Minuteness, the Atmosphere, Meteoric Stones, Popular Fallacies, Weather Prognostics, the Thermometer, the Barometer, Sound, &c. 85 Illustrations, cloth gilt, 2s. 6d.

Steam and its Uses. Including the Steam Engine, the Locomotive, and Steam Navigation. 89 Illustrations, cloth gilt, 2s.

Popular Astronomy. Containing How to observe the Heavens—The Earth, Sun, Moon, Planets, Light, Comets, Eclipses, Astronomical Influences, &c. 182 Illustrations, 4s. 6d.

The Bee and White Ants: Their Manners and Habits. With Illustrations of Animal Instinct and Intelligence. 135 Illustrations, cloth gilt, 2s.

The Electric Telegraph Popularized. To render intelligible to all who can Read, irrespective of any previous Scientific Acquirements, the various forms of Telegraphy in Actual Operation. 100 Illustrations, cloth gilt, 1s. 6d.

COUNTING-HOUSE WORK, TABLES, etc.

Accounts for Manufacturers.
FACTORY ACCOUNTS: Their Principles and Practice. A Handbook for Accountants and Manufacturers, with Appendices on the Nomenclature of Machine Details; the Income Tax Acts; the Rating of Factories; Fire and Boiler Insurance; the Factory and Workshop Acts, &c., including also a Glossary of Terms and a large number of Specimen Rulings. By EMILE GARCKE and J. M. FELLS. Third Edition. Demy 8vo, 250 pages, price 6s. strongly bound. [*Just published.*
"A very interesting description of the requirements of Factory Accounts. . . . the principle of assimilating the Factory Accounts to the general commercial books is one which we thoroughly agree with."—*Accountants' Journal.*
"Characterised by extreme thoroughness. There are few owners of Factories who would not derive great benefit from the perusal of this most admirable work."—*Local Government Chronicle.*

Foreign Commercial Correspondence.
THE FOREIGN COMMERCIAL CORRESPONDENT: Being Aids to Commercial Correspondence in Five Languages—English, French, German, Italian and Spanish. By CONRAD E. BAKER. Second Edition, Revised. Crown 8vo, 3s. 6d. cloth. [*Just published.*
"Whoever wishes to correspond in all the languages mentioned by Mr. Baker cannot do better than study this work, the materials of which are excellent and conveniently arranged. They consist not of entire specimen letters, but what are far more useful—short passages, sentences, or phrases expressing the same general idea in various forms."—*Athenæum.*
"A careful examination has convinced us that it is unusually complete, well arranged and reliable. The book is a thoroughly good one."—*Schoolmaster.*

Intuitive Calculations.
THE COMPENDIOUS CALCULATOR; or, Easy and Concise Methods of Performing the various Arithmetical Operations required in Commercial and Business Transactions, together with Useful Tables. By DANIEL O'GORMAN. Corrected and Extended by J. R. YOUNG, formerly Professor of Mathematics at Belfast College. Twenty-seventh Edition, carefully Revised by C. NORRIS. Fcap. 8vo, 3s. 6d. strongly half-bound in leather.
"It would be difficult to exaggerate the usefulness of a book like this to everyone engaged in commerce or manufacturing industry. It is crammed full of rules and formulæ for shortening and employing calculations."—*Knowledge.*
"Supplies special and rapid methods for all kinds of calculations. Of great utility to persons engaged in any kind of commercial transactions."—*Scotsman.*

Modern Metrical Units and Systems.
MODERN METROLOGY: A Manual of the Metrical Units and Systems of the Present Century. With an Appendix containing a proposed English System. By LOWIS D'A. JACKSON, A.M.Inst.C.E., Author of "Aid to Survey Practice," &c. Large crown 8vo, 12s. 6d. cloth.
"The author has brought together much valuable and interesting information. . . . We cannot but recommend the work to the consideration of all interested in the practical reform of our weights and measures."—*Nature.*
"For exhaustive tables of equivalent weights and measures of all sorts, and for clear demonstrations of the effects of the various systems that have been proposed or adopted, Mr. Jackson's treatise is without a rival."—*Academy.*

The Metric System and the British Standards.
A SERIES OF METRIC TABLES, in which the British Standard Measures and Weights are compared with those of the Metric System at present in Use on the Continent. By C. H. DOWLING, C.E. 8vo, 10s. 6d. strongly bound.
"Their accuracy has been certified by Professor Airy, the Astronomer-Royal."—*Builder.*
"Mr. Dowling's Tables are well put together as a ready-reckoner for the conversion of one system into the other."—*Athenæum.*

Iron and Metal Trades' Calculator.
THE IRON AND METAL TRADES' COMPANION. For expeditiously ascertaining the Value of any Goods bought or sold by Weight, from 1s. per cwt. to 112s. per cwt., and from one farthing per pound to one shilling per pound. Each Table extends from one pound to 100 tons. To which are appended Rules on Decimals, Square and Cube Root, Mensuration of Superficies and Solids, &c.; Tables of Weights of Materials, and other Useful Memoranda. By THOS. DOWNIE. 396 pp., 9s. Strongly bound in leather.
"A most useful set of tables, and will supply a want, for nothing like them before existed."—*Building News.*
"Although specially adapted to the iron and metal trades, the tables will be found useful in every other business in which merchandise is bought and sold by weight."—*Railway News.*

Calculator for Numbers and Weights Combined.

THE NUMBER AND WEIGHT CALCULATOR. Containing upwards of 250,000 Separate Calculations, showing at a glance the value at 421 different rates, ranging from $\frac{1}{16}$th of a Penny to 20s. each, or per cwt., and £20 per ton, of any number of articles consecutively, from 1 to 470.—Any number of cwts., qrs., and lbs., from 1 cwt. to 470 cwts.—Any number of tons, cwts., qrs., and lbs., from 1 to 23½ tons. By WILLIAM CHADWICK, Public Accountant. Second Edition, Revised and Improved, and specially adapted for the Apportionment of Mileage Charges for Railway Traffic. 8vo, price 18s., strongly bound for Office wear and tear. [*Just published.*

☞ *This comprehensive and entirely unique and original Calculator is adapted for the use of Accountants and Auditors, Railway Companies, Canal Companies, Shippers, Shipping Agents, General Carriers, etc. Ironfounders, Brassfounders, Metal Merchants, Iron Manufacturers, Ironmongers, Engineers, Machinists, Boiler Makers, Millwrights, Roofing, Bridge and Girder Makers, Colliery Proprietors, etc. Timber Merchants, Builders, Contractors, Architects, Surveyors, Auctioneers, Valuers, Brokers, Mill Owners and Manufacturers, Mill Furnishers, Merchants and General Wholesale Tradesmen.*

*** OPINIONS OF THE PRESS.

"The book contains the answers to questions, and not simply a set of ingenious puzzle methods of arriving at results. It is as easy of reference for any answer or any number of answers as a dictionary, and the references are even more quickly made. For making up accounts or estimates, the book must prove invaluable to all who have any considerable quantity of calculations involving price and measure in any combination to do."—*Engineer.*

"The most complete and practical ready reckoner which it has been our fortune yet to see. It is difficult to imagine a trade or occupation in which it could not be of the greatest use, either in saving human labour or in checking work. The Publishers have placed within the reach of every commercial man an invaluable and unfailing assistant."—*The Miller.*

"The most perfect work of the kind yet prepared."—*Glasgow Herald.*

Comprehensive Weight Calculator.

THE WEIGHT CALCULATOR. Being a Series of Tables upon a New and Comprehensive Plan, exhibiting at One Reference the exact Value of any Weight from 1 lb. to 15 tons, at 300 Progressive Rates, from 1d. to 168s. per cwt., and containing 186,000 Direct Answers, which, with their Combinations, consisting of a single addition (mostly to be performed at sight), will afford an aggregate of 10,266,000 Answers; the whole being calculated and designed to ensure correctness and promote despatch. By HENRY HARBEN, Accountant. Fourth Edition, carefully Corrected. Royal 8vo, strongly half-bound, £1 5s.

"A practical and useful work of reference for men of business generally; it is the best of the kind we have seen."—*Ironmonger.*

"Of priceless value to business men. It is a necessary book in all mercantile offices."—*Sheffield Independent.*

Comprehensive Discount Guide.

THE DISCOUNT GUIDE. Comprising several Series of Tables for the use of Merchants, Manufacturers, Ironmongers, and others, by which may be ascertained the exact Profit arising from any mode of using Discounts, either in the Purchase or Sale of Goods, and the method of either Altering a Rate of Discount or Advancing a Price, so as to produce, by one operation, a sum that will realise any required profit after allowing one or more Discounts: to which are added Tables of Profit or Advance from 1¼ to 90 per cent., Tables of Discount from 1¼ to 98¾ per cent., and Tables of Commission, &c., from ⅛ to 10 per cent. By HENRY HARBEN, Accountant, Author of "The Weight Calculator." New Edition, carefully Revised and Corrected. Demy 8vo, 544 pp. half-bound, £1 5s.

"A book such as this can only be appreciated by business men, to whom the saving of time means saving of money. We have the high authority of Professor J. R. Young that the tables throughout the work are constructed upon strictly accurate principles. The work is a model of typographical clearness, and must prove of great value to merchants, manufacturers, and general traders."—*British Trade Journal.*

Iron Shipbuilders' and Merchants' Weight Tables.

IRON-PLATE WEIGHT TABLES: For Iron Shipbuilders, *Engineers and Iron Merchants.* Containing the Calculated Weights of upwards of 150,000 different sizes of Iron Plates, from 1 foot by 6 in. by ¼ in. to 10 feet by 5 feet by 1 in. Worked out on the basis of 40 lbs. to the square foot of Iron of 1 inch in thickness. Carefully compiled and thoroughly Revised by H. BURLINSON and W. H. SIMPSON. Oblong 4to, 25s. half-bound.

"This work will be found of great utility. The authors have had much practical experience of what is wanting in making estimates; and the use of the book will save much time in making elaborate calculations. —*English Mechanic.*

INDUSTRIAL AND USEFUL ARTS.

Soap-making.

THE ART OF SOAP-MAKING: A Practical Handbook of the Manufacture of Hard and Soft Soaps, Toilet Soaps, etc. Including many New Processes, and a Chapter on the Recovery of Glycerine from Waste Leys. By ALEXANDER WATT, Author of "Electro-Metallurgy Practically Treated," &c. With numerous Illustrations. Third Edition, Revised. Crown 8vo, 7s. 6d. cloth.

"The work will prove very useful, not merely to the technological student, but to the practical soap-boiler who wishes to understand the theory of his art."—*Chemical News.*

"Really an excellent example of a technical manual, entering, as it does, thoroughly and exhaustively both into the theory and practice of soap manufacture. The book is well and honestly done, and deserves the considerable circulation with which it will doubtless meet."—*Knowledge.*

"Mr. Watt's book is a thoroughly practical treatise on an art which has almost no literature in our language. We congratulate the author on the success of his endeavour to fill a void in English technical literature."—*Nature.*

Paper Making.

THE ART OF PAPER MANUFACTURE: A Practical Handbook of the Manufacture of Paper from Rags, Esparto, Wood and other Fibres. By ALEXANDER WATT, Author of "The Art of Soap-Making," "The Art of Leather Manufacture," &c. With numerous Illustrations. Cr. 8vo. [*In the press.*

Leather Manufacture.

THE ART OF LEATHER MANUFACTURE. Being a Practical Handbook, in which the Operations of Tanning, Currying, and Leather Dressing are fully Described, and the Principles of Tanning Explained, and many Recent Processes introduced; as also Methods for the Estimation of Tannin, and a Description of the Arts of Glue Boiling, Gut Dressing, &c. By ALEXANDER WATT, Author of "Soap-Making," "Electro-Metallurgy," &c. With numerous Illustrations. Second Edition. Crown 8vo, 9s. cloth.

"A sound, comprehensive treatise on tanning and its accessories. . . . An eminently valuable production, which redounds to the credit of both author and publishers."—*Chemical Review.*

"This volume is technical without being tedious, comprehensive and complete without being prosy, and it bears on every page the impress of a master hand. We have never come across a better trade treatise, nor one that so thoroughly supplied an absolute want."—*Shoe and Leather Trades' Chronicle.*

Boot and Shoe Making.

THE ART OF BOOT AND SHOE-MAKING. A Practical Handbook, including Measurement, Last-Fitting, Cutting-Out, Closing and Making, with a Description of the most approved Machinery employed. By JOHN B. LENO, late Editor of *St. Crispin*, and *The Boot and Shoe-Maker*. With numerous Illustrations. Third Edition. 12mo, 2s. cloth limp.

"This excellent treatise is by far the best work ever written on the subject. A new work, embracing all modern improvements, was much wanted. This want is now satisfied. The chapter on clicking, which shows how waste may be prevented, will save fifty times the price of the book."—*Scottish Leather Trader.*

"This volume is replete with matter well worthy the perusal of boot and shoe manufacturers and experienced craftsmen, and instructive and valuable in the highest degree to all young beginners and craftsmen in the trade of which it treats."—*Leather Trades' Circular.*

Dentistry.

MECHANICAL DENTISTRY: A Practical Treatise on the Construction of the various kinds of Artificial Dentures. Comprising also Useful Formulæ, Tables and Receipts for Gold Plate, Clasps, Solders, &c. &c. By CHARLES HUNTER. Third Edition, Revised. With upwards of 100 Wood Engravings. Crown 8vo, 3s. 6d. cloth.

"The work is very practical."—*Monthly Review of Dental Surgery.*

"We can strongly recommend Mr. Hunter's treatise to all students preparing for the profession of dentistry, as well as to every mechanical dentist."—*Dublin Journal of Medical Science.*

"A work in a concise form that few could read without gaining information from."—*British Journal of Dental Science.*

Wood Engraving.

A PRACTICAL MANUAL OF WOOD ENGRAVING. With a Brief Account of the History of the Art. By WILLIAM NORMAN BROWN. With numerous Illustrations. Crown 8vo, 2s. cloth.

"The author deals with the subject in a thoroughly practical and easy series of representative lessons."—*Paper and Printing Trades' Journal.*

"The book is clear and complete, and will be useful to anyone wanting to understand the first elements of the beautiful art of wood engraving."—*Graphic.*

HANDYBOOKS FOR HANDICRAFTS. By PAUL N. HASLUCK.

☞ These Handybooks are written to supply Handicraftsmen with information on workshop practice, and are intended to convey, in plain language, technical knowledge of the several crafts. Workshop terms are used, and workshop practice described, the text being freely illustrated with drawings of modern tools, appliances and processes.

N.B. The following Volumes are already published, and others are in preparation.

Metal Turning.

THE METAL TURNER'S HANDYBOOK. A Practical Manual for Workers at the Foot-Lathe: Embracing Information on the Tools, Appliances and Processes employed in Metal Turning. By PAUL N. HASLUCK, Author of "Lathe-Work." With upwards of One Hundred Illustrations. Second Edition, Revised. Crown 8vo, 2s. cloth.

"Altogether admirably adapted to initiate students into the art of turning."—*Leicester Post.*
"Clearly and concisely written, excellent in every way, we heartily commend it to all interested in metal turning."—*Mechanical World.*

Wood Turning.

THE WOOD TURNER'S HANDYBOOK. A Practical Manual for Workers at the Lathe: Embracing Information on the Tools, Appliances and Processes Employed in Wood Turning. By PAUL N. HASLUCK. With upwards of One Hundred Illustrations. Crown 8vo, 2s. cloth.

"We recommend the book to young turners and amateurs. A multitude of workmen have hitherto sought in vain for a manual of this special industry."—*Mechanical World.*

Watch Repairing.

THE WATCH JOBBER'S HANDYBOOK. A Practical Manual on Cleaning, Repairing and Adjusting. Embracing Information on the Tools, Materials, Appliances and Processes Employed in Watchwork. By PAUL N. HASLUCK. With upwards of One Hundred Illustrations. Cr. 8vo, 2s. cloth.

"All young persons connected with the trade should acquire and study this excellent, and at the same time, inexpensive work."—*Clerkenwell Chronicle.*

Pattern Making.

THE PATTERN MAKER'S HANDYBOOK. A Practical Manual, embracing Information on the Tools, Materials and Appliances employed in Constructing Patterns for Founders. By PAUL N. HASLUCK. With One Hundred Illustrations. Crown 8vo, 2s. cloth.

"We commend it to all who are interested in the counsels it so ably gives."—*Colliery Guardian.*
"This handy volume contains sound information of considerable value to students and artificers."—*Hardware Trades Journal.*

Mechanical Manipulation.

THE MECHANIC'S WORKSHOP HANDYBOOK. A Practical Manual on Mechanical Manipulation. Embracing Information on various Handicraft Processes, with Useful Notes and Miscellaneous Memoranda. By PAUL N. HASLUCK, Crown 8vo, 2s. cloth.

"It is a book which should be found in every workshop, as it is one which will be continually referred to for a very great amount of standard information."—*Saturday Review.*

Model Engineering.

THE MODEL ENGINEER'S HANDYBOOK: A Practical Manual on Model Steam Engines. Embracing Information on the Tools, Materials and Processes Employed in their Construction. By PAUL N. HASLUCK. With upwards of 100 Illustrations. Crown 8vo, 2s. cloth.

"Mr. Hasluck's latest volume is of greater importance than would at first appear; and indeed he has produced a very good little book."—*Builder.*
"By carefully going through the work, amateurs may pick up an excellent notion of the construction of full-sized steam engines."—*Telegraphic Journal.*

Clock Repairing.

THE CLOCK JOBBER'S HANDYBOOK: A Practical Manual on Cleaning, Repairing and Adjusting. Embracing Information on the Tools, Materials, Appliances and Processes Employed in Clockwork. By PAUL N. HASLUCK. With upwards of 100 Illustrations. Cr. 8vo. 2s. cloth. [*Just ready.*

INDUSTRIAL AND USEFUL ARTS. 33

Electrolysis of Gold, Silver, Copper, etc.
ELECTRO-DEPOSITION : A Practical Treatise on the Electrolysis of Gold, Silver, Copper, Nickel, and other Metals and Alloys. With descriptions of Voltaic Batteries, Magneto and Dynamo-Electric Machines, Thermopiles, and of the Materials and Processes used in every Department of the Art, and several Chapters on Electro-Metallurgy. By ALEXANDER WATT, Author of "Electro-Metallurgy," &c. With numerous Illustrations. Third Edition, Revised and Enlarged. Crown 8vo, 9s. cloth.
"Eminently a book for the practical worker in electro-deposition. It contains minute and practical descriptions of methods, processes and materials as actually pursued and used in the workshop. Mr. Watt's book recommends itself to all interested in its subjects."—*Engineer.*

Electro-Metallurgy.
ELECTRO-METALLURGY; Practically Treated. By ALEXANDER WATT, Author of "Electro-Deposition," &c. Ninth Edition, including the most recent Processes. 12mo, 4s. cloth boards.
"From this book both amateur and artisan may learn everything necessary for the successful prosecution of electroplating."—*Iron.*

Electroplating.
ELECTROPLATING : A Practical Handbook on the Deposition of Copper, Silver, Nickel, Gold, Aluminium, Brass, Platinum, &c. &c. With Descriptions of the Chemicals, Materials, Batteries and Dynamo Machines used in the Art. By J. W. URQUHART, C.E., Author of "Electric Light," &c. Second Edition, Revised, with Additions. Numerous Illustrations. Crown 8vo, 5s. cloth.
"An excellent practical manual."—*Engineering.*
"This book will show any person how to become an expert in electro-deposition."—*Builder.*
"An excellent work, giving the newest information."—*Horological Journal.*

Electrotyping.
ELECTROTYPING : The Reproduction and Multiplication of Printing Surfaces and Works of Art by the Electro-deposition of Metals. By J. W. URQUHART, C.E. Crown 8vo, 5s. cloth.
"The book is thoroughly practical. The reader is, therefore, conducted through the leading laws of electricity, then through the metals used by electrotypers, the apparatus, and the depositing processes, up to the final preparation of the work."—*Art Journal.*

Goldsmiths' Work.
THE GOLDSMITH'S HANDBOOK. By GEORGE E. GEE, Jeweller, &c. Third Edition, considerably Enlarged. 12mo, 3s. 6d. cloth.
"A good, sound, technical educator, and will be generally accepted as an authority."—*Horological Journal.*
"A standard book which few will care to be without."—*Jeweller and Metalworker.*

Silversmiths' Work.
THE SILVERSMITH'S HANDBOOK. By GEORGE E. GEE, Jeweller, &c. Second Edition, Revised, with Illustrations. 12mo, 3s. 6d. cloth.
"The chief merit of the work is its practical character. . . . The workers in the trade will speedily discover its merits when they sit down to study it."—*English Mechanic.*
* *The above two works together, strongly half-bound, price 7s.*

Bread and Biscuit Baking.
THE BREAD AND BISCUIT BAKER'S AND SUGAR-BOILER'S ASSISTANT. Including a large variety of Modern Recipes. With Remarks on the Art of Bread-making. By ROBERT WELLS, Practical Baker. Crown 8vo, 2s. cloth. [*Just published.*
"A large number of wrinkles for the ordinary cook, as well as the baker."—*Saturday Review.*
"A book of instruction for learners and for daily reference in the bakehouse."—*Bakers' Times.*

Confectionery.
THE PASTRYCOOK AND CONFECTIONER'S GUIDE. For Hotels, Restaurants and the Trade in general, adapted also for Family Use. By ROBERT WELLS, Author of "The Bread and Biscuit Baker's and Sugar Boiler's Assistant." Crown 8vo, 2s. cloth. [*Just published.*
"We cannot speak too highly of this really excellent work. In these days of keen competition our readers cannot do better than purchase this book."—*Baker's Times.*
"Will be found as serviceable by private families as by restaurant chefs and victuallers in general."—*Miller.*

Laundry Work.
A HANDBOOK OF LAUNDRY MANAGEMENT. For Use in Steam and Hand-Power Laundries and Private Houses. By the Editor of THE LAUNDRY JOURNAL. Crown 8vo, 2s. 6d. cloth. [*Just published.*

Horology.

A TREATISE ON MODERN HOROLOGY, in Theory and Practice. Translated from the French of CLAUDIUS SAUNIER, ex-Director of the School of Horology at Macon, by JULIEN TRIPPLIN, F.R.A.S., Besancon, Watch Manufacturer, and EDWARD RIGG, M.A., Assayer in the Royal Mint. With Seventy-eight Woodcuts and Twenty-two Coloured Copper Plates. Second Edition. Super-royal 8vo, £2 2s. cloth; £2 10s. half-calf.

"There is no horological work in the English language at all to be compared to this production of M. Saunier's for clearness and completeness. It is alike good as a guide for the student and as a reference for the experienced horologist and skilled workman."—*Horological Journal*.
"The latest, the most complete, and the most reliable of those literary productions to which continental watchmakers are indebted for the mechanical superiority over their English brethren—in fact, the Book of Books, is M. Saunier's 'Treatise.'"—*Watchmaker, Jeweller and Silversmith*.

Watchmaking.

THE WATCHMAKER'S HANDBOOK. Translated from the French of CLAUDIUS SAUNIER, and considerably Enlarged by JULIEN TRIPPLIN, F.R.A.S., Vice-President of the Horological Institute, and EDWARD RIGG, M.A., Assayer in the Royal Mint. With Numerous Woodcuts and Fourteen Copper Plates. Second Edition, Revised. With Appendix. Cr. 8vo, 9s. cloth.

"Each part is truly a treatise in itself. The arrangement is good and the language is clear and concise. It is an admirable guide for the young watchmaker."—*Engineering*.
"It is impossible to speak too highly of its excellence. It fulfils every requirement in a handbook intended for the use of a workman. Should be found in every workshop."—*Watch and Clockmaker*.

CHEMICAL MANUFACTURES & COMMERCE.

Alkali Trade, Manufacture of Sulphuric Acid, etc.

A MANUAL OF THE ALKALI TRADE, including the Manufacture of Sulphuric Acid, Sulphate of Soda, and Bleaching Powder. By JOHN LOMAS, Alkali Manufacturer, Newcastle-upon-Tyne and London. With 232 Illustrations and Working Drawings, and containing 390 pages of Text. Second Edition, with Additions. Super-royal 8vo, £1 10s. cloth.

"This book is written by a manufacturer for manufacturers. The working details of the most approved forms of apparatus are given, and these are accompanied by no less than 232 wood engravings, all of which may be used for the purposes of construction. Every step in the manufacture is very fully described in this manual, and each improvement explained."—*Athenæum*.
"We find here not merely a sound and luminous explanation of the chemical principles of the trade, but a notice of numerous matters which have a most important bearing on the successful conduct of alkali works, but which are generally overlooked by even experienced technological authors."—*Chemical Review*.

Brewing.

A HANDBOOK FOR YOUNG BREWERS. By HERBERT EDWARDS WRIGHT, B.A. Crown 8vo, 3s. 6d. cloth.

"This little volume, containing such a large amount of good sense in so small a compass, ought to recommend itself to every brewery pupil, and many who have passed that stage."—*Brewers' Guardian*.
"The book is very clearly written, and the author has successfully brought his scientific knowledge to bear upon the various processes and details of brewing."—*Brewer*.

Commercial Chemical Analysis.

THE COMMERCIAL HANDBOOK OF CHEMICAL ANALYSIS; or, Practical Instructions for the determination of the Intrinsic or Commercial Value of Substances used in Manufactures, in Trades, and in the Arts. By A. NORMANDY, Editor of Rose's "Treatise on Chemical Analysis." New Edition, to a great extent Re-written by HENRY M. NOAD, Ph.D., F.R.S. With numerous Illustrations. Crown 8vo, 12s. 6d. cloth.

"We strongly recommend this book to our readers as a guide, alike indispensable to the housewife as to the pharmaceutical practitioner."—*Medical Times*.
"Essential to the analysts appointed under the new Act. The most recent results are given, and the work is well edited and carefully written."—*Nature*.

Explosives.

A HANDBOOK OF MODERN EXPLOSIVES. Being a Practical Treatise on the Manufacture and Application of Dynamite, Gun-Cotton, Nitro-Glycerine, and other Explosive Compounds. By M. EISSLER, Mining Engineer, Author of "The Metallurgy of Gold," "The Metallurgy of Silver," &c. With about 100 Illustrations. Crown 8vo. [*In the press*.

AGRICULTURE, FARMING, GARDENING, etc. 35

Dye-Wares and Colours.

THE MANUAL OF COLOURS AND DYE-WARES : Their Properties, Applications, Valuation, Impurities, and Sophistications. For the use of Dyers, Printers, Drysalters, Brokers, &c. By J. W. SLATER. Second Edition, Revised and greatly Enlarged. Crown 8vo, 7s. 6d. cloth.

"A complete encyclopædia of the *materia tinctoria*. The information given respecting each article is full and precise, and the methods of determining the value of articles such as these, so liable to sophistication, are given with clearness, and are practical as well as valuable."—*Chemist and Druggist.*

"There is no other work which covers precisely the same ground. To students preparing for examinations in dyeing and printing it will prove exceedingly useful."—*Chemical News.*

Pigments.

THE ARTIST'S MANUAL OF PIGMENTS. Showing their Composition, Conditions of Permanency, Non-Permanency, and Adulterations; Effects in Combination with Each Other and with Vehicles ; and the most Reliable Tests of Purity. Together with the Science and Arts Department's Examination Questions on Painting. By H. C. STANDAGE. Second Edition, Revised. Small crown 8vo, 2s. 6d. cloth.

"This work is indeed *multum-in-parvo*, and we can, with good conscience, recommend it to all who come in contact with pigments, whether as makers, dealers or users."—*Chemical Review.*

"This manual cannot fail to be a very valuable aid to all painters who wish their work to endure and be of a sound character; it is complete and comprehensive."—*Spectator.*

"The author supplies a great deal of very valuable information and memoranda as to the chemical qualities and artistic effect of the principal pigments used by painters."—*Builder.*

Gauging. Tables and Rules for Revenue Officers, Brewers, etc.

A POCKET BOOK OF MENSURATION AND GAUGING : Containing Tables, Rules and Memoranda for Revenue Officers, Brewers, Spirit Merchants, &c. By J. B. MANT (Inland Revenue). Oblong 18mo, 4s. leather, with elastic band.

"This handy and useful book is adapted to the requirements of the Inland Revenue Department, and will be a favourite book of reference. The range of subjects is comprehensive, and the arrangement simple and clear."—*Civilian.*

"A most useful book. It should be in the hands of every practical brewer."—*Brewers' Journal.*

AGRICULTURE, FARMING, GARDENING, etc.

Agricultural Facts and Figures.

NOTE-BOOK OF AGRICULTURAL FACTS AND FIGURES FOR FARMERS AND FARM STUDENTS. By PRIMROSE MCCONNELL, Fellow of the Highland and Agricultural Society ; late Professor of Agriculture, Glasgow Veterinary College. Third Edition. Royal 32mo, full roan, gilt edges, with elastic band, 4s.

"The most complete and comprehensive Note-book for Farmers and Farm Students that we have seen. It literally teems with information, and we can cordially recommend it to all connected with agriculture."—*North British Agriculturist.*

Youatt and Burn's Complete Grazier.

THE COMPLETE GRAZIER, and FARMER'S and CATTLE-BREEDER'S ASSISTANT. A Compendium of Husbandry; especially in the departments connected with the Breeding, Rearing, Feeding, and General Management of Stock; the Management of the Dairy, &c. With Directions for the Culture and Management of Grass Land, of Grain and Root Crops, the Arrangement of Farm Offices, the use of Implements and Machines, and on Draining, Irrigation, Warping, &c. ; and the Application and Relative Value of Manures. By WILLIAM YOUATT, Esq., V.S. Twelfth Edition, Enlarged by ROBERT SCOTT BURN, Author of "Outlines of Modern Farming," "Systematic Small Farming," &c. One large 8vo volume, 860 pp., with 244 Illustrations, £1 1s. half-bound.

"The standard and text-book with the farmer and grazier."—*Farmer's Magazine.*

"A treatise which will remain a standard work on the subject as long as British agriculture endures."—*Mark Lane Express* (First Notice).

"The book deals with all departments of agriculture, and contains an immense amount of valuable information. It is, in fact, an encyclopædia of agriculture put into readable form, and it is the only work equally comprehensive brought down to present date. It is excellently printed on thick paper, and strongly bound, and deserves a place in the library of every agriculturist."—*Mark Lane Express* (Second Notice).

"This esteemed work is well worthy of a place in the libraries of agriculturists."—*North British Agriculturist.*

Flour Manufacture, Milling, etc.

FLOUR MANUFACTURE: A Treatise on Milling Science and Practice. By FRIEDRICH KICK, Imperial Regierungsrath, Professor of Mechanical Technology in the Imperial German Polytechnic Institute, Prague. Translated from the Second Enlarged and Revised Edition with Supplement. By H. H. P. POWLES, A.M.I.C.E. Nearly 400 pp. Illustrated with 28 Folding Plates, and 167 Woodcuts. Royal 8vo, 25s. cloth.

"This valuable work is, and will remain, the standard authority on the science of milling. . . The miller who has read and digested this work will have laid the foundation, so to speak, of a successful career; he will have acquired a number of general principles which he can proceed to apply. In this handsome volume we at last have the accepted text-book of modern milling in good, sound English, which has little, if any, trace of the German idiom."—*The Miller.*

"The appearance of this celebrated work in English is very opportune, and British millers will, we are sure, not be slow in availing themselves of its pages."—*Millers' Gazette.*

Small Farming.

SYSTEMATIC SMALL FARMING; or, The Lessons of my Farm. Being an Introduction to Modern Farm Practice for Small Farmers. By ROBERT SCOTT BURN, Author of "Outlines of Modern Farming." With numerous Illustrations, crown 8vo, 6s. cloth.

"This is the completest book of its class we have seen, and one which every amateur farmer will read with pleasure and accept as a guide."—*Field.*

"The volume contains a vast amount of useful information. No branch of farming is left untouched, from the labour to be done to the results achieved. It may be safely recommended to all who think they will be in paradise when they buy or rent a three-acre farm."—*Glasgow Herald.*

Modern Farming.

OUTLINES OF MODERN FARMING. By R. SCOTT BURN. Soils, Manures, and Crops—Farming and Farming Economy—Cattle, Sheep, and Horses — Management of Dairy, Pigs and Poultry — Utilisation of Town-Sewage, Irrigation, &c. Sixth Edition. In One Vol., 1,250 pp., half-bound, profusely Illustrated, 12s.

"The aim of the author has been to make his work at once comprehensive and trustworthy, and in this aim he has succeeded to a degree which entitles him to much credit."—*Morning Advertiser.* "No farmer should be without this book."—*Banbury Guardian.*

Agricultural Engineering.

FARM ENGINEERING, THE COMPLETE TEXT-BOOK OF. Comprising Draining and Embanking; Irrigation and Water Supply; Farm Roads, Fences, and Gates; Farm Buildings, their Arrangement and Construction, with Plans and Estimates; Barn Implements and Machines; Field Implements and Machines; Agricultural Surveying, Levelling, &c. By Prof. JOHN SCOTT, Professor of Agriculture at the Royal Agricultural College, Cirencester, &c. In One Vol., 1,150 pages, half-bound, 600 Illustrations, 12s.

"Written with great care, as well as with knowledge and ability. The author has done his work well; we have found him a very trustworthy guide wherever we have tested his statements. The volume will be of great value to agricultural students."—*Mark Lane Express.*

"For a young agriculturist we know of no handy volume so likely to be more usefully studied."—*Bell's Weekly Messenger.*

English Agriculture.

THE FIELDS OF GREAT BRITAIN : A Text-Book of Agriculture, adapted to the Syllabus of the Science and Art Department. For Elementary and Advanced Students. By HUGH CLEMENTS (Board of Trade). Second Edition, Revised and Enlarged. 18mo, 2s. 6d. cloth.

"A most comprehensive volume, giving a mass of information."—*Agricultural Economist.*

"It is a long time since we have seen a book which has pleased us more, or which contains such a vast and useful fund of knowledge."—*Educational Times.*

New Pocket Book for Farmers.

TABLES, MEMORANDA, AND CALCULATED RESULTS for Farmers, Graziers, Agricultural Students, Surveyors, Land Agents Auctioneers, etc. With a New System of Farm Book-keeping. Selected and Arranged by SIDNEY FRANCIS. Second Edition, Revised. 272 pp., waistcoat-pocket size, 1s. 6d., limp leather. [*Just published.*

"Weighing less than 1 oz., and occupying no more space than a match box, it contains a mass of facts and calculations which has never before, in such handy form, been obtainable. Every operation on the farm is dealt with. The work may be taken as thoroughly accurate, having been revised by Dr. Fream. We cordially recommend it."—*Bell's Weekly Messenger.*

"A marvellous little book. . . . The agriculturist who possesses himself of it will not be disappointed with his investment."—*The Farm.*

AGRICULTURE, FARMING, GARDENING, etc. 37

Farm and Estate Book-keeping.
BOOK-KEEPING FOR FARMERS & ESTATE OWNERS. A Practical Treatise, presenting, in Three Plans, a System adapted to all Classes of Farms. By JOHNSON M. WOODMAN, Chartered Accountant. Second Edition, Revised. Crown 8vo, 3s. 6d. cloth boards; or 2s. 6d. cloth limp.
"The volume is a capital study of a most important subject."—*Agricultural Gazette.*
"Will be found of great assistance by those who intend to commence a system of book-keeping, the author's examples being clear and explicit, and his explanations, while full and accurate, being to a large extent free from technicalities."—*Live Stock Journal.*

Farm Account Book.
WOODMAN'S YEARLY FARM ACCOUNT BOOK. Giving a Weekly Labour Account and Diary, and showing the Income and Expenditure under each Department of Crops, Live Stock, Dairy, &c. &c. With Valuation, Profit and Loss Account, and Balance Sheet at the end of the Year, and an Appendix of Forms. Ruled and Headed for Entering a Complete Record of the Farming Operations. By JOHNSON M. WOODMAN, Chartered Accountant, Author of "Book-keeping for Farmers." Folio, 7s. 6d. half bound. [*culture.*
"Contains every requisite form for keeping farm accounts readily and accurately."—*Agri-*

Early Fruits, Flowers and Vegetables.
THE FORCING GARDEN; or, How to Grow Early Fruits, Flowers, and Vegetables. With Plans and Estimates for Building Glasshouses, Pits and Frames. Containing also Original Plans for Double Glazing, a New Method of Growing the Gooseberry under Glass, &c. &c., and on Ventilation, &c. With Illustrations. By SAMUEL WOOD. Crown 8vo, 3s. 6d. cloth.
"A good book, and fairly fills a place that was in some degree vacant. The book is written with great care, and contains a great deal of valuable teaching."—*Gardeners' Magazine.*
"Mr. Wood's book is an original and exhaustive answer to the question 'How to Grow Early Fruits, Flowers and Vegetables?'"—*Land and Water.*

Good Gardening.
A PLAIN GUIDE TO GOOD GARDENING; or, How to Grow Vegetables, Fruits, and Flowers. With Practical Notes on Soils, Manures, Seeds, Planting, Laying-out of Gardens and Grounds, &c. By S. WOOD. Third Edition, with considerable Additions, &c., and numerous Illustrations. Crown 8vo, 5s. cloth.
"A very good book, and one to be highly recommended as a practical guide. The practical directions are excellent."—*Athenæum.*
"May be recommended to young gardeners, cottagers and amateurs, for the plain and trustworthy information it gives on common matters too often neglected."—*Gardeners' Chronicle.*

Gainful Gardening.
MULTUM-IN-PARVO GARDENING; or, How to make One Acre of Land produce £620 a-year by the Cultivation of Fruits and Vegetables; also, How to Grow Flowers in Three Glass Houses, so as to realise £176 per annum clear Profit. By SAMUEL WOOD, Author of "Good Gardening," &c. Fourth and cheaper Edition, Revised, with Additions. Crown 8vo, 1s. sewed.
"We are bound to recommend it as not only suited to the case of the amateur and gentleman's gardener, but to the market grower."—*Gardeners' Magazine.*

Gardening for Ladies.
THE LADIES' MULTUM-IN-PARVO FLOWER GARDEN, and Amateurs' Complete Guide. By S. WOOD. Crown 8vo, 3s. 6d. cloth.
"This volume contains a good deal of sound, common sense instruction."—*Florist.*
"Full of shrewd hints and useful instructions, based on a lifetime of experience."—*Scotsman.*

Receipts for Gardeners.
GARDEN RECEIPTS. Edited by CHARLES W. QUIN. 12mo, 1s. 6d. cloth limp.
"A useful and handy book, containing a good deal of valuable information."—*Athenæum.*

Market Gardening.
MARKET AND KITCHEN GARDENING. By Contributors to "The Garden." Compiled by C. W. SHAW, late Editor of "Gardening Illustrated." 12mo, 3s. 6d. cloth boards. [*Just published.*
"The most valuable compendium of kitchen and market-garden work published."—*Farmer.*

Cottage Gardening.
COTTAGE GARDENING; or, Flowers, Fruits, and Vegetables for Small Gardens. By E. HOBDAY. 12mo, 1s. 6d. cloth limp.
"Contains much useful information at a small charge."—*Glasgow Herald.*

38   CROSBY LOCKWOOD & SON'S CATALOGUE.

ESTATE MANAGEMENT, AUCTIONEERING, LAW, etc.

Hudson's Land Valuer's Pocket-Book.
THE LAND VALUER'S BEST ASSISTANT: Being Tables on a very much Improved Plan, for Calculating the Value of Estates. With Tables for reducing Scotch, Irish, and Provincial Customary Acres to Statute Measure, &c. By R. HUDSON, C.E. New Edition. Royal 32mo, leather, elastic band, 4s.

"This new edition includes tables or ascertaining the value of leases for any term of years; and for showing how to lay out plots of ground of certain acres in forms, square, round, &c., with valuable rules for ascertaining the probable worth of standing timber to any amount; and is of incalculable value to the country gentleman and professional man."—*Farmers' Journal.*

Ewart's Land Improver's Pocket-Book.
THE LAND IMPROVER'S POCKET-BOOK OF FORMULÆ, TABLES and MEMORANDA required in any Computation relating to the Permanent Improvement of Landed Property. By JOHN EWART, Land Surveyor and Agricultural Engineer. Second Edition, Revised. Royal 32mo, oblong, leather, gilt edges, with elastic band, 4s.

"A compendious and handy little volume."—*Spectator.*

Complete Agricultural Surveyor's Pocket-Book.
THE LAND VALUER'S AND LAND IMPROVER'S COMPLETE POCKET-BOOK. Consisting of the above Two Works bound together. Leather, gilt edges, with strap, 7s. 6d.

"Hudson's book is the best ready-reckoner on matters relating to the valuation of land and crops, and its combination with Mr. Ewart's work greatly enhances the value and usefulness of the latter-mentioned. . . . It is most useful as a manual for reference."—*North of England Farmer.*

Auctioneer's Assistant.
THE APPRAISER, AUCTIONEER, BROKER, HOUSE AND ESTATE AGENT AND VALUER'S POCKET ASSISTANT, for the Valuation for Purchase, Sale, or Renewal of Leases, Annuities and Reversions, and of property generally; with Prices for Inventories, &c. By JOHN WHEELER, Valuer, &c. Fifth Edition, re-written and greatly extended by C. NORRIS, Surveyor, Valuer, &c. Royal 32mo, 5s. cloth.

"A neat and concise book of reference, containing an admirable and clearly-arranged list of prices for inventories, and a very practical guide to determine the value of furniture, &c."—*Standard.*

"Contains a large quantity of varied and useful information as to the valuation for purchase, sale, or renewal of leases, annuities and reversions, and of property generally, with prices for inventories, and a guide to determine the value of interior fittings and other effects."—*Builder.*

Auctioneering.
AUCTIONEERS: *Their Duties and Liabilities.* By ROBERT SQUIBBS, Auctioneer. Demy 8vo, 10s. 6d. cloth.

"The position and duties of auctioneers treated compendiously and clearly."—*Builder.*
"Every auctioneer ought to possess a copy of this excellent work."—*Ironmonger.*
"Of great value to the profession. . . . We readily welcome this book from the fact that it treats the subject in a manner somewhat new to the profession."—*Estates Gazette.*

Legal Guide for Pawnbrokers.
THE PAWNBROKERS', FACTORS' AND MERCHANTS' GUIDE TO THE LAW OF LOANS AND PLEDGES. With the Statutes and a Digest of Cases on Rights and Liabilities, Civil and Criminal, as to Loans and Pledges of Goods, Debentures, Mercantile and other Securities. By H. C. FOLKARD, Esq., Barrister-at-Law, Author of "The Law of Slander and Libel," &c. With Additions and Corrections. Fcap. 8vo, 3s. 6d. cloth.

"This work contains simply everything that requires to be known concerning the department of the law of which it treats. We can safely commend the book as unique and very nearly perfect."—*Iron.*

"The task undertaken by Mr. Folkard has been very satisfactorily performed. . . . Such explanations as are needful have been supplied with great clearness and with due regard to brevity."—*City Press.*

ESTATE MANAGEMENT, AUCTIONEERING, LAW, etc. 39

How to Invest.
HINTS FOR INVESTORS : Being an Explanation of the Mode of Transacting Business on the Stock Exchange. To which are added Comments on the Fluctuations and Table of Quarterly Average prices of Consols since 1759. Also a Copy of the London Daily Stock and Share List. By WALTER M. PLAYFORD, Sworn Broker. Crown 8vo, 2s. cloth.
"An invaluable guide to investors and speculators."—*Bullionist*

Metropolitan Rating Appeals.
REPORTS OF APPEALS HEARD BEFORE THE COURT OF GENERAL ASSESSMENT SESSIONS, from the Year 1871 to 1885. By EDWARD RYDE and ARTHUR LYON RYDE. Fourth Edition, brought down to the Present Date, with an Introduction to the Valuation (Metropolis) Act, 1869, and an Appendix by WALTER C. RYDE, of the Inner Temple, Barrister-at-Law. 8vo, 16s. cloth.
"A useful work, occupying a place mid-way between a handbook for a lawyer and a guide to the surveyor. It is compiled by a gentleman eminent in his profession as a land agent, whose specialty, it is acknowledged, lies in the direction of assessing property for rating purposes."—*Land Agents' Record.*

House Property.
HANDBOOK OF HOUSE PROPERTY. A Popular and Practical Guide to the Purchase, Mortgage, Tenancy, and Compulsory Sale of Houses and Land, including the Law of Dilapidations and Fixtures; with Examples of all kinds of Valuations, Useful Information on Buildings, and Suggestive Elucidations of Fine Art. By E. L. TARBUCK, Architect and Surveyor. Fourth Edition, Enlarged. 12mo, 5s. cloth.
"The advice is thoroughly practical."—*Law Journal.*
"For all who have dealings with house property, this is an indispensable guide."—*Decoration.*
"Carefully brought up to date, and much improved by the addition of a division on fine art."
" A well-written and thoughtful work."—*Land Agents Record.*

Inwood's Estate Tables.
TABLES FOR THE PURCHASING OF ESTATES, Freehold, Copyhold, or Leasehold ; Annuities, Advowsons, etc., and for the Renewing of Leases held under Cathedral Churches, Colleges, or other Corporate bodies, for Terms of Years certain, and for Lives ; also for Valuing Reversionary Estates, Deferred Annuities, Next Presentations, &c. ; together with SMART'S Five Tables of Compound Interest, and an Extension of the same to Lower and Intermediate Rates. By W. INWOOD. 23rd Edition, with considerable Additions, and new and valuable Tables of Logarithms for the more Difficult Computations of the Interest of Money, Discount, Annuities, &c., by M. FEDOR THOMAN, of the Société Crédit Mobilier of Paris. Crown 8vo, 8s. cloth.
"Those interested in the purchase and sale of estates, and in the adjustment of compensation cases, as well as in transactions in annuities, life insurances, &c., will find the present edition of eminent service."—*Engineering.*
"'Inwood's Tables' still maintain a most enviable reputation. The new issue has been enriched by large additional contributions by M. Fedor Thoman, whose carefully arranged Tables cannot fail to be of the utmost utility."—*Mining Journal.*

Agricultural and Tenant-Right Valuation.
THE AGRICULTURAL AND TENANT-RIGHT-VALUER'S ASSISTANT. A Practical Handbook on Measuring and Estimating the Contents, Weights and Values of Agricultural Produce and Timber, the Values of Estates and Agricultural Labour, the Forms of Tenant-Right-Valuations, Scales of Compensation under the Agricultural Holdings Act, 1883, &c. &c. By TOM BRIGHT, Agricultural Surveyor. Crown 8vo, 3s. 6d. cloth.
"Full of tables and examples in connection with the valuation of tenant-right, estates, labour, contents, and weights of timber, and farm produce of all kinds."—*Agricultural Gazette.*
"An eminently practical handbook, full of practical tables and data of undoubted interest and value to surveyors and auctioneers in preparing valuations of all kinds."—*Farmer.*

Plantations and Underwoods.
POLE PLANTATIONS AND UNDERWOODS: A Practical Handbook on Estimating the Cost of Forming, Renovating, Improving and Grubbing Plantations and Underwoods, their Valuation for Purposes of Transfer, Rental, Sale or Assessment. By TOM BRIGHT, F.S.Sc., Author of "The Agricultural and Tenant-Right-Valuer's Assistant," &c. Crown 8vo, 3s. 6d. cloth. [*Just published.*
"Very useful to those actually engaged in managing wood."—*Bell's Weekly Messenger.*
"To valuers, foresters and agents it will be a welcome aid."—*North British Agriculturist.*
"Well calculated to assist the valuer in the discharge of his duties, and of undoubted interest and use both to surveyors and auctioneers in preparing valuations of all kinds.'—*Kent Herald.*

A Complete Epitome of the Laws of this Country.

EVERY MAN'S OWN LAWYER: A Handy-Book of the Principles of Law and Equity. By A BARRISTER. Twenty-sixth Edition. Reconstructed, Thoroughly Revised, and much Enlarged. Including the Legislation of the Two Sessions of 1888, and including careful digests of *The Local Government Act*, 1888; *County Electors Act*, 1888; *County Courts Act*, 1888; *Glebe Lands Act*, 1888; *Law of Libel Amendment Act*, 1888; *Patents, Designs and Trade Marks Act*, 1888; *Solicitors Act*, 1888; *Preferential Payments in Bankruptcy Act*, 1888; *Land Charges Registration and Searches Act*, 1888; *Trustees Act*, 1888, &c. Crown 8vo, 688 pp., price 6s. 8d. (saved at every consultation!), strongly bound in cloth. [*Just published.*

*** THE BOOK WILL BE FOUND TO COMPRISE (AMONGST OTHER MATTER)—

THE RIGHTS AND WRONGS OF INDIVIDUALS—MERCANTILE AND COMMERCIAL LAW—PARTNERSHIPS, CONTRACTS AND AGREEMENTS—GUARANTEES, PRINCIPALS AND AGENTS—CRIMINAL LAW—PARISH LAW—COUNTY COURT LAW—GAME AND FISHERY LAWS—POOR MEN'S LAWSUITS—LAWS OF BANKRUPTCY—WAGERS—CHEQUES, BILLS AND NOTES—COPYRIGHT—ELECTIONS AND REGISTRATION—INSURANCE—LIBEL AND SLANDER—MARRIAGE AND DIVORCE—MERCHANT SHIPPING—MORTGAGES—SETTLEMENTS—STOCK EXCHANGE PRACTICE—TRADE MARKS AND PATENTS—TRESPASS—NUISANCES—TRANSFER OF LAND—WILLS, &c. &c. Also LAW FOR LANDLORD AND TENANT—MASTER AND SERVANT—HEIRS—DEVISEES AND LEGATEES—HUSBAND AND WIFE—EXECUTORS AND TRUSTEES—GUARDIAN AND WARD—MARRIED WOMEN AND INFANTS—LENDER, BORROWER AND SURETIES—DEBTOR AND CREDITOR—PURCHASER AND VENDOR—COMPANIES—FRIENDLY SOCIETIES—CLERGYMEN—CHURCHWARDENS—MEDICAL PRACTITIONERS—BANKERS—FARMERS—CONTRACTORS—STOCK BROKERS—SPORTSMEN—GAMEKEEPERS—FARRIERS—HORSE DEALERS—AUCTIONEERS—HOUSE AGENTS—INNKEEPERS—BAKERS—MILLERS—PAWNBROKERS—SURVEYORS—RAILWAYS AND CARRIERS—CONSTABLES—SEAMEN—SOLDIERS, &c. &c.

☞ *The following subjects may be mentioned as amongst those which have received special attention during the revision in question:*—Marriage of British Subjects Abroad; Police Constables; Pawnbrokers; Intoxicating Liquors; Licensing; Domestic Servants; Landlord and Tenant; Vendors and Purchasers; Municipal Elections; Local Elections; Corrupt Practices at Elections; Public Health and Nuisances; Highways; Churchwardens; Legal and Illegal Ritual; Vestry Meetings; Rates.

It is believed that the extensions and amplifications of the present edition, while intended to meet the requirements of the ordinary Englishman, will also have the effect of rendering the book useful to the legal practitioner in the country.

One result of the reconstruction and revision, with the extensive additions thereby necessitated, has been *the enlargement of the book by nearly a hundred and fifty pages*, while the price remains as before.

The PUBLISHERS feel every confidence, therefore, that this standard work will continue to be regarded, as hitherto, as an absolute necessity FOR EVERY MAN OF BUSINESS AS WELL AS EVERY HEAD OF A FAMILY.

*** OPINIONS OF THE PRESS.

"It is a complete code of English Law, written in plain language, which all can understand. . . . Should be in the hands of every business man, and all who wish to abolish lawyers' bills."—*Weekly Times*.

"A useful and concise epitome of the law, compiled with considerable care. —*Law Magazine*.

"A concise, cheap and complete epitome of the English law. So plainly written that he who runs may read, and he who reads may understand."—*Figaro*.

"A dictionary of legal facts well put together. The book is a very useful one."—*Spectator*.

"A work which has long been wanted, which is thoroughly well done, and which we most cordially recommend."—*Sunday Times*.

Private Bill Legislation and Provisional Orders.

HANDBOOK FOR THE USE OF SOLICITORS AND ENGINEERS Engaged in Promoting Private Acts of Parliament and Provisional Orders, for the Authorization of Railways, Tramways, Works for the Supply of Gas and Water, and other undertakings of a like character. By L. LIVINGSTON MACASSEY, of the Middle Temple, Barrister-at-Law, and Member of the Institution of Civil Engineers; Author of "Hints on Water Supply." Demy 8vo, 950 pp., price 25s. cloth.

"The volume is a desideratum on a subject which can be only acquired by practical experience, and the order of procedure in Private Bill Legislation and Provisional Orders is followed. The author's suggestions and notes will be found of great value to engineers and others professionally engaged in this class of practice."—*Building News*.

"The author's double experience as an engineer and barrister has eminently qualified him for the task, and enabled him to approach the subject alike from an engineering and legal point of view. The volume will be found a great help both to engineers and lawyers engaged in promoting Private Acts of Parliament and Provisional Orders."—*Local Government Chronicle*.

OGDEN, SMALE AND CO. LIMITED, PRINTERS, GREAT SAFFRON HILL, E.C.

Weale's Rudimentary Series.

LONDON, 1862.
THE PRIZE MEDAL
Was awarded to the Publishers of
"WEALE'S SERIES."

A NEW LIST OF

WEALE'S SERIES
RUDIMENTARY SCIENTIFIC, EDUCATIONAL, AND CLASSICAL.

Comprising nearly Three Hundred and Fifty distinct works in almost every department of Science, Art, and Education, recommended to the notice of Engineers, Architects, Builders, Artisans, and Students generally, as well as to those interested in Workmen's Libraries, Literary and Scientific Institutions, Colleges, Schools, Science Classes, &c., &c.

☞ " WEALE'S SERIES includes Text-Books on almost every branch of Science and Industry, comprising such subjects as Agriculture, Architecture and Building, Civil Engineering, Fine Arts, Mechanics and Mechanical Engineering, Physical and Chemical Science, and many miscellaneous Treatises. The whole are constantly undergoing revision, and new editions, brought up to the latest discoveries in scientific research, are constantly issued. The prices at which they are sold are as low as their excellence is assured."—*American Literary Gazette.*

" Amongst the literature of technical education, WEALE'S SERIES has ever enjoyed a high reputation, and the additions being made by Messrs. CROSBY LOCKWOOD & SON render the series more complete, and bring the information upon the several subjects down to the present time."—*Mining Journal.*

" It is not too much to say that no books have ever proved more popular with, or more useful to, young engineers and others than the excellent treatises comprised in WEALE'S SERIES."—*Engineer.*

" The excellence of WEALE'S SERIES is now so well appreciated, that it would be wasting our space to enlarge upon their general usefulness and value."—*Builder.*

" The volumes of WEALE'S SERIES form one of the best collections of elementary technical books in any language."—*Architect.*

" WEALE'S SERIES has become a standard as well as an unrivalled collection of treatises in all branches of art and science."—*Public Opinion.*

PHILADELPHIA, 1876.
THE PRIZE MEDAL
Was awarded to the Publishers for
Books: Rudimentary, Scientific,
"WEALE'S SERIES," ETC.

CROSBY LOCKWOOD & SON,
7, STATIONERS' HALL COURT, LUDGATE HILL, LONDON, E.C.

WEALE'S RUDIMENTARY SCIENTIFIC SERIES.

*** The volumes of this Series are freely Illustrated with Woodcuts, or otherwise, where requisite. Throughout the following List it must be understood that the books are bound in limp cloth, unless otherwise stated; *but the volumes marked with a ‡ may also be had strongly bound in cloth boards for 6d. extra.*

N.B.—In ordering from this List it is recommended, as a means of facilitating business and obviating error, to quote the numbers affixed to the volumes, as well as the titles and prices.

CIVIL ENGINEERING, SURVEYING, ETC.

No.
31. *WELLS AND WELL-SINKING*. By JOHN GEO. SWINDELL, A.R.I.B.A., and G. R. BURNELL, C.E. Revised Edition. With a New Appendix on the Qualities of Water. Illustrated. 2s.
35. *THE BLASTING AND QUARRYING OF STONE*, for Building and other Purposes. By Gen. Sir J. BURGOYNE, Bart. 1s. 6d.
43. *TUBULAR, AND OTHER IRON GIRDER BRIDGES*, particularly describing the Britannia and Conway Tubular Bridges. By G. DRYSDALE DEMPSEY, C.E. Fourth Edition. 2s.
44. *FOUNDATIONS AND CONCRETE WORKS*, with Practical Remarks on Footings, Sand, Concrete, Béton, Pile-driving, Caissons, and Cofferdams, &c. By E. DOBSON. Fifth Edition. 1s. 6d.
60. *LAND AND ENGINEERING SURVEYING*. By T. BAKER, C.E. Fourteenth Edition, revised by Professor J. R. YOUNG. 2s.‡
80*. *EMBANKING LANDS FROM THE SEA*. With examples and Particulars of actual Embankments, &c. By J. WIGGINS, F.G.S. 2s.
81. *WATER WORKS*, for the Supply of Cities and Towns. With a Description of the Principal Geological Formations of England as influencing Supplies of Water, &c. By S. HUGHES, C.E. New Edition. 4s.‡
118. *CIVIL ENGINEERING IN NORTH AMERICA*, a Sketch of. By DAVID STEVENSON, F.R.S.E., &c. Plates and Diagrams. 3s.
167. *IRON BRIDGES, GIRDERS, ROOFS, AND OTHER WORKS*. By FRANCIS CAMPIN, C.E. 2s. 6d.‡
197. *ROADS AND STREETS*. By H. LAW, C.E., revised and enlarged by D. K. CLARK, C.E., including pavements of Stone, Wood, Asphalte, &c. 4s. 6d.‡
203. *SANITARY WORK IN THE SMALLER TOWNS AND IN VILLAGES*. By C. SLAGG, A.M.I.C.E. Revised Edition. 3s.‡
212. *GAS-WORKS, THEIR CONSTRUCTION AND ARRANGEMENT*; and the Manufacture and Distribution of Coal Gas. Originally written by SAMUEL HUGHES, C.E. Re-written and enlarged by WILLIAM RICHARDS, C.E. Seventh Edition, with important additions. 5s. 6d.‡
213. *PIONEER ENGINEERING*. A Treatise on the Engineering Operations connected with the Settlement of Waste Lands in New Countries. By EDWARD DOBSON, Assoc. Inst. C.E. 4s. 6d.‡
216. *MATERIALS AND CONSTRUCTION*; A Theoretical and Practical Treatise on the Strains, Designing, and Erection of Works of Construction. By FRANCIS CAMPIN, C.E. Second Edition, revised. 3s.‡
219. *CIVIL ENGINEERING*. By HENRY LAW, M.Inst. C.E. Including HYDRAULIC ENGINEERING by GEO. R. BURNELL, M.Inst. C.E. Seventh Edition, revised, with large additions by D. KINNEAR CLARK, M.Inst. C.E. 6s. 6d., Cloth boards, 7s. 6d.
268. *THE DRAINAGE OF LANDS, TOWNS, & BUILDINGS*. By G. D. DEMPSEY, C.E. Revised, with large Additions on Recent Practice in Drainage Engineering, by D. KINNEAR CLARK, M.I.C.E. Second Edition, Corrected. 4s. 6d.‡ [*Just published.*

☞ *The ‡ indicates that these vols. may be had strongly bound at 6d. extra.*

LONDON: CROSBY LOCKWOOD AND SON,

MECHANICAL ENGINEERING, ETC.

33. *CRANES*, the Construction of, and other Machinery for Raising Heavy Bodies. By JOSEPH GLYNN, F.R.S. Illustrated. 1s. 6d.
34. *THE STEAM ENGINE*. By Dr. LARDNER. Illustrated. 1s. 6d.
59. *STEAM BOILERS:* their Construction and Management. By R. ARMSTRONG, C.E. Illustrated. 1s. 6d.
82. *THE POWER OF WATER*, as applied to drive Flour Mills, and to give motion to Turbines, &c. By JOSEPH GLYNN, F.R.S. 2s.‡
98. *PRACTICAL MECHANISM*, the Elements of; and Machine Tools. By T. BAKER, C.E. With Additions by J. NASMYTH, C.E. 2s. 6d.‡
139. *THE STEAM ENGINE*, a Treatise on the Mathematical Theory of, with Rules and Examples for Practical Men. By T. BAKER, C.E. 1s. 6d.
164. *MODERN WORKSHOP PRACTICE*, as applied to Steam Engines, Bridges, Ship-building, Cranes, &c. By J. G. WINTON. Fourth Edition, much enlarged and carefully revised. 3s. 6d.‡ [*Just published.*
165. *IRON AND HEAT*, exhibiting the Principles concerned in the Construction of Iron Beams, Pillars, and Girders. By J. ARMOUR. 2s. 6d.‡
166. *POWER IN MOTION:* Horse-Power, Toothed-Wheel Gearing, Long and Short Driving Bands, and Angular Forces. By J. ARMOUR, 2s.‡
171. *THE WORKMAN'S MANUAL OF ENGINEERING* DRAWING. By J. MAXTON. 6th Edn. With 7 Plates and 350 Cuts. 3s. 6d.‡
190. *STEAM AND THE STEAM ENGINE*, Stationary and Portable. Being an Extension of the Elementary Treatise on the Steam Engine of Mr. JOHN SEWELL. By D. K. CLARK, M.I.C.E. 3s. 6d.‡
200. *FUEL*, its Combustion and Economy. By C. W. WILLIAMS With Recent Practice in the Combustion and Economy of Fuel—Coal, Coke Wood, Peat, Petroleum, &c.—by D. K. CLARK, M.I.C.E. 3s. 6d.‡
202. *LOCOMOTIVE ENGINES*. By G. D. DEMPSEY, C.E.; with large additions by D. KINNEAR CLARK, M.I.C.E. 3s.‡
211. *THE BOILERMAKER'S ASSISTANT* in Drawing, Templating, and Calculating Boiler and Tank Work. By JOHN COURTNEY, Practical Boiler Maker. Edited by D. K. CLARK, C.E. 100 Illustrations. 2s.
217. *SEWING MACHINERY:* Its Construction, History, &c., with full Technical Directions for Adjusting, &c. By J. W. URQUHART, C.E. 2s.‡
223. *MECHANICAL ENGINEERING*. Comprising Metallurgy, Moulding, Casting, Forging, Tools, Workshop Machinery, Manufacture of the Steam Engine, &c. By FRANCIS CAMPIN, C.E. Second Edition. 2s. 6d.‡
236. *DETAILS OF MACHINERY*. Comprising Instructions for the Execution of various Works in Iron. By FRANCIS CAMPIN, C.E. 3s.‡
237. *THE SMITHY AND FORGE;* including the Farrier's Art and Coach Smithing. By W. J. E. CRANE. Illustrated. 2s. 6d.‡
238. *THE SHEET-METAL WORKER'S GUIDE;* a Practical Handbook for Tinsmiths, Coppersmiths, Zincworkers, &c. With 94 Diagrams and Working Patterns. By W. J. E. CRANE. Second Edition, revised. 1s. 5d.
251. *STEAM AND MACHINERY MANAGEMENT:* with Hints on Construction and Selection. By M. POWIS BALE, M.I.M.E. 2s. 6d.‡
254. *THE BOILERMAKER'S READY-RECKONER*. By J. COURTNEY. Edited by D. K. CLARK, C.E. 4s., limp; 5s., half-bound.
255. *LOCOMOTIVE ENGINE-DRIVING*. A Practical Manual for Engineers in charge of Locomotive Engines. By MICHAEL REYNOLDS, M.S.E. Eighth Edition. 3s. 6d., limp; 4s. 6d. cloth boards.
256. *STATIONARY ENGINE-DRIVING*. A Practical Manual for Engineers in charge of Stationary Engines. By MICHAEL REYNOLDS, M.S.E. Third Edition. 3s. 6d., limp; 4s. 6d. cloth boards.
260. *IRON BRIDGES OF MODERATE SPAN:* their Construction and Erection. By HAMILTON W. PENDRED, C.E. 2s.

☞ *The ‡ indicates that these vols. may be had strongly bound at 6d. extra.*

7, STATIONERS' HALL COURT, LUDGATE HILL, E.C.

MINING, METALLURGY, ETC.

4. *MINERALOGY*, Rudiments of; a concise View of the General Properties of Minerals. By A. RAMSAY, F.G.S., F.R.G.S., &c. Third Edition, revised and enlarged. Illustrated. 3s. 6d.‡

117. *SUBTERRANEOUS SURVEYING*, with and without the Magnetic Needle. By T. FENWICK and T. BAKER, C.E. Illustrated. 2s. 6d.‡

135. *ELECTRO-METALLURGY*; Practically Treated. By ALEXANDER WATT. Ninth Edition, enlarged and revised, with additional Illustrations, and including the most recent Processes. 3s. 6d.‡

172. *MINING TOOLS*, Manual of. For the Use of Mine Managers, Agents, Students, &c. By WILLIAM MORGANS. 2s. 6d.

172*. *MINING TOOLS, ATLAS* of Engravings to Illustrate the above, containing 235 Illustrations, drawn to Scale. 4to. 4s. 6d.

176. *METALLURGY OF IRON*. Containing History of Iron Manufacture, Methods of Assay, and Analyses of Iron Ores, Processes of Manufacture of Iron and Steel, &c. By H. BAUERMAN, F.G.S. Sixth Edition, revised and enlarged. 5s.‡ [*Just published.*

180. *COAL AND COAL MINING*. By the late Sir WARINGTON W. SMYTH, M.A., F.R.S. Seventh Edition, revised. 3s. 6d.‡ [*Just published.*

195. *THE MINERAL SURVEYOR AND VALUER'S COMPLETE GUIDE*. By W. LINTERN, M.E. Third Edition, including Magnetic and Angular Surveying. With Four Plates. 3s. 6d.‡

214. *SLATE AND SLATE QUARRYING*, Scientific, Practical, and Commercial. By D. C. DAVIES, F.G.S., Mining Engineer, &c. 3s.‡

264. *A FIRST BOOK OF MINING AND QUARRYING*, with the Sciences connected therewith, for Primary Schools and Self Instruction. By J. H. COLLINS, F.G.S. Second Edition, with additions. 1s. 6d.

ARCHITECTURE, BUILDING, ETC.

16. *ARCHITECTURE—ORDERS*—The Orders and their Æsthetic Principles. By W. H. LEEDS. Illustrated. 1s. 6d.

17. *ARCHITECTURE—STYLES*—The History and Description of the Styles of Architecture of Various Countries, from the Earliest to the Present Period. By T. TALBOT BURY, F.R.I.B.A., &c. Illustrated. 2s.
** ORDERS AND STYLES OF ARCHITECTURE, in One Vol., 3s. 6d.

18. *ARCHITECTURE—DESIGN*—The Principles of Design in Architecture, as deducible from Nature and exemplified in the Works of the Greek and Gothic Architects. By E. L. GARBETT, Architect. Illustrated. 2s.6d.
** The three preceding Works, in One handsome Vol., half bound, entitled "MODERN ARCHITECTURE," price 6s.

22. *THE ART OF BUILDING*, Rudiments of. General Principles of Construction, Materials used in Building, Strength and Use of Materials, Working Drawings, Specifications, and Estimates. By E. DOBSON, 2s.‡

25. *MASONRY AND STONECUTTING*: Rudimentary Treatise on the Principles of Masonic Projection and their application to Construction. By EDWARD DOBSON, M.R.I.B.A., &c. 2s. 6d.‡

42. *COTTAGE BUILDING*. By C. BRUCE ALLEN, Architect. Tenth Edition, revised and enlarged. With a Chapter on Economic Cottages for Allotments, by EDWARD E. ALLEN, C.E. 2s.

45. *LIMES, CEMENTS, MORTARS, CONCRETES, MASTICS*, PLASTERING, &c. By G. R. BURNELL, C.E. Thirteenth Edition. 1s. 6d.

57. *WARMING AND VENTILATION*. An Exposition of the General Principles as applied to Domestic and Public Buildings, Mines, Lighthouses, Ships, &c. By C. TOMLINSON, F.R.S., &c. Illustrated. 3s.

111. *ARCHES, PIERS, BUTTRESSES, &c.*: Experimental Essays on the Principles of Construction. By W. BLAND. Illustrated. 1s. 6d.

☞ The ‡ indicates that these vols. may be had strongly bound at 6d. extra.

LONDON: CROSBY LOCKWOOD AND SON,

WEALE'S RUDIMENTARY SERIES. 5

Architecture, Building, etc., *continued.*

116. *THE ACOUSTICS OF PUBLIC BUILDINGS;* or, The Principles of the Science of Sound applied to the purposes of the Architect and Builder. By T. ROGER SMITH, M.R.I.B.A., Architect. Illustrated. 1s. 6d.
127. *ARCHITECTURAL MODELLING IN PAPER,* the Art of. By T. A. RICHARDSON, Architect. Illustrated. 1s. 6d.
128. *VITRUVIUS — THE ARCHITECTURE OF MARCUS VITRUVIUS POLLO.* In Ten Books. Translated from the Latin by JOSEPH GWILT, F.S.A., F.R.A.S. With 23 Plates. 5s.
130. *GRECIAN ARCHITECTURE,* An Inquiry into the Principles of Beauty in; with an Historical View of the Rise and Progress of the Art in Greece. By the EARL OF ABERDEEN. 1s.
** *The two preceding Works in One handsome Vol., half bound, entitled* "ANCIENT ARCHITECTURE," *price* 6s.
132. *THE ERECTION OF DWELLING-HOUSES.* Illustrated by a Perspective View, Plans, Elevations, and Sections of a pair of Semi-detached Villas, with the Specification, Quantities, and Estimates, &c. By S. H. BROOKS. New Edition, with Plates. 2s. 6d.‡
156. *QUANTITIES & MEASUREMENTS* in Bricklayers', Masons', Plasterers', Plumbers', Painters', Paperhangers', Gilders', Smiths', Carpenters' and Joiners' Work. By A. C. BEATON, Surveyor. New Edition. 1s. 6d.
175. *LOCKWOOD'S BUILDER'S PRICE BOOK FOR* 1891. A Comprehensive Handbook of the Latest Prices and Data for Builders, Architects, Engineers, and Contractors. Re-constructed, Re-written, and greatly Enlarged. By FRANCIS T. W. MILLER, A.R.I.B.A. 650 pages. 3s. 6d.; cloth boards, 4s. [*Just Published.*
182. *CARPENTRY AND JOINERY*—THE ELEMENTARY PRIN-CIPLES OF CARPENTRY. Chiefly composed from the Standard Work of THOMAS TREDGOLD, C.E. With a TREATISE ON JOINERY by E. WYNDHAM TARN, M.A. Fifth Edition, Revised. 3s. 6d.‡
182*. *CARPENTRY AND JOINERY. ATLAS* of 35 Plates to accompany the above. With Descriptive Letterpress. 4to. 6s.
185. *THE COMPLETE MEASURER;* the Measurement of Boards, Glass, &c.; Unequal-sided, Square-sided, Octagonal-sided, Round Timber and Stone, and Standing Timber, &c. By RICHARD HORTON. Fifth Edition. 4s.; strongly bound in leather, 5s.
187. *HINTS TO YOUNG ARCHITECTS.* By G. WIGHTWICK. New Edition. By G. H. GUILLAUME. Illustrated. 3s. 6d.‡
188. *HOUSE PAINTING, GRAINING, MARBLING, AND SIGN WRITING:* with a Course of Elementary Drawing for House-Painters, Sign-Writers, &c., and a Collection of Useful Receipts. By ELLIS A. DAVIDSON. Fifth Edition. With Coloured Plates. 5s. cloth limp; 6s. cloth boards.
189. *THE RUDIMENTS OF PRACTICAL BRICKLAYING.* In Six Sections: General Principles; Arch Drawing, Cutting, and Setting; Pointing; Paving, Tiling, Materials; Slating and Plastering; Practical Geometry, Mensuration, &c. By ADAM HAMMOND. Seventh Edition. 1s. 6d.
191. *PLUMBING.* A Text-Book to the Practice of the Art or Craft of the Plumber. With Chapters upon House Drainage and Ventilation. Fifth Edition. With 380 Illustrations. By W. P. BUCHAN. 3s. 6d.‡
192. *THE TIMBER IMPORTER'S, TIMBER MERCHANT'S,* and BUILDER'S STANDARD GUIDE. By R. E. GRANDY. 2s.
206. *A BOOK ON BUILDING, Civil and Ecclesiastical,* including CHURCH RESTORATION. With the Theory of Domes and the Great Pyramid, &c. By Sir EDMUND BECKETT, Bart., LL.D., Q.C., F.R.A.S. 4s. 6d.‡
226. *THE JOINTS MADE AND USED BY BUILDERS* in the Construction of various kinds of Engineering and Architectural Works. By WYVILL J. CHRISTY, Architect. With upwards of 160 Engravings on Wood. 3s.‡
228. *THE CONSTRUCTION OF ROOFS OF WOOD AND IRON.* By E. WYNDHAM TARN, M.A., Architect. Second Edition, revised. 1s. 6d.

☞ *The ‡ indicates that these vols. may be had strongly bound at 6d. extra.*

7, STATIONERS' HALL COURT, LUDGATE HILL, E.C.

Architecture, Building, etc., *continued*.

229. *ELEMENTARY DECORATION:* as applied to the Interior and Exterior Decoration of Dwelling-Houses, &c. By J. W. FACEY. 2s.
257. *PRACTICAL HOUSE DECORATION.* A Guide to the Art of Ornamental Painting. By JAMES W. FACEY. 2s. 6d.
*** The two preceding Works, in One handsome Vol., half-bound, entitled "HOUSE DECORATION, ELEMENTARY AND PRACTICAL," *price* 5s.
230. *HANDRAILING.* Showing New and Simple Methods for finding the Pitch of the Plank, Drawing the Moulds, Bevelling, Jointing-up, and Squaring the Wreath. By GEORGE COLLINGS. Second Edition, Revised including A TREATISE ON STAIRBUILDING. Plates and Diagrams. 2s. 6d.
247. *BUILDING ESTATES:* a Rudimentary Treatise on the Development, Sale, Purchase, and General Management of Building Land. By FOWLER MAITLAND, Surveyor. Second Edition, revised. 2s.
248. *PORTLAND CEMENT FOR USERS.* By HENRY FAIJA, Assoc. M. Inst. C.E. Third Edition, corrected. Illustrated. 2s.
252. *BRICKWORK:* a Practical Treatise, embodying the General and Higher Principles of Bricklaying, Cutting and Setting, &c. By F. WALKER. Second Edition, Revised and Enlarged. 1s. 6d.
23. *THE PRACTICAL BRICK AND TILE BOOK.* Comprising: 189. BRICK AND TILE MAKING, by E. DOBSON, A.I.C.E.; PRACTICAL BRICKLAYING, by A. HAMMOND; BRICKCUTTING AND SETTING, by A. HAMMOND. 534 pp. with 270 Illustrations. 6s. Strongly half-bound.
253. *THE TIMBER MERCHANT'S, SAW-MILLER'S, AND IMPORTER'S FREIGHT-BOOK AND ASSISTANT.* By WM. RICHARDSON. With a Chapter on Speeds of Saw-Mill Machinery, &c. By M. POWIS BALE, A.M.Inst.C.E. 3s.‡
258. *CIRCULAR WORK IN CARPENTRY AND JOINERY.* A Practical Treatise on Circular Work of Single and Double Curvature. By GEORGE COLLINGS. Second Edition, 2s. 6d.
259. *GAS FITTING:* A Practical Handbook treating of every Description of Gas Laying and Fitting. By JOHN BLACK. With 122 Illustrations. 2s. 6d.‡
261. *SHORING AND ITS APPLICATION:* A Handbook for the Use of Students. By GEORGE H. BLAGROVE. 1s. 6d. [*Just published.*
265. *THE ART OF PRACTICAL BRICK CUTTING & SETTING.* By ADAM HAMMOND. With 90 Engravings. 1s. 6d. [*Just published.*
267. *THE SCIENCE OF BUILDING:* An Elementary Treatise on the Principles of Construction. Adapted to the Requirements of Architectural Students. By E. WYNDHAM TARN, M.A. Lond. Third Edition, Revised and Enlarged. With 59 Wood Engravings. 3s. 6d.‡ [*Just published.*
271. *VENTILATION:* a Text-book to the Practice of the Art of Ventilating Buildings, with a Supplementary Chapter upon Air Testing. By WILLIAM PATON BUCHAN, R.P., Sanitary and Ventilating Engineer, Author of "Plumbing," &c. 3s. 6d.‡ [*Just published.*

SHIPBUILDING, NAVIGATION, MARINE ENGINEERING, ETC.

51. *NAVAL ARCHITECTURE.* An Exposition of the Elementary Principles of the Science, and their Practical Application to Naval Construction. By J. PEAKE. Fifth Edition, with Plates and Diagrams. 3s. 6d.‡
53*. *SHIPS FOR OCEAN & RIVER SERVICE*, Elementary and Practical Principles of the Construction of. By H. A. SOMMERFELDT. 1s. 6d.
53**. *AN ATLAS OF ENGRAVINGS* to Illustrate the above. Twelve large folding plates. Royal 4to, cloth. 7s. 6d.
54. *MASTING, MAST-MAKING, AND RIGGING OF SHIPS,* Also Tables of Spars, Rigging, Blocks; Chain, Wire, and Hemp Ropes, &c., relative to every class of vessels. By ROBERT KIPPING, N.A. 2s.

☞ *The* ‡ *indicates that these vols. may be had strongly bound at* 6d. *extra.*

LONDON : CROSBY LOCKWOOD AND SON,

WEALE'S RUDIMENTARY SERIES.

Shipbuilding, Navigation, Marine Engineering, etc., *cont.*
54*. *IRON SHIP-BUILDING.* With Practical Examples and Details.
By JOHN GRANTHAM, C.E. Fifth Edition. 4s.
55. *THE SAILOR'S SEA BOOK:* a Rudimentary Treatise on
Navigation. By JAMES GREENWOOD, B.A. With numerous Woodcuts and
Coloured Plates. New and enlarged edition. By W. H. ROSSER. 2s. 6d.‡
80. *MARINE ENGINES AND STEAM VESSELS.* By ROBERT
MURRAY, C.E. Eighth Edition, thoroughly Revised, with Additions by the
Author and by GEORGE CARLISLE, C.E. 4s. 6d. limp; 5s. cloth boards.
83*bis*. *THE FORMS OF SHIPS AND BOATS.* By W. BLAND.
Seventh Edition, Revised, with numerous Illustrations and Models. 1s. 6d.
99. *NAVIGATION AND NAUTICAL ASTRONOMY,* in Theory
and Practice. By Prof. J. R. YOUNG. New Edition. 2s. 6d.
106. *SHIPS' ANCHORS,* a Treatise on. By G. COTSELL, N.A. 1s. 6d.
149. *SAILS AND SAIL-MAKING.* With Draughting, and the Centre
of Effort of the Sails; Weights and Sizes of Ropes; Masting, Rigging,
and Sails of Steam Vessels, &c. 12th Edition. By R. KIPPING, N.A., 2s. 6d.‡
155. *ENGINEER'S GUIDE TO THE ROYAL & MERCANTILE
NAVIES.* By a PRACTICAL ENGINEER. Revised by D. F. M'CARTHY. 3s.
55 *PRACTICAL NAVIGATION.* Consisting of The Sailor's
& Sea-Book. By JAMES GREENWOOD and W. H. ROSSER. Together with
204. the requisite Mathematical and Nautical Tables for the Working of the
Problems. By H. LAW, C.E., and Prof. J. R. YOUNG. 7s. Half-bound.

AGRICULTURE, GARDENING, ETC.
61*. *A COMPLETE READY RECKONER FOR THE ADMEA-
SUREMENT OF LAND,* &c. By A. ARMAN. Third Edition, revised
and extended by C. NORRIS, Surveyor, Valuer, &c. 2s.
131. *MILLER'S, CORN MERCHANT'S, AND FARMER'S
READY RECKONER.* Second Edition, with a Price List of Modern
Flour-Mill Machinery, by W. S. HUTTON, C.E. 2s.
140. *SOILS, MANURES, AND CROPS.* (Vol. 1. OUTLINES OF
MODERN FARMING.) By R. SCOTT BURN. Woodcuts. 2s.
141. *FARMING & FARMING ECONOMY,* Notes, Historical and
Practical, on. (Vol. 2. OUTLINES OF MODERN FARMING.) By R. SCOTT BURN. 3s.
142. *STOCK; CATTLE, SHEEP, AND HORSES.* (Vol. 3.
OUTLINES OF MODERN FARMING.) By R. SCOTT BURN. Woodcuts. 2s. 6d.
145. *DAIRY, PIGS, AND POULTRY,* Management of the. By
R. SCOTT BURN. (Vol. 4. OUTLINES OF MODERN FARMING.) 2s.
146. *UTILIZATION OF SEWAGE, IRRIGATION, AND
RECLAMATION OF WASTE LAND.* (Vol. 5. OUTLINES OF MODERN
FARMING.) By R. SCOTT BURN. Woodcuts. 2s. 6d.
** *Nos.* 140-1-2-5-6, *in One Vol., handsomely half-bound, entitled* "OUTLINES OF
MODERN FARMING." By ROBERT SCOTT BURN. *Price* 12s.
177. *FRUIT TREES,* The Scientific and Profitable Culture of. From
the French of DU BREUIL. Revised by GEO. GLENNY. 187 Woodcuts. 3s. 6d.‡
198. *SHEEP; THE HISTORY, STRUCTURE, ECONOMY, AND
DISEASES OF.* By W. C. SPOONER, M.R.V.C., &c. Fifth Edition,
enlarged, including Specimens ot New and Improved Breeds. 3s. 6d.‡
201. *KITCHEN GARDENING MADE EASY.* By GEORGE M. F.
GLENNY. Illustrated. 1s. 6d.‡
207. *OUTLINES OF FARM MANAGEMENT,* and the Organi-
zation of Farm Labour. By R. SCOTT BURN. 2s. 6d.‡
208. *OUTLINES OF LANDED ESTATES MANAGEMENT.*
By R. SCOTT BURN. 2s. 6d.‡
** *Nos.* 207 & 208 *in One Vol., handsomely half-bound, entitled* "OUTLINES OF
LANDED ESTATES AND FARM MANAGEMENT." By R. SCOTT BURN. *Price* 6s.

☞ *The ‡ indicates that these vols. may be had strongly bound at* 6d. *extra.*

7, STATIONERS' HALL COURT, LUDGATE HILL, E.C.

Agriculture, Gardening, etc., *continued.*

209. **THE TREE PLANTER AND PLANT PROPAGATOR.**
A Practical Manual on the Propagation of Forest Trees, Fruit Trees, Flowering Shrubs, Flowering Plants, &c. By SAMUEL WOOD. 2s.

210. **THE TREE PRUNER.** A Practical Manual on the Pruning of Fruit Trees, including also their Training and Renovation; also the Pruning of Shrubs, Climbers, and Flowering Plants. By SAMUEL WOOD. 1s. 6d.

∗ *Nos.* 209 & 210 *in One Vol., handsomely half-bound, entitled* "THE TREE PLANTER, PROPAGATOR, AND PRUNER." By SAMUEL WOOD. *Price* 3s. 6d.

218. **THE HAY AND STRAW MEASURER:** Being New Tables for the Use of Auctioneers, Valuers, Farmers, Hay and Straw Dealers, &c. By JOHN STEELE. Fourth Edition. 2s.

222. **SUBURBAN FARMING.** The Laying-out and Cultivation of Farms, adapted to the Produce of Milk, Butter, and Cheese, Eggs, Poultry, and Pigs. By Prof. JOHN DONALDSON and R. SCOTT BURN. 3s. 6d.‡

231. **THE ART OF GRAFTING AND BUDDING.** By CHARLES BALTET. With Illustrations. 2s. 6d.‡

232. **COTTAGE GARDENING;** or, Flowers, Fruits, and Vegetables for Small Gardens. By E. HOBDAY. 1s. 6d.

233. **GARDEN RECEIPTS.** Edited by CHARLES W. QUIN. 1s. 6d.

234. **MARKET AND KITCHEN GARDENING.** By C. W. SHAW, late Editor of "Gardening Illustrated." 3s.‡ [*Just published.*

239. **DRAINING AND EMBANKING.** A Practical Treatise, embodying the most recent experience in the Application of Improved Methods. By JOHN SCOTT, late Professor of Agriculture and Rural Economy at the Royal Agricultural College, Cirencester. With 68 Illustrations. 1s. 6d.

240. **IRRIGATION AND WATER SUPPLY.** A Treatise on Water Meadows, Sewage Irrigation, and Warping; the Construction of Wells, Ponds, and Reservoirs, &c. By Prof. JOHN SCOTT. With 34 Illus. 1s. 6d.

241. **FARM ROADS, FENCES, AND GATES.** A Practical Treatise on the Roads, Tramways, and Waterways of the Farm; the Principles of Enclosures; and the different kinds of Fences, Gates, and Stiles. By Professor JOHN SCOTT. With 75 Illustrations. 1s. 6d.

242. **FARM BUILDINGS.** A Practical Treatise on the Buildings necessary for various kinds of Farms, their Arrangement and Construction, with Plans and Estimates. By Prof. JOHN SCOTT. With 105 Illus. 2s.

243. **BARN IMPLEMENTS AND MACHINES.** A Practical Treatise on the Application of Power to the Operations of Agriculture; and on various Machines used in the Threshing-barn, in the Stock-yard, and in the Dairy, &c. By Prof. J. SCOTT. With 123 Illustrations. 2s.

244. **FIELD IMPLEMENTS AND MACHINES.** A Practical Treatise on the Varieties now in use, with Principles and Details of Construction, their Points of Excellence, and Management. By Professor JOHN SCOTT. With 138 Illustrations. 2s.

245. **AGRICULTURAL SURVEYING.** A Practical Treatise on Land Surveying, Levelling, and Setting-out; and on Measuring and Estimating Quantities, Weights, and Values of Materials, Produce, Stock, &c. By Prof. JOHN SCOTT. With 62 Illustrations. 1s. 6d.

∗ *Nos.* 239 *to* 245 *in One Vol., handsomely half-bound, entitled* "THE COMPLETE TEXT-BOOK OF FARM ENGINEERING." By Professor JOHN SCOTT. *Price* 12s.

250. **MEAT PRODUCTION.** A Manual for Producers, Distributors, &c. By JOHN EWART. 2s. 6d.‡

266. **BOOK-KEEPING FOR FARMERS & ESTATE OWNERS.** By J. M. WOODMAN, Chartered Accountant. 2s. 6d. cloth limp; 3s. 6d. cloth boards. [*Just published.*

☞ *The* ‡ *indicates that these vols. may be had strongly bound at 6d. extra.*

LONDON : CROSBY LOCKWOOD AND SON,

MATHEMATICS, ARITHMETIC, ETC.

32. *MATHEMATICAL INSTRUMENTS*, a Treatise on; Their Construction, Adjustment, Testing, and Use concisely Explained. By J. F. HEATHER, M.A. Fourteenth Edition, revised, with additions, by A. T. WALMISLEY, M.I.C.E., Fellow of the Surveyors' Institution. Original Edition, in 1 vol., Illustrated. 2s.‡ [*Just published.*
⁎ *In ordering the above, be careful to say, " Original Edition " (No. 32), to distinguish it from the Enlarged Edition in 3 vols. (Nos. 168-9-70.)*

76. *DESCRIPTIVE GEOMETRY*, an Elementary Treatise on; with a Theory of Shadows and of Perspective, extracted from the French of G. MONGE. To which is added, a description of the Principles and Practice of Isometrical Projection. By J. F. HEATHER, M.A. With 14 Plates. 2s.

178. *PRACTICAL PLANE GEOMETRY:* giving the Simplest Modes of Constructing Figures contained in one Plane and Geometrical Construction of the Ground. By J. F. HEATHER, M.A. With 215 Woodcuts. 2s.

83. *COMMERCIAL BOOK-KEEPING.* With Commercial Phrases and Forms in English, French, Italian, and German. By JAMES HADDON, M.A., Arithmetical Master of King's College School, London. 1s. 6d.

84. *ARITHMETIC*, a Rudimentary Treatise on: with full Explanations of its Theoretical Principles, and numerous Examples for Practice. By Professor J. R. YOUNG. Eleventh Edition. 1s. 6d.

84*. A KEY to the above, containing Solutions in full to the Exercises, together with Comments, Explanations, and Improved Processes, for the Use of Teachers and Unassisted Learners. By J. R. YOUNG. 1s. 6d.

85. *EQUATIONAL ARITHMETIC*, applied to Questions of Interest, Annuities, Life Assurance, and General Commerce; with various Tables by which all Calculations may be greatly facilitated. By W. HIPSLEY. 2s.

86. *ALGEBRA*, the Elements of. By JAMES HADDON, M.A. With Appendix, containing miscellaneous Investigations, and a Collection of Problems in various parts of Algebra. 2s.

86*. A KEY AND COMPANION to the above Book, forming an extensive repository of Solved Examples and Problems in Illustration of the various Expedients necessary in Algebraical Operations. By J. R. YOUNG. 1s. 6d.

88. *EUCLID*, THE ELEMENTS OF : with many additional Propositions
89. and Explanatory Notes: to which is prefixed, an Introductory Essay on Logic. By HENRY LAW, C.E. 2s. 6d.‡
⁎ *Sold also separately, viz.:—*
88. EUCLID, The First Three Books. By HENRY LAW, C.E. 1s. 6d.
89. EUCLID, Books 4, 5, 6, 11, 12. By HENRY LAW, C.E. 1s. 6d.

90. *ANALYTICAL GEOMETRY AND CONIC SECTIONS*, By JAMES HANN. A New Edition, by Professor J. R. YOUNG. 2s.‡

91. *PLANE TRIGONOMETRY*, the Elements of. By JAMES HANN, formerly Mathematical Master of King's College, London. 1s. 6d.

92. *SPHERICAL TRIGONOMETRY*, the Elements of. By JAMES HANN. Revised by CHARLES H. DOWLING, C.E. 1s.
⁎ *Or with " The Elements of Plane Trigonometry," in One Volume, 2s. 6d.*

93. *MENSURATION AND MEASURING.* With the Mensuration and Levelling of Land for the Purposes of Modern Engineering. By T. BAKER, C.E. New Edition by E. NUGENT, C.E. Illustrated. 1s. 6d.

101. *DIFFERENTIAL CALCULUS*, Elements of the. By W. S. B. WOOLHOUSE, F.R.A.S., &c. 1s. 6d.

102. *INTEGRAL CALCULUS*, Rudimentary Treatise on the. By HOMERSHAM COX, B.A. Illustrated. 1s.

136. *ARITHMETIC*, Rudimentary, for the Use of Schools and Self-Instruction. By JAMES HADDON, M.A. Revised by A. ARMAN. 1s. 6d.

137. A KEY TO HADDON'S RUDIMENTARY ARITHMETIC. By A. ARMAN. 1s. 6d.

☞ *The ‡ indicates that these vols. may be had strongly bound at 6d. extra.*

7, STATIONERS' HALL COURT, LUDGATE HILL, E.C.

WEALE'S RUDIMENTARY SERIES.

Mathematics, Arithmetic, etc., *continued.*
168. *DRAWING AND MEASURING INSTRUMENTS.* Including—I. Instruments employed in Geometrical and Mechanical Drawing, and in the Construction, Copying, and Measurement of Maps and Plans. II. Instruments used for the purposes of Accurate Measurement, and for Arithmetical Computations. By J. F. HEATHER, M.A. Illustrated. 1s. 6d.
169. *OPTICAL INSTRUMENTS.* Including (more especially) Telescopes, Microscopes, and Apparatus for producing copies of Maps and Plans by Photography. By J. F. HEATHER, M.A. Illustrated. 1s. 6d.
170. *SURVEYING AND ASTRONOMICAL INSTRUMENTS.* Including—I. Instruments Used for Determining the Geometrical Features of a portion of Ground. II. Instruments Employed in Astronomical Observations. By J. F. HEATHER, M.A. Illustrated. 1s. 6d.
*** *The above three volumes form an enlargement of the Author's original work "Mathematical Instruments." (See No. 32 in the Series.).*
168.⎫ *MATHEMATICAL INSTRUMENTS.* By J. F. HEATHER,
169. ⎬ M.A. Enlarged Edition, for the most part entirely re-written. The 3 Parts as
170.⎭ above, in One thick Volume. With numerous Illustrations. 4s. 6d.‡
158. *THE SLIDE RULE, AND HOW TO USE IT;* containing full, easy, and simple Instructions to perform all Business Calculations with unexampled rapidity and accuracy. By CHARLES HOARE, C.E. Fifth Edition. With a Slide Rule in tuck of cover. 2s. 6d.‡
196. *THEORY OF COMPOUND INTEREST AND ANNUITIES;* with Tables of Logarithms for the more Difficult Computations of Interest, Discount, Annuities, &c. By FÉDOR THOMAN. 4s.‡
199. *THE COMPENDIOUS CALCULATOR;* or, Easy and Concise Methods of Performing the various Arithmetical Operations required in Commercial and Business Transactions; together with Useful Tables. By D. O'GORMAN. Twenty-seventh Edition, carefully revised by C. NORRIS. 2s. 6d., cloth limp; 3s. 6d., strongly half-bound in leather.
204. *MATHEMATICAL TABLES,* for Trigonometrical, Astronomical, and Nautical Calculations; to which is prefixed a Treatise on Logarithms. By HENRY LAW, C.E. Together with a Series of Tables for Navigation and Nautical Astronomy. By Prof. J. R. YOUNG. New Edition. 4s.
204*. *LOGARITHMS.* With Mathematical Tables for Trigonometrical, Astronomical, and Nautical Calculations. By HENRY LAW, M.Inst.C.E. New and Revised Edition. (Forming part of the above Work). 3s.
221. *MEASURES, WEIGHTS, AND MONEYS OF ALL NATIONS,* and an Analysis of the Christian, Hebrew, and Mahometan Calendars. By W. S. B. WOOLHOUSE, F.R.A.S., F.S.S. Sixth Edition. 2s.‡
227. *MATHEMATICS AS APPLIED TO THE CONSTRUCTIVE ARTS.* Illustrating the various processes of Mathematical Investigation, by means of Arithmetical and Simple Algebraical Equations and Practical Examples. By FRANCIS CAMPIN, C.E. Second Edition. 3s.‡

PHYSICAL SCIENCE, NATURAL PHILOSOPHY, ETC.

1. *CHEMISTRY.* By Professor GEORGE FOWNES, F.R.S. With an Appendix on the Application of Chemistry to Agriculture. 1s.
2. *NATURAL PHILOSOPHY,* Introduction to the Study of. By C. TOMLINSON. Woodcuts. 1s. 6d.
6. *MECHANICS,* Rudimentary Treatise on. By CHARLES TOMLINSON. Illustrated. 1s. 6d.
7. *ELECTRICITY;* showing the General Principles of Electrical Science, and the purposes to which it has been applied. By Sir W. SNOW HARRIS, F.R.S., &c. With Additions by R. SABINE, C.E., F.S.A. 1s. 6d.
7*. *GALVANISM.* By Sir W. SNOW HARRIS. New Edition by ROBERT SABINE, C.E., F.S.A. 1s. 6d.
8. *MAGNETISM;* being a concise Exposition of the General Principles of Magnetical Science. By Sir W. SNOW HARRIS. New Edition, revised by H. M. NOAD, Ph.D. With 165 Woodcuts. 3s. 6d.‡

☞ *The ‡ indicates that these vols. may be had strongly bound at 6d. extra.*

LONDON: CROSBY LOCKWOOD AND SON,

Physical Science, Natural Philosophy, etc., *continued*.

11. *THE ELECTRIC TELEGRAPH;* its History and Progress; with Descriptions of some of the Apparatus. By R. SABINE, C.E., F.S.A. 3s.
12. *PNEUMATICS,* including Acoustics and the Phenomena of Wind Currents. for the Use of Beginners By CHARLES TOMLINSON, F.R.S. Fourth Edition, enlarged. Illustrated. 1s. 6d. [*Just published.*
72. *MANUAL OF THE MOLLUSCA;* a Treatise on Recent and Fossil Shells. By Dr. S. P. WOODWARD, A.L.S. Fourth Edition. With Plates and 300 Woodcuts. 7s. 6d., cloth.
96. *ASTRONOMY.* By the late Rev. ROBERT MAIN, M.A. Third Edition, by WILLIAM THYNNE LYNN, B.A., F.R.A.S. 2s.
97. *STATICS AND DYNAMICS,* the Principles and Practice of; embracing also a clear development of Hydrostatics, Hydrodynamics, and Central Forces. By T. BAKER, C.E. Fourth Edition. 1s. 6d.
173. *PHYSICAL GEOLOGY,* partly based on Major-General PORTLOCK's "Rudiments of Geology." By RALPH TATE, A.L.S., &c. Woodcuts. 2s.
174. *HISTORICAL GEOLOGY,* partly based on Major-General PORTLOCK's "Rudiments." By RALPH TATE, A.L.S., &c. Woodcuts. 2s. 6d.
173 & 174. *RUDIMENTARY TREATISE ON GEOLOGY,* Physical and Historical. Partly based on Major-General PORTLOCK's " Rudiments of Geology." By RALPH TATE, A.L.S., F.G.S., &c. In One Volume. 4s. 6d.‡
183 & 184. *ANIMAL PHYSICS,* Handbook of. By Dr. LARDNER, D.C.L., formerly Professor of Natural Philosophy and Astronomy in University College, Lond. With 520 Illustrations. In One Vol. 7s. 6d., cloth boards.
*** *Sold also in Two Parts, as follows :—*
183. ANIMAL PHYSICS. By Dr. LARDNER. Part I., Chapters I.—VII. 4s.
184. ANIMAL PHYSICS. By Dr. LARDNER. Part II., Chapters VIII.—XVIII. 3s.
269. *LIGHT:* an Introduction to the Science of Optics, for the Use of Students of Architecture, Engineering, and other Applied Sciences. By E. WYNDHAM TARN, M.A. 1s. 6d. [*Just published.*

FINE ARTS.

20. *PERSPECTIVE FOR BEGINNERS.* Adapted to Young Students and Amateurs in Architecture, Painting, &c. By GEORGE PYNE. 2s.
40 *GLASS STAINING, AND THE ART OF PAINTING ON GLASS.* From the German of Dr. GESSERT and EMANUEL OTTO FROMBERG. With an Appendix on THE ART OF ENAMELLING. 2s. 6d.
69. *MUSIC,* A Rudimentary and Practical Treatise on. With numerous Examples. By CHARLES CHILD SPENCER. 2s. 6d.
71. *PIANOFORTE,* The Art of Playing the. With numerous Exercises & Lessons from the Best Masters. By CHARLES CHILD SPENCER. 1s.6d.
69-71. *MUSIC & THE PIANOFORTE.* In one vol. Half bound, 5s.
181. *PAINTING POPULARLY EXPLAINED,* including Fresco, Oil, Mosaic, Water Colour, Water-Glass, Tempera, Encaustic, Miniature, Painting on Ivory, Vellum, Pottery, Enamel, Glass, &c. With Historical Sketches of the Progress of the Art by THOMAS JOHN GULLICK, assisted by JOHN TIMBS, F.S.A. Fifth Edition, revised and enlarged. 5s.‡
186. *A GRAMMAR OF COLOURING,* applied to Decorative Painting and the Arts. By GEORGE FIELD. New Edition, enlarged and adapted to the Use of the Ornamental Painter and Designer. By ELLIS A. DAVIDSON. With two new Coloured Diagrams, &c. 3s.‡
246. *A DICTIONARY OF PAINTERS, AND HANDBOOK FOR PICTURE AMATEURS;* including Methods of Painting, Cleaning, Relining and Restoring, Schools of Painting, &c. With Notes on the Copyists and Imitators of each Master. By PHILIPPE DARYL. 2s. 6d.‡

The ‡ indicates that these vols. may be had strongly bound at 6d. extra.

INDUSTRIAL AND USEFUL ARTS.

23. **BRICKS AND TILES**, Rudimentary Treatise on the Manufacture of. By E. Dobson, M.R.I.B.A. Illustrated, 3s.‡
67. **CLOCKS, WATCHES, AND BELLS**, a Rudimentary Treatise on. By Sir Edmund Beckett, LL.D., Q.C. Seventh Edition, revised and enlarged. 4s. 6d. limp; 5s. 6d. cloth boards.
83**. **CONSTRUCTION OF DOOR LOCKS**. Compiled from the Papers of A. C. Hobbs, and Edited by Charles Tomlinson, F.R.S. 2s. 6d.
162. **THE BRASS FOUNDER'S MANUAL**; Instructions for Modelling, Pattern-Making, Moulding, Turning, Filing, Burnishing, Bronzing, &c. With copious Receipts, &c. By Walter Graham. 2s.‡
205. **THE ART OF LETTER PAINTING MADE EASY**. By J. G. Badenoch. Illustrated with 12 full-page Engravings of Examples. 1s. 6d.
215. **THE GOLDSMITH'S HANDBOOK**, containing full Instructions for the Alloying and Working of Gold. By George E. Gee, 3s.‡
225. **THE SILVERSMITH'S HANDBOOK**, containing full Instructions for the Alloying and Working of Silver. By George E. Gee. 3s.‡
⁎ The two preceding Works, in One handsome Vol., half-bound, entitled "The Goldsmith's & Silversmith's Complete Handbook," 7s.
249. **THE HALL-MARKING OF JEWELLERY PRACTICALLY CONSIDERED**. By George E. Gee. 3s.‡
224. **COACH BUILDING**, A Practical Treatise, Historical and Descriptive. By J. W. Burgess. 2s. 6d.‡
235. **PRACTICAL ORGAN BUILDING**. By W. E. Dickson, M.A., Precentor of Ely Cathedral. Illustrated. 2s. 6d.‡
262. **THE ART OF BOOT AND SHOEMAKING**. By John Bedford Leno. Numerous Illustrations. Third Edition. 2s.
263. **MECHANICAL DENTISTRY**: A Practical Treatise on the Construction of the Various Kinds of Artificial Dentures, with Formulæ, Tables, Receipts, &c. By Charles Hunter. Third Edition. 3s.‡
270. **WOOD ENGRAVING**: A Practical and Easy Introduction to the Study of the Art. By W. N. Brown. 1s. 6d.

MISCELLANEOUS VOLUMES.

36. *A DICTIONARY OF TERMS used in ARCHITECTURE, BUILDING, ENGINEERING, MINING, METALLURGY, ARCHÆOLOGY, the FINE ARTS, &c.* By John Weale. Fifth Edition. Revised by Robert Hunt, F.R.S. Illustrated. 5s. limp; 6s. cloth boards.
50. **THE LAW OF CONTRACTS FOR WORKS AND SERVICES**. By David Gibbons. Third Edition, enlarged. 3s.‡
112. **MANUAL OF DOMESTIC MEDICINE**. By R. Gooding, B.A., M.D. A Family Guide in all Cases of Accident and Emergency. 2s.‡
112*. **MANAGEMENT OF HEALTH**. A Manual of Home and Personal Hygiene. By the Rev. James Baird, B.A. 1s.
150. **LOGIC**, Pure and Applied. By S. H. Emmens. 1s. 6d.
153. **SELECTIONS FROM LOCKE'S ESSAYS ON THE HUMAN UNDERSTANDING**. With Notes by S. H. Emmens. 2s.
154. **GENERAL HINTS TO EMIGRANTS**. 2s.
157. **THE EMIGRANT'S GUIDE TO NATAL**. By Robert James Mann, F.R.A.S., F.M.S. Second Edition. Map. 2s.
193. **HANDBOOK OF FIELD FORTIFICATION**. By Major W. W. Knollys, F.R.G.S. With 161 Woodcuts. 3s.‡
194. **THE HOUSE MANAGER**: Being a Guide to Housekeeping, Practical Cookery, Pickling and Preserving, Household Work, Dairy Management, &c. By An Old Housekeeper. 3s. 6d.‡
194, **HOUSE BOOK** (The). Comprising :—I. The House Manager. 112 & By an Old Housekeeper. II. Domestic Medicine. By R. Gooding, M.D. 112*. III. Management of Health. By J. Baird. In One Vol., half-bound, 6s.

☞ *The ‡ indicates that these vols may be had strongly bound at 6d. extra.*

LONDON : CROSBY LOCKWOOD AND SON,

EDUCATIONAL AND CLASSICAL SERIES.

HISTORY.

1. **England, Outlines of the History of;** more especially with reference to the Origin and Progress of the English Constitution. By WILLIAM DOUGLAS HAMILTON, F.S.A., of Her Majesty's Public Record Office. 4th Edition, revised. 5s.; cloth boards, 6s.
5. **Greece, Outlines of the History of;** in connection with the Rise of the Arts and Civilization in Europe. By W. DOUGLAS HAMILTON, of University College, London, and EDWARD LEVIEN, M.A., of Balliol College, Oxford. 2s. 6d.; cloth boards, 3s. 6d.
7. **Rome, Outlines of the History of:** from the Earliest Period to the Christian Era and the Commencement of the Decline of the Empire. By EDWARD LEVIEN, of Balliol College, Oxford. Map, 2s. 6d.; cl. bds. 3s. 6d.
9. **Chronology of History, Art, Literature, and Progress,** from the Creation of the World to the Present Time. The Continuation by W. D. HAMILTON, F.S.A. 3s.; cloth boards, 3s. 6d.
50. **Dates and Events in English History,** for the use of Candidates in Public and Private Examinations. By the Rev. E. RAND. 1s.

ENGLISH LANGUAGE AND MISCELLANEOUS.

11. **Grammar of the English Tongue,** Spoken and Written. With an Introduction to the Study of Comparative Philology. By HYDE CLARKE, D.C.L. Fourth Edition. 1s. 6d.
12. **Dictionary of the English Language,** as Spoken and Written. Containing above 100,000 Words. By HYDE CLARKE, D.C.L. 3s. 6d.; cloth boards, 4s. 6d.; complete with the GRAMMAR, cloth bds., 5s. 6d.
48. **Composition and Punctuation,** familiarly Explained for those who have neglected the Study of Grammar. By JUSTIN BRENAN. 18th Edition. 1s. 6d.
49. **Derivative Spelling-Book:** Giving the Origin of Every Word from the Greek, Latin, Saxon, German, Teutonic, Dutch, French, Spanish, and other Languages; with their present Acceptation and Pronunciation. By J. ROWBOTHAM, F.R.A.S. Improved Edition. 1s. 6d.
51. **The Art of Extempore Speaking:** Hints for the Pulpit, the Senate, and the Bar. By M. BAUTAIN, Vicar-General and Professor at the Sorbonne. Translated from the French. 8th Edition, carefully corrected. 2s. 6d.
53. **Places and Facts in Political and Physical Geography,** for Candidates in Examinations. By the Rev. EDGAR RAND, B.A. 1s.
54. **Analytical Chemistry,** Qualitative and Quantitative, a Course of. To which is prefixed, a Brief Treatise upon Modern Chemical Nomenclature and Notation. By WM. W. PINK and GEORGE E. WEBSTER. 2s.

THE SCHOOL MANAGERS' SERIES OF READING BOOKS,

Edited by the Rev. A. R. GRANT, Rector of Hitcham, and Honorary Canon of Ely; formerly H.M. Inspector of Schools.

INTRODUCTORY PRIMER, 3d.

	s. d.		s. d.
FIRST STANDARD	0 6	FOURTH STANDARD	1 2
SECOND ,,	0 10	FIFTH ,,	1 6
THIRD ,,	1 0	SIXTH ,,	1 6

LESSONS FROM THE BIBLE. Part I. Old Testament. 1s.
LESSONS FROM THE BIBLE. Part II. New Testament, to which is added THE GEOGRAPHY OF THE BIBLE, for very young Children. By Rev. C. THORNTON FORSTER. 1s. 2d. *₊* Or the Two Parts in One Volume. 2s.

7, STATIONERS' HALL COURT, LUDGATE HILL, E.C.

WEALE'S EDUCATIONAL AND CLASSICAL SERIES.

FRENCH.

24. **French Grammar.** With Complete and Concise Rules on the Genders of French Nouns. By G. L. STRAUSS, Ph.D. 1s. 6d.
25. **French-English Dictionary.** Comprising a large number of New Terms used in Engineering, Mining, &c. By ALFRED ELWES. 1s. 6d.
26. **English-French Dictionary.** By ALFRED ELWES. 2s.
25,26. **French Dictionary** (as above). Complete, in One Vol., 3s.; cloth boards, 3s. 6d. *** Or with the GRAMMAR, cloth boards, 4s. 6d.
47. **French and English Phrase Book:** containing Introductory Lessons, with Translations, several Vocabularies of Words, a Collection of suitable Phrases, and Easy Familiar Dialogues. 1s. 6d.

GERMAN.

39. **German Grammar.** Adapted for English Students, from Heyse's Theoretical and Practical Grammar, by Dr. G. L. STRAUSS. 1s. 6d.
40. **German Reader:** A Series of Extracts, carefully culled from the most approved Authors of Germany; with Notes, Philological and Explanatory. By G. L. STRAUSS, Ph.D. 1s.
41-43. **German Triglot Dictionary.** By N. E. S. A. HAMILTON. In Three Parts. Part I. German-French-English. Part II. English-German-French. Part III. French-German-English. 3s., or cloth boards, 4s.
41-43 & 39. **German Triglot Dictionary** (as above), together with German Grammar (No. 39), in One Volume, cloth boards, 5s.

ITALIAN.

27. **Italian Grammar,** arranged in Twenty Lessons, with a Course of Exercises. By ALFRED ELWES. 1s. 6d.
28. **Italian Triglot Dictionary,** wherein the Genders of all the Italian and French Nouns are carefully noted down. By ALFRED ELWES. Vol. 1. Italian-English-French. 2s. 6d.
30. **Italian Triglot Dictionary.** By A. ELWES. Vol. 2. English-French-Italian. 2s. 6d.
32. **Italian Triglot Dictionary.** By ALFRED ELWES. Vol. 3. French-Italian-English. 2s. 6d.
28,30,32. **Italian Triglot Dictionary** (as above). In One Vol., 7s. 6d. Cloth boards.

SPANISH AND PORTUGUESE.

34. **Spanish Grammar,** in a Simple and Practical Form. With a Course of Exercises. By ALFRED ELWES. 1s. 6d.
35. **Spanish-English and English-Spanish Dictionary.** Including a large number of Technical Terms used in Mining, Engineering, &c. with the proper Accents and the Gender of every Noun. By ALFRED ELWES 4s.; cloth boards, 5s. *** Or with the GRAMMAR, cloth boards, 6s.
55. **Portuguese Grammar,** in a Simple and Practical Form. With a Course of Exercises. By ALFRED ELWES. 1s. 6d.
56. **Portuguese-English and English-Portuguese Dictionary.** Including a large number of Technical Terms used in Mining, Engineering, &c., with the proper Accents and the Gender of every Noun. By ALFRED ELWES. Second Edition, Revised, 5s.; cloth boards, 6s. *** Or with the GRAMMAR, cloth boards, 7s.

HEBREW.

46*. **Hebrew Grammar.** By Dr. BRESSLAU. 1s. 6d.
44. **Hebrew and English Dictionary,** Biblical and Rabbinical; containing the Hebrew and Chaldee Roots of the Old Testament Post-Rabbinical Writings. By Dr. BRESSLAU. 6s.
46. **English and Hebrew Dictionary.** By Dr. BRESSLAU. 3s.
44,46. **Hebrew Dictionary** (as above), in Two Vols., complete, with 46*. the GRAMMAR, cloth boards, 12s.

LONDON: CROSBY LOCKWOOD AND SON,

LATIN.

19. **Latin Grammar.** Containing the Inflections and Elementary Principles of Translation and Construction. By the Rev. THOMAS GOODWIN, M.A., Head Master of the Greenwich Proprietary School. 1s. 6d.
20. **Latin-English Dictionary.** By the Rev. THOMAS GOODWIN, M.A. 2s.
22. **English-Latin Dictionary;** together with an Appendix of French and Italian Words which have their origin from the Latin. By the Rev. THOMAS GOODWIN, M.A. 1s. 6d.
20,22. **Latin Dictionary** (as above). Complete in One Vol., 3s. 6d. cloth boards, 4s. 6d. *** Or with the GRAMMAR, cloth boards, 5s. 6d.

LATIN CLASSICS. With Explanatory Notes in English.
1. **Latin Delectus.** Containing Extracts from Classical Authors, with Genealogical Vocabularies and Explanatory Notes, by H. YOUNG. 1s. 6d.
2. **Cæsaris Commentarii de Bello Gallico.** Notes, and a Geographical Register for the Use of Schools, by H. YOUNG. 2s.
3. **Cornelius Nepos.** With Notes. By H. YOUNG. 1s.
4. **Virgilii Maronis Bucolica et Georgica.** With Notes on the Bucolics by W. RUSHTON, M.A., and on the Georgics by H. YOUNG. 1s. 6d.
5. **Virgilii Maronis Æneis.** With Notes, Critical and Explanatory, by H. YOUNG. New Edition, revised and improved. With copious Additional Notes by Rev. T. H. L. LEARY, D.C.L., formerly Scholar of Brasenose College, Oxford. 3s.
5* ——— Part 1. Books i.—vi., 1s. 6d.
5** ——— Part 2. Books vii.—xii., 2s.
6. **Horace;** Odes, Epode, and Carmen Sæculare. Notes by H. YOUNG. 1s. 6d.
7. **Horace;** Satires, Epistles, and Ars Poetica. Notes by W. BROWNRIGG SMITH, M.A., F.R.G.S. 1s. 6d.
8. **Sallustii Crispi Catalina et Bellum Jugurthinum.** Notes, Critical and Explanatory, by W. M. DONNE, B.A., Trin. Coll., Cam. 1s. 6d.
9. **Terentii Andria et Heautontimorumenos.** With Notes, Critical and Explanatory, by the Rev. JAMES DAVIES, M.A. 1s. 6d.
10. **Terentii Adelphi, Hecyra, Phormio.** Edited, with Notes, Critical and Explanatory, by the Rev. JAMES DAVIES, M.A. 2s.
11. **Terentii Eunuchus, Comœdia.** Notes, by Rev. J. DAVIES, M.A. 1s. 6d.
12. **Ciceronis Oratio pro Sexto Roscio Amerino.** Edited, with an Introduction, Analysis, and Notes, Explanatory and Critical, by the Rev. JAMES DAVIES, M.A. 1s. 6d.
13. **Ciceronis Orationes in Catilinam, Verrem, et pro Archia.** With Introduction, Analysis, and Notes, Explanatory and Critical, by Rev. T. H. L. LEARY, D.C.L. formerly Scholar of Brasenose College, Oxford. 1s. 6d.
14. **Ciceronis Cato Major, Lælius, Brutus, sive de Senectute, de Amicitia, de Claris Oratoribus Dialogi.** With Notes by W. BROWNRIGG SMITH, M.A., F.R.G.S. 2s.
16. **Livy:** History of Rome. Notes by H. YOUNG and W. B. SMITH, M.A. Part 1. Books i., ii., 1s. 6d.
16*. ——— Part 2. Books iii., iv., v., 1s. 6d.
17. ——— Part 3. Books xxi., xxii., 1s. 6d.
19. **Latin Verse Selections,** from Catullus, Tibullus, Propertius, and Ovid. Notes by W. B. DONNE, M.A., Trinity College, Cambridge. 2s.
20. **Latin Prose Selections,** from Varro, Columella, Vitruvius, Seneca, Quintilian, Florus, Velleius Paterculus, Valerius Maximus Suetonius, Apuleius, &c. Notes by W. B. DONNE, M.A. 2s.
21. **Juvenalis Satiræ.** With Prolegomena and Notes by T. H. S. ESCOTT, B.A., Lecturer on Logic at King's College, London. 2s.

7, STATIONERS' HALL COURT, LUDGATE HILL, E.C.

GREEK.

14. **Greek Grammar,** in accordance with the Principles and Philological Researches of the most eminent Scholars of our own day. By HANS CLAUDE HAMILTON. 1s. 6d.

15,17. **Greek Lexicon.** Containing all the Words in General Use, with their Significations, Inflections, and Doubtful Quantities. By HENRY R. HAMILTON. Vol. 1. Greek-English, 2s. 6d.; Vol. 2. English-Greek, 2s. Or the Two Vols. in One, 4s. 6d.: cloth boards, 5s.

14,15. **Greek Lexicon** (as above). Complete, with the GRAMMAR, in
17. One Vol., cloth boards, 6s.

GREEK CLASSICS. With Explanatory Notes in English.

1. **Greek Delectus.** Containing Extracts from Classical Authors, with Genealogical Vocabularies and Explanatory Notes, by H. YOUNG. New Edition, with an improved and enlarged Supplementary Vocabulary, by JOHN HUTCHISON, M.A., of the High School, Glasgow. 1s. 6d.

2, 3. **Xenophon's Anabasis;** or, The Retreat of the Ten Thousand. Notes and a Geographical Register, by H. YOUNG. Part 1. Books i. to iii., 1s. Part 2. Books iv. to vii., 1s.

4. **Lucian's Select Dialogues.** The Text carefully revised, with Grammatical and Explanatory Notes, by H. YOUNG. 1s. 6d.

5-12. **Homer, The Works of.** According to the Text of BAEUMLEIN. With Notes, Critical and Explanatory, drawn from the best and latest Authorities, with Preliminary Observations and Appendices, by T. H. L. LEARY, M.A., D.C.L.

THE ILIAD:	Part 1. Books i. to vi., 1s. 6d.	Part 3. Books xiii. to xviii., 1s. 6d.
	Part 2. Books vii. to xii., 1s. 6d.	Part 4. Books xix. to xxiv., 1s. 6d.
THE ODYSSEY:	Part 1. Books i. to vi., 1s. 6d.	Part 3. Books xiii. to xviii., 1s. 6d.
	Part 2. Books vii. to xii., 1s. 6d.	Part 4. Books xix. to xxiv., and Hymns, 2s.

13. **Plato's Dialogues:** The Apology of Socrates, the Crito, and the Phædo. From the Text of C. F. HERMANN. Edited with Notes, Critical and Explanatory, by the Rev. JAMES DAVIES, M.A. 2s.

14-17. **Herodotus, The History of,** chiefly after the Text of GAISFORD. With Preliminary Observations and Appendices, and Notes, Critical and Explanatory, by T. H. L. LEARY, M.A., D.C.L.
Part 1. Books i., ii. (The Clio and Euterpe), 2s.
Part 2. Books iii., iv. (The Thalia and Melpomene), 2s.
Part 3. Books v.-vii. (The Terpsichore, Erato, and Polymnia), 2s.
Part 4. Books viii., ix. (The Urania and Calliope) and Index, 1s. 6d.

18. **Sophocles:** Œdipus Tyrannus. Notes by H. YOUNG. 1s.

20. **Sophocles:** Antigone. From the Text of DINDORF. Notes, Critical and Explanatory, by the Rev. JOHN MILNER, B.A. 2s.

23. **Euripides:** Hecuba and Medea. Chiefly from the Text of DINDORF. With Notes, Critical and Explanatory, by W. BROWNRIGG SMITH, M.A., F.R.G.S. 1s. 6d.

26. **Euripides:** Alcestis. Chiefly from the Text of DINDORF. With Notes, Critical and Explanatory, by JOHN MILNER, B.A. 1s. 6d.

30. **Æschylus:** Prometheus Vinctus: The Prometheus Bound. From the Text of DINDORF. Edited, with English Notes, Critical and Explanatory, by the Rev. JAMES DAVIES, M.A. 1s.

32. **Æschylus:** Septem Contra Thebes: The Seven against Thebes. From the Text of DINDORF. Edited, with English Notes, Critical and Explanatory, by the Rev. JAMES DAVIES, M.A. 1s.

40. **Aristophanes:** Acharnians. Chiefly from the Text of C. H. WEISE. With Notes, by C. S. T. TOWNSHEND, M.A. 1s. 6d.

41. **Thucydides:** History of the Peloponnesian War. Notes by H. YOUNG. Book 1. 1s. 6d.

42. **Xenophon's Panegyric on Agesilaus.** Notes and Introduction by LL. F. W. JEWITT. 1s. 6d.

43. **Demosthenes.** The Oration on the Crown and the Philippics. With English Notes. By Rev. T. H. L. LEARY, D.C.L., formerly Scholar of Brasenose College, Oxford. 1s. 6d.

www.ingramcontent.com/pod-product-compliance
Lightning Source LLC
Chambersburg PA
CBHW022110230426
43672CB00008B/1330